D1219490

FEMTOCELLS

FEMTOCELLS: TECHNOLOGIES AND DEPLOYMENT

Jie Zhang
University of Bedfordshire, UK

Guillaume de la Roche
University of Bedfordshire, UK

WILEY
A John Wiley and Sons, Ltd., Publication

This edition first published 2010

Registered office
John Wiley & Sons Ltd, The Atrium, Southern Gate, Chichester, West Sussex, PO19 8SQ, United Kingdom

For details of our global editorial offices, for customer services and for information about how to apply for permission to reuse the copyright material in this book please see our website at www.wiley.com.

Library of Congress Cataloguing-in-Publication Data
Zhang, Jie, 1967-
Femtocells : technologies and deployment / Jie Zhang, Guillaume de la Roche.
p. cm.
Includes bibliographical references and index.
ISBN 978-0-470-74298-3 (cloth)
1. Femtocells. 2. Wireless LANs – Equipment and supplies. 3. Cellular telephone systems – Equipment and supplies. 4. Radio relay systems. 5. Telephone repeaters. I. De la Roche, Guillaume. II. Title.
TK5103.2.Z524 2010
621.382′1–dc22

2009039766

A catalogue record for this book is available from the British Library.

ISBN: 978-0-470-74298-3 (H/B)

Set in 10/12 Times by Laserwords Private Limited, Chennai, India
Printed and bound in Singapore by Markono Print Media Pte Ltd

for
MALCOM FOSTER
for his tireless reviews

Contents

About the Authors

Jie Zhang is a professor of wireless communications and networks and the director of CWiND (Centre for Wireless Network Design, www.cwind.org) at the DCST (Department of Computer Science and Technology) of UoB (University of Bedfordshire). He joined UoB as a Senior Lecturer in 2002, becoming professor in 2006.

He received his PhD in industrial automation from East China University of Science and Technology (www.ecust.edu.cn), Shanghai, China, in 1995. From 1997 to 2001, he was a postdoctoral research fellow with University College London, Imperial College London, and Oxford University.

Since 2003, he has been awarded more than 12 projects worth over 4 million Euros (his share). In addition, Professor Zhang is responsible for projects worth a few million Euros with his industrial partners. These projects are centred on new radio propagation models, UMTS/HSPA/ WiMAX/LTE simulation, planning and optimization, indoor radio network design and femtocells.

He is an evaluator for both EPSRC (Engineering and Physical Science Research Council) and the EU Framework Program.

He has published over 100 refereed journal and conference papers. He is the chair of a femtocell panel titled Femtocells: Deployment and Applications at IEEE ICC 2009. He has been a panellist at IEEE Globecom and IEEE PIMRC.

Prof. Zhang is an Associate Editor of Telecommunication Systems (Springer) and is in the editorial board of Computer Communications (Elsevier).

Guillaume de la Roche has been working as a research fellow at the Centre for Wireless Network Design (UK) since 2007. He received the Dipl-Ing in Telecommunication from the School of Chemistry Physics and Electronics (CPE Lyon), France, an MSc degree in Signal Processing (2003) and PhD in Wireless Communication (2007) from the National Institute of Applied Sciences (INSA Lyon), France.

From 2001 to 2002 he was a research engineer with Siemens-Infineon in Munich, Germany. From 2003 to 2004 he worked in a small French company where he deployed and optimized 802.11 wireless networks. He was responsible for a team developing a WiFi planning tool. From 2004 to 2007 he was with the CITI Laboratory at INSA Lyon, France. His research was about radio propagation in indoor environments and WiFi network planning and optimization.

He has supervised a number of students and taught laboratory courses in GSM network planning. He still teaches object programming and Java at Lyon 1 University. He has been involved in EU and UK funded projects, and is currently the principal investigator for an FP7 project (CWNetPlan) related to the coexistence and the optimization of indoor and outdoor wireless networks. His current research includes femtocells, channel modelling, and wireless network planning and optimization.

Alvaro Valcarce obtained his MEng in telecommunications engineering from the University of Vigo (Spain) in 2006. During 2005 he worked at 'Telefonica I+D' in Madrid (Spain), integrating an applications-streaming platform into an 'ATG Dynamo Server', as well as developing a system for 'applications-on-demand'.

In 2006, he worked as a researcher at the WiSAAR consortium in Saarbruecken (Germany), where he performed several WiMAX field trials, radio propagation measurements, data analysis and study of radio performance. The outcome of this project was an empirical propagation model especially designed for WiMAX coverage prediction at 3.5 GHz in urban environments.

Alvaro joined the Centre for Wireless Network Design (CWiND) of the University of Bedfordshire (UK) in 2007 with the support of a European Marie Curie Host Fellowship for Early Stage Research Training (EST). During 2008, he was one of the main researchers of the first British EPSRC-funded research project on femtocells – 'The feasibility study of WiMAX-based femtocells for indoor coverage' (EP/F067364/1). His PhD is integrated in the FP6 RANPLAN-HEC project (MEST-CT-2005-020958): 'Automatic 3G/4G Radio Access Network Planning and Optimization – A High End Computing Approach'. This project studies, among other wireless topics, the indoor-to-outdoor wireless channel and its applicability to network planning. Alvaro's main research interests currently include radio channel modelling, multicarrier systems, finite-difference algorithms, wireless networks planning and optimization methods and femtocells.

David López-Pérez received his bachelor and master degrees in telecommunication from Miguel Hernandez University, Elche, Alicante (Spain) in 2003 and 2006, respectively. He joined Vodafone Spain in 2005, working at the Radio Frequency Department in the area of network planning and optimization. He participated in the development of the Vodafone Automatic Frequency Planning tool for GSM and DCS networks.

He took up a research PhD scholarship at the Cork Institute of Technology in Ireland in 2006 for a year where he worked on a project called 'UbiOne The Multi-Modal WiFi Positioning System'. This project proposed a multi-modal positioning system utilizing off-the-shelf WiFi based equipment for a cost-effective solution, providing accurate location data in office/open building and campus environments.

Nowadays, he is a Marie-Curie fellow at the Centre for Wireless Network Design at the University of Bedfordshire, and his research is supported by the 'FP6 Marie Curie RANPLAN-HEC project' (MESTCT-2005-020958). He also participates as a guest researcher in the first EPSRC-funded research project on femtocells – 'The feasibility study of WiMAX based femtocells for indoor coverage' (EP/F067364/1) in the UK. His research is focused on the study of 2G/3G/LTE/WiMAX network planning and optimization, and self-organization for macrocells and femtocells two-tier networks. He is also interested on cooperative communications, optimization and simulation techniques.

Enjie Liu is a Senior Lecturer at the Department of Computer Science and Technology of the University of Bedfordshire. She joined UoB in 2003. She is a member of the networking teaching group in the department and responsible for delivering both wired and wireless modules to undergraduates as well as post graduates. She received her PhD from Queen Mary College, University of London in 2002. Then she worked as research fellow with the Centre for Communication Systems Research (CCSR), the University of Surrey. She was granted a Newly Appointed Lecturers Award by The Nuffield Foundation. She was a co-investigator of EU FP6 RANPLAN-HEC project that oversees 3G/4G radio network planning and optimization. She was the principal investigator of the first EPSRC funded research project on femtocells – 'The feasibility study of WiMAX based femtocell for indoor coverage' (EP/F067364/1). Before her PhD, she had more than 10 years of industrial experience in telecommunications with Nortel Networks. She first worked with Nortel in China on installation, on-site testing and maintenance of wireless networks such as GSM and CDMA. She also worked with the Nortel Networks Harlow Laboratory in the UK.

Hui Song is a PhD student and research associate at the Center for Wireless Network Design (CWiND), University of Bedfordshire. His interest is network planning and optimization technologies. His current focus is on modeling OFDM fading channels. Before joining CWiND, he was the manager of the technology department at Bynear Telecom Software Ltd, Shanghai, China. There he was responsible for developing and maintaining the nation-first network planning and optimization suite (including GSM, WCDMA and TD-SCDMA). Song holds a mathematics degree from Fudan University, Shanghai, China. He currently resides in the United Kindom.

Preface

In cellular networks, it is estimated that $\frac{2}{3}$ of calls and over 90% of data services occur indoors. However, some surveys show that many households and businesses experience a poor indoor coverage problem. It has been identified that poor coverage is the main reason for churn, which is very costly for operators in saturated markets. How to provide good indoor coverage cost effectively is thus a demanding challenge for operators.

The recent development of femtocells provides a fresh opportunity for operators to address the poor indoor coverage problem. Femtocells represent a more cost-effective solution than do other indoor solutions such as DAS (Distributed Antenna Systems) and picocells in indoor scenarios such as home and SOHO (Small Office and Home Office). Many operators such as Vodafone and AT&T have expressed a strong interest in femtocells and announced commercial launches of femtocells within their mobile networks, starting in the second half of 2009.

The deployment of a large number of femtocells (in particular, spectrum-efficient co-channel deployment) will have an impact on the macrocell layer. This impact and the performance of both macrocell and femtocell layers have to be fully evaluated before large-scale deployment. The evaluation can be done either through trials or simulation-based approaches. There is currently a lack of documentation that provides a comprehensive and organized explanation for the challenging issues arising from the deployment of femtocells in an existing macrocell network. We therefore believe that there is an urgent need for a book that covers femtocell technologies (such as femtocell architecture and air interface technologies) and the issues arising from femtocell deployment (such as interference modelling and mitigation, self-optimization, mobility management, etc.).

In recent years, CWiND has been funded by the EPSRC (Engineering and Physical Science Research Council) and the European Commission to research femtocells and indoor radio network design. These projects equipped us with a good understanding of femtocell technologies and the challenging issues arising from deployment of femtocells. It is also fortunate that CWiND could dedicate a large amount of human resources from February to June 2009 to the writing of this very much needed book.

In this book, the method used to study femtocell deployment is computer-aided simulation rather than trial based. This method is more convenient and more cost effective.

The book is written in a tutorial style. We believe that it suits a wide range of readers, e.g., RF engineers from operators, R&D engineers from telecom vendors, academics and researchers from universities, consultants for wireless networks and employees from regulatory bodies.

This book is organized as follows:

In Chapter 1 (Introduction), an introduction to femtocell concepts and the book is given. The advantages and disadvantages of using femtocells, the standardization and business models are also briefly touched on.

In Chapter 2 (Indoor Coverage Techniques), an overview of the different indoor coverage techniques is given. As femtocell is mainly used for indoors, we think a brief introduction to other indoor coverage techniques might be useful for readers. In this chapter, the evolution from macrocell to femtocell is presented and the different methods are compared. Advantages and drawbacks of the different techniques, like Distributed Antenna System (DAS), repeaters, and picocells are also given and the main challenges related to femtocells are introduced. It needs to be pointed out that femtocells can also be used outdoors, provided that backhaul connections are available or can be easily established.

In Chapter 3 (Access Network Architecture), the evolution of femtocell architecture from 3GPP Release 8 and different options to ensure the connectivity of the femtocell to the core network are described. Functional split between HNB and HNB-GW, new interfaces such as Iuh are also described. Security aspects are also touched on.

In Chapter 4 (Air Interface Technologies), different air interface technologies for femtocells are presented. In particular, femtocell specific features in the discussed air interfaces are described. The technologies covered in this chapter include Global System for Mobile communication (GSM), Universal Mobile Telecommunication System (UMTS), High Speed Packet Access (HSPA), Wireless Interoperability for Microwave Access (WiMAX) and Long Term Evolution (LTE).

In Chapter 5 (System-Level Simulation for Femtocell Scenarios), the methodology of how to simulate femtocells is detailed. The development of a femtocell simulation tool is presented, from the radio coverage level, to the system level. Simulation methods, including both static and dynamic approaches, are illustrated with some femtocell deployment examples. Coverage and capacity analysis is given for the given scenarios of a hybrid femtocell/macrocell WiMAX network.

In Chapter 6 (Interference in the Presence of Femtocells), interference between femtocell and macrocell (so-called cross-layer interference), as well as between neighbouring femtocells (so-called co-layer interference) are analysed for both CDMA and OFDMA based femto/macro networks. The performance of a UMTS macrocell network in the presence of femtocells is also given. Moreover, some interference cancellation and avoidance techniques are also presented in this chapter.

In Chapter 7 (Mobility Management), issues related to mobility management such as cell selection/reselection and handovers in two-tier femto/macro networks for various access methods (CSG, open access and hybrid access) are discussed in detail. Mobility management is a major issue and presents a big challenge for hybrid femto/macro network.

In Chapter 8 (Self-Organization), issues related to femtocell self-organization are presented. Self-organization includes self-configuration, self-optimisation and self-healing. With self-conguration, the initial femtocell parameters are automatically selected (such as PCI, neighbouring list, channel and power). Self-optimization kicks in when FAPs are operational and optimize the FAP parameters taking into account the fluctuations of the channel and resources available. In order to achieve self-organisation, FAPs should know their radio environments; hence, radio channel sensing techniques such as those using message exchange and measurement report are also described in this chapter. Femtocells are plug-and-play devices, self-organization capability is key to the successful deployment of femtocells.

In Chapter 9 (Further Femtocell Issues), some other important challenges that have to be solved are presented, these include ensuring the timing accuracy, the security and the identification of location of the femtocell devices. In addition, access methods, femtocell applications and health issues are also discussed in this chapter.

No book is perfect and this one is no exception. In order to provide a remedy to this fact, we will present further materials related to this book at the following website: www.deployfemtocell.com. We also plan to create a discussion board at this site, so that the interactions between the authors and readers and between readers themselves can be facilitated. Finally, we hope that you will like this book and give us feedback so that we can improve the book for the next edition.

Acknowledgements

We would like to thank our publishers, Tiina Ruonamaa, Anna Smart, Sarah Tilley, Brett Wells and the rest of the wireless team at Wiley. They produce the largest collection of best wireless books! We are grateful for their encouragement, enthusiasm and vision about this book, as well as for their professionalism. We believe they are all great assets for Wiley! We learned a lot from them. We thank Brett Wells and Dhanya Ramesh for their excellent work at the production stage.

We thank anonymous reviewers for their helpful comments that have improved the quality of this book.

Jie Zhang would like to thank Simon Saunders for the invitation to the Femto Forum meeting in Dallas in December 2008. This gave Jie an overview of the Femto Forum activities. The Femto Forum white paper on WCDMA interference management was also useful for this book.

The authors would like to thank Holger Claussen and Malek Shahid from Alcatel-Lucent. They both have a great understanding of femtocells. The discussions with them were very helpful.

The authors would like to thank De Chen and Eric (Linfeng) Xia from Huawei Technologies. The discussions with them improved our understanding of LTE femtocells.

We would like to thank John Malcolm Foster, a great friend of ours, for his wisdom and endless corrections of research proposals, research papers and book chapters in the last 7 years. Malcolm corrected the English for all the chapters of this book. We all learned a lot from him in the last few years.

We would also like to thank other CWiND members whose research might be directly/indirectly useful for this book, such as Alpár Jüttner, Raymond Kwan, Ákos Ladányi and Zhihua Lai (according to alphabetic order of surnames). We are really proud of working with so many talented, self-motivated and extremely able young researchers. Together, we have made CWiND a special place with so many achievements in a very short time.

We express our thanks to the EPSRC (Engineering and Physical Science Research Council) and the European Commission for their support of our research on femtocells and indoor radio network design. We would like to extend our thanks to the project partners on these projects Ranplan Wireless Network Design Ltd (in particular, Joyce Wu) and INSA-Lyon (in particular, Jean-Marie Gorce).

We thank all our teachers/supervisors who illuminated us during our studies from the primary school to the PhD. In many ways, they lit up our dreams.

We express our gratitude to all our families for their support throughout the years. We know that without their support, we can not even live in this world.

Jie Zhang would like to thank Joyce for all the work she does at home and, in particular, for her delivery of Jie's biggest achievements Jennifer and James. Jie is grateful for Jennifer's love of engineering and believes that she will do better than him in engineering.

Acronyms

3GPP	3rd Generation Partnership Project
AAA	Authentication, Authorization and Accounting
ACIR	Adjacent Channel Interference Rejection
ACL	Allowed CSG List
ACLR	Adjacent Channel Leakage Ratio
ACPR	Adjacent Channel Power Ratio
ACS	Adjacent Channel Selectivity
ADSL	Asymmetric Digital Subscriber Line
AGCH	Access Grant Channel
AH	Authentication Header
AKA	Authentication and Key Agreement
AMC	Adaptive Modulation and Coding
API	Application Programming Interface
ARPU	Average Revenue Per Unit
AS	Access Stratum
ASE	Area Spectral Efficiency
ASN	Access Service Network
ATM	Asynchronous Transfer Mode
AUC	Authentication Centre
AWGN	Additive White Gaussian Noise
BCCH	Broadcast Control Channel
BCH	Broadcast Channel
BE	Best Effort
BER	Bit Error Rate
BLER	BLock Error Rate
BPSK	Binary Phase-Shift Keying
BS	Base Station
BSC	Base Station Controller
BSIC	Base Station Identity Code
BSS	Base Station Subsystem
BTS	Base Transceiver Station
CAC	Call Admission Control
CAPEX	CAPital EXpenditure
CAZAC	Constant Amplitude Zero Auto-Correlation
CCCH	Common Control Channel
CCPCH	Common Control Physical Channel

CCTrCH	Coded Composite Transport Channel
CDMA	Code Division Multiple Access
CGI	Cell Global Identity
CN	Core Network
CPCH	Common Packet Channel
CPE	Customer Premises Equipment
CPICH	Common Pilot Channel
CQI	Channel Quality Indicator
CRC	Cyclic Redundance Check
CSG	Closed Subscriber Group
CSG ID	CSG Identity
CSI	Channel State Information
CTCH	Common Traffic Channel
CWiND	Centre for Wireless Network Design
DAS	Distributed Antenna System
DCCH	Dedicated Control Channel
DCH	Dedicated Channel
DCS	Digital Communication System
DFT	Discrete Fourier Transform
DL	DownLink
DoS	Denial of Service
DPCCH	Dedicated Physical Control Channel
DPDCH	Dedicated Physical Data Channel
DRX	Discontinuous Reception
DSCH	Downlink Shared Channel
DSL	Digital Subscriber Line
DTCH	Dedicated Traffic Channel
DXF	Drawing Interchange Format
EAGCH	Enhanced uplink Absolute Grant Channel
EAP	Extensible Authentication Protocol
ECRM	Effective Code Rate Map
EDCH	Enhanced Dedicated Channel
EESM	Exponential Effective SINR Mapping
EHICH	EDCH HARQ Indicator Channel
EIR	Equipment Identity Register
EMS	Enhanced Messaging Srvice
EPC	Enhanced Packet Core
EPLMN	Equivalent PLMN
ERGCH	Enhanced uplink Relative Grant Channel
ertPS	extended real time Polling Service
ESP	Encapsulating Security Payload
EUTRA	Evolved UTRA
EUTRAN	Evolved UTRAN
EVDO	Evolution-Data Optimized
FACCH	Fast Associated Control Channel
FACH	Forward Access Channel

FAP	Femtocell Access Point
FCC	Federal Communications Commission
FCCH	Frequency-Correlation Channel
FCH	Frame Control Header
FDD	Frequency Division Duplexing
FDTD	Finite-Difference Time-Domain
FFT	Fast Fourier Transform
FGW	Femto Gateway
FIFO	First In–First Out
FMC	Fixed Mobile Convergence
FTP	File Transfer Protocol
FUSC	Full Usage of Subchannels
GAN	Generic Access Network
GANC	Generic Access Network Controller
GERAN	GSM EDGE Radio Access Network
GGSN	Gateway GPRS Support Node
GMSC	Gateway Mobile Switching Centre
GPRS	General Packet Radio Service
GPS	Global Positioning System
GPU	Graphics Processing Unit
GSM	Global System for Mobile communication
HARQ	Hybrid Automatic Repeat reQuest
HBS	Home Base Station
HCS	Hierarchical Cell Structure
HeNB	Home eNodeB
HLR	Home Location Register
HNB	Home NodeB
HNBAP	Home NodeB Application Protocol
HNBGW	Home NodeB Gateway
HPLMN	Home PLMN
HSDPA	High Speed Downlink Packet Access
HSDSCH	High-Speed DSCH
HSPA	High Speed Packet Access
HSS	Home Subscriber Server
HSUPA	High Speed Uplink Packet Access
HUA	Home User Agent
IC	Interference Cancellation
ICI	Intercarrier Interference
ICNIRP	International Commission on Non-Ionizing Radiation Protection
ICS	IMS centralized service
IDFT	Inverse Discrete Fourier Transform
IETF	Internet Engineering Task Force
IFFT	Inverse Fast Fourier Transform
IKE	Internet Key Exchange
IKEv2	Internet Key Exchange version 2
IMEI	International Mobile Equipment Identity

IMS	IP Multimedia Subsystem
IMSI	International Mobile Subscriber Identity
IP	Internet Protocol
IPsec	Internet Protocol Security
ISI	Intersymbol Interference
IWF	IMS Interworking Function
Iub	UMTS Interface between RNC and Node B
Iuh	Iu Home
KPI	Key Performance Indicator
LA	Location Area
LAC	Location Area Code
LAI	Location Area Identity
LAU	Location Area Update
LLS	Link-Level Simulation
LOS	Line Of Sight
LTE	Long Term Evolution
LUT	Look Up Table
MAC	Medium Access Control
MAP	Media Access Protocol
MBMS	Multimedia Broadcast Multicast Service
MBS	Macrocell Base Station
MBSFN	Multi-media Broadcast over a Single-Frequency Network
MC	Modulation and Coding
MGW	Media Gateway
MIB	Master Information Block
MIC	Mean Instantaneous Capacity
MIMO	Multiple Input–Multiple Output
MM	Mobility Management
MME	Mobility Management Entity
MMSE	Minimum Mean Square Error
MNC	Mobile Network Code
MNO	Mobile Network Operator
MR	Measurement Report
MS	Mobile Station
MSC	Mobile Switching Centre
MSISDN	Mobile Subscriber Integrated Services Digital Network Number
NAS	Non-Access Stratum
NCL	Neighbour Cell List
NGMN	Next Generation Mobile Networks
NIR	Non Ionization radiation
NLOS	Non-Line Of Sight
nrtPS	non-real-time Polling Service
NSS	Network Switching Subsystem
NTP	Network Time Protocol
NWG	Network Working Group
OAM&P	Operation, Administration, Maintenance and Provisioning

OC	Optimum Combining
OCXO	Oven Controlled Oscillator
OFDM	Orthogonal Frequency Division Multiplexing
OFDMA	Orthogonal Frequency Division Multiple Access
OPEX	OPerational EXpenditure
OSI	Open Systems Interconnection
OSS	Operation Support Subsystem
P2P	Point to Point
PAPR	Peak-to-Average Power Ratio
PC	Power Control
PCCH	Paging Control Channel
PCCPCH	Primary Common Control Physical Channel
PCH	Paging Channel
PCI	Physical Cell Identity
PCPCH	Physical Common Packet Channel
PCPICH	Primary Common Pilot Channel
PDSCH	Physical Downlink Shared Channel
PDU	Packet Data Unit
PF	Proportional Fair
PHY	Physical
PIC	Parallel Interference Cancellation
PKI	Public Key Infrastructure
PLMN	Public Land Mobile Network
PLMN ID	PLMN Identity
PN	Pseudorandom Noise
PRACH	Physical Random Access Channel
PSC	Primary Scrambling Code
PSTN	Public Switched Telephone Network
PUSC	Partial Usage of Subchannels
QAM	Quadrature Amplitude Modulation
QoS	Quality of Service
QPSK	Quadrature Phase Shift Keying
RAB	Radio Access Bearer
RACH	Random Access Channel
RADIUS	Remote Authentication Dial-In User Services
RAN	Radio Access Network
RANAP	Radio Access Network Application Part
RAT	Radio Access Technology
RF	Radio Frequency
RLC	Radio Link Control
RMSE	Root Mean Square Error
RNC	Radio Network Controller
RPLMN	Registered PLMN
RRM	Radio Resource Management
RTP	Real-time Transport Protocol
rtPS	real-time Polling Service

RUA	RANAP User Adaptation
SACCH	Slow Associated Control Channel
SAIC	Single Antenna Interference Cancellation
SAP	Service Access Point
SCCPCH	Secondary Common Control Physical Channel
SCFDMA	Single Carrier FDMA
SCH	Synchronization Channel
SCTP	Stream Control Transmission Protocol
SDCCH	Standalone Dedicated Control Channel
SDU	Service Data Unit
SG	Signalling Gateway
SGSN	Serving GPRS Support Node
SI	State Insertion
SIB	System Information Block
SIC	Successive Interference Cancellation
SIGTRAN	Signalling Transport
SIM	Subscriber Identity Module
SINR	Signal to Interference plus Noise Ratio
SIP	Session Initiation Protocol
SLS	System-Level Simulation
SMS	Short Message Service
SNMP	Simple Network Management Protocol
SOHO	Small Office/Home Office
SON	Self-Organizing Network
SSL	Secure Socket Layer
TAI	Tracking Area Identity
TAU	Tracking Area Update
TCH	Traffic Channel
TCXO	Temperature Controlled Oscillator
TDD	Time Division Duplex
TDMA	Time Division Multiple Access
TLS	Transport Layer Security
TPM	Trusted Platform Module
TSG	Technical Specification Group
TTG	Transmit/Receive Transition Gap
TTI	Transmission Time Interval
TV	Television
UARFCN	UTRA Absolute Radio Frequency Channel Number
UDP	User Datagram Protocol
UE	User Equipment
UGS	Unsolicited Grant Service
UICC	Universal Integrated Circuit Card
UL	UpLink
UMA	Unlicensed Mobile Access
UMTS	Universal Mobile Telecommunication System
USIM	Universal Subscriber Identity Module

UTRA UMTS Terrestrial Radio Access
UTRAN UMTS Terrestrial Radio Access Network
UWB Ultra Wide Band
VLR Visitor Location Register
VoIP Voice-Over IP
VPLMN Visited PLMN
WCDMA Wideband Code Division Multiple Access
WAP WiFi Access Point
WEP Wired Equivalent Privacy
WG Working Group
WHO World Health Organization
WiFi Wireless Fidelity
WiMAX Wireless Interoperability for Microwave Access

1

Introduction

Jie Zhang, Guillaume de la Roche and Enjie Liu

1.1 The Indoor Coverage Challenge

In cellular networks, it is estimated that 2/3 of calls and over 90% of data services occur indoors. Hence, it is extremely important for cellular operators to provide good indoor coverage for not only voice but also video and high speed data services, which are becoming increasingly important. However, some surveys show that 45% of households and 30% of businesses [1] experience poor indoor coverage problem. Good indoor coverage and service quality will generate more revenues for operators, enhance subscriber loyalty and reduce churn. On the other hand, poor indoor coverage will do exactly the opposite. Hence, how to provide good indoor coverage, in particular, for high speed data services, is a big challenge for operators.

A typical approach to providing indoor coverage is to use outdoor macrocells. This approach has a number of drawbacks:

- It is very expensive to provide indoor coverage using an 'outside in' approach. For example, in UMTS, an indoor user will require higher power drain from the base station in order to overcome high penetration loss. This will result in less power to be used by other users and lead to reduced cell throughput. This is because the power used by indoor users is not efficient in terms of generating capacity and in UMTS capacity is linked to power. Hence, the cost per Mb of using 'outside in' approach will become higher and more expensive than using indoor solutions.
- A high capacity network needs a lot of outdoor base station sites, the acquisition of which has become very challenging in densely populated areas.
- It is less likely that a high capacity network using such an approach will be built, due to the interference and higher power drain from base stations to serve indoor users from outdoor macrocells, etc.

Femtocells: Technologies and Deployment Jie Zhang and Guillaume de la Roche

- As the cell sites become denser, the network planning and optimization becomes a big challenge in such networks. For example, in GSM/GPRS/EDGE networks, the frequency planning and in CDMA based networks, the planning of soft handover regions, etc.
- 3G and beyond networks will normally work at 2 GHz or above, the building penetration is a challenge for networks operating above 2 GHz.
- The network performance (e.g., throughput) indoors can not be guaranteed, in particular, in the side not facing the macrocell sites. In order to achieve higher data rates, higher modulation and coding schemes are needed. The higher modulation and coding schemes in HSDPA, WiMAX and LTE require better channel conditions, which can only be met near those windows facing macrocell sites.

Hence, indoor solutions such as DAS (Distributed Antenna Systems) and picocells become an attractive and viable business proposition in hotspots such as large business centres, office buildings and shopping malls. These indoor systems are deployed by operators. The indoor solutions will improve in-building coverage, offload traffic from outdoor macrocells, enhance service quality and facilitate high data rate services due to the improved performance of radio links. With indoor solutions, in UMTS, the orthogonality can be improved, which will result in high throughput. In HSPA/LTE or WiMAX, the better channel conditions will enable high modulation and coding scheme to be used and thus deliver richer services that further drive demand.

Even though the above mentioned indoor solutions are more cost effective than using outdoor macrocells to provide indoor coverage for voice and high speed data services, such solutions are still too expensive to be used in some scenarios such as SOHO (Small Office and Home Office) and home users (for personal communications and entertaining, etc.). The scale of SOHO and home use normally does not represent a viable business proposition for operators. Recently, the development of femtocells provides a good opportunity for low cost indoor solutions for such scenarios. Unlike picocells, femtocells are deployed by users.

1.2 Concepts of Femtocells

1.2.1 What is a Femtocell?

Femtocells, also known as 'home base station', are cellular network access points that connect standard mobile devices to a mobile operator's network using residential DSL, cable broadband connections, optical fibres or wireless last-mile technologies.

1.2.2 A Brief History

The concept of 'home base station' was first studied by Bell Labs of Alcatel-Lucent in 1999. In 2002, Motorola announced the first 3G-based home base station product. However, it was not until 2005 that the 'home base station' concept started to gain a wider acceptance. In 2006, 'femtocell' as a term was coined. In February 2007, a number of companies demonstrated femtocells at the 3GSM World Congress (Barcelona), with operators announcing trials. In July 2007, the Femto Forum [2] was founded to promote femtocell standardization and deployment worldwide. As of December 2008, the forum includes over 100 telecom hardware and software vendors, mobile operators, content providers and start-ups. In 2008. Home NodeB (HNB) and Home eNodeB (HeNB) were first introduced in 3rd Generation Partnership Project (3GPP) *Release 8*, signalling that it had become a mainstream wireless access technology. Large scale femtocell deployment is expected in 2010. It is likely that the roll-out of Long Term Evolution (LTE) networks will include both outdoor macrocells and indoor femtocells from the early stage of network deployment. Femtocells are also very promising for enterprise applications.

1.2.3 What is Included in a Femtocell Access Point?

The femtocell unit incorporates the functionality of a typical base station (Node-B in UMTS). A femtocell unit looks like a WiFi access point, see Figure 1.1. However, it also contains RNC (Radio Network Controller; in the case of GSM, BSC) and all the core network elements. Thus, it does not require a cellular core network, requiring only a data connection to the DSL or cable to the Internet, through which it is then connected to the mobile operator's core network, see Figure 1.1. In this book, we use femtocell access point (FAP) to stand for the femtocell unit that contains base station and core network functionalities, and use femtocell to refer to the service area covered by the FAP. A FAP looks like a WiFi access point (WAP). However, inside, they are fundamentally different. WAP implements WiFi technologies such as IEEE 802.11b, 802.11g, and 802.11n. FAP implements cellular technologies such as GSM/GPRS/EDGE, UMTS/HSPA/LTE and mobile WiMAX (IEEE 802.16e). A comprehensive comparison of WiFi and cellular technologies is beyond the scope of this chapter.

1.2.4 FAP Technologies

The technologies behind femtocell are cellular technologies. As the key driver of femtocell is the demand for higher and higher data rates indoors, UMTS/HSPA FAPs are the current main focus. However, FAPs can also be based on GSM/GPRS/EDGE. 2G/3G based femtocells have been developed by various vendors. The development of WiMAX and LTE based femtocells is also under way.

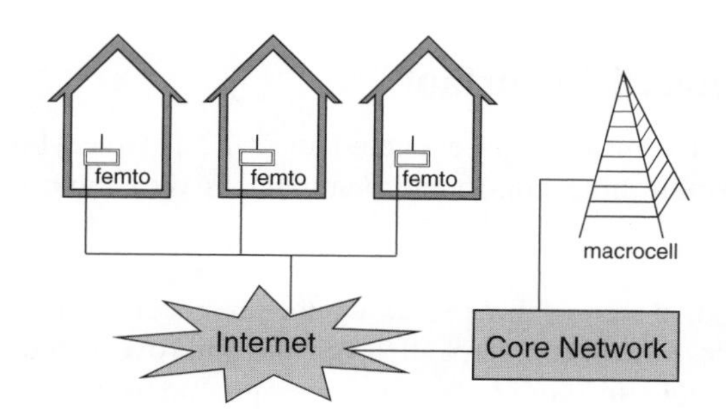

Figure 1.1 Typical femtocell and macrocell scenario

1.2.5 FAP Deployment

Unlike picocells, FAPs are self-deployed by users rather than operators. They should be regarded as consumer electronics. In order to generate minimum interference to outdoor macrocells and neighboring femtocells, a FAP must be able to configure itself automatically. Automatic configuration of FAP can be divided into a sensing phase, in which radio environment will be assessed, and an auto-tuning phase, in which FAP parameters (e.g., downlink Tx power and sub-channel allocation, etc.) will be automatically configured. Automatic configuration of FAP is key to the successful deployment of femtocells. Before FAPs can actually be self-deployed by users, operators must test typical femtocell deployment scenarios by trials and/or simulation. The main purpose of the simulation and trials is to find out the impact of femtocell deployment on the macrocell layer. In addition, how femtocells will affect each other, as well as the performance of both femto- and macro-cell layers will also need to be investigated. A femtocell deployment tool that incorporates system level simulation for various RATs (Radio Access Technologies), accurate radio propagation models (e.g., using 3D ray tracing or FDTD), 3D modelling and visualisation of building structure (for example, to read from AutoCAD .dxf file to generate 3D viewer of building floor structures) and optimization engine will be highly desirable for this purpose because compared with trials, it is much cheaper and convenient. CWiND's industrial partner Ranplan Wireless Network Design Ltd (www.ranplan.co.uk) is developing such a tool and will be ready for commercial offering at the beginning of 2010.

1.2.6 FAP Classification

According to their capacity, FAP can be classified into two categories, namely home FAP, which can support 3–5 simultaneous users, and enterprise FAP, which can support 8–16 users. The key drive of FAP is to provide high data rate services for the residential sector. There is a low probability that all the subscribers will simultaneously use the femtocell, which is why home femtocells supporting more than five simultaneous users would be too useless compared with the real demand. In addition, this is also restricted by the bandwidth limitation of the uplink ADSL. According to the cellular technologies used, FAP can be classified into UMTS FAP, GSM FAP, WiMAX FAP, and so on. There is a trend to combine different air interfaces into one FAP.

1.3 Why is Femtocell Important?

A large deployment of femtocells is expected in 2012 [3] (see Figure 1.2 and Figure 1.3), but why is this small thing important? Femtocell is very important for the following reasons:

- It can provide indoor coverage for places where macrocells cannot.
- It can offload traffic from the macrocell layer and improve macrocell capacity (in the case of using macrocells to provide indoor coverage, more power from the base station will be needed to compensate for high penetration loss, resulting in a decrease in macrocell capacity).

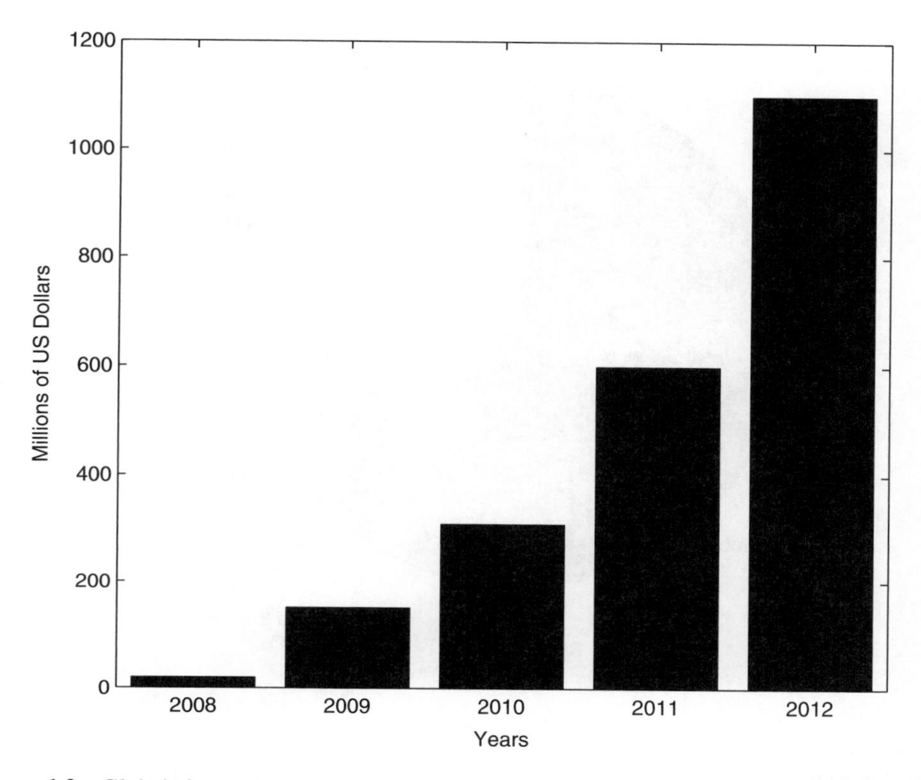

Figure 1.2 Global femto base station infrastructure equipment market forecast (Data from www.wirelessweek.com)

- Assume that good isolation (hence, the signal leakage from indoor to outdoor will be small) can be achieved, the addition of a femtocell layer will significantly improve the total network capacity by reusing radio spectrum indoors.
- There is a growing demand for higher and higher data rates. Due to the high penetration loss, high data rate services can not be provided to indoors apart from those areas near windows that are facing a macrocell site. This is because high data rate requires high performance RF links. High data rate services such as those facilitated by HSDPA are the key drive of femtocells.
- Femtocells can provide significant power saving to UEs. The path loss to indoor FAP is much smaller than that to the outdoor macrocell base station, and so is the required transmitting power from UE to the FAP. Battery life is one of the biggest bottlenecks for providing high speed data services to mobile terminals.
- As FAPs only need to be switched on when the users are at home (for home femtocells) or at work (for enterprise femtocells), the use of femtocell is 'greener' than macrocells. The power consumption of base stations accounts for a considerable amount of an operator's OPEX. In the UK, the power to run base stations is over 3 watts per subscriber. In some developing countries, the power consumption accounts for some $\frac{2}{3}$ of the OPEX. A base station consumes far more power than that is used for transmitting

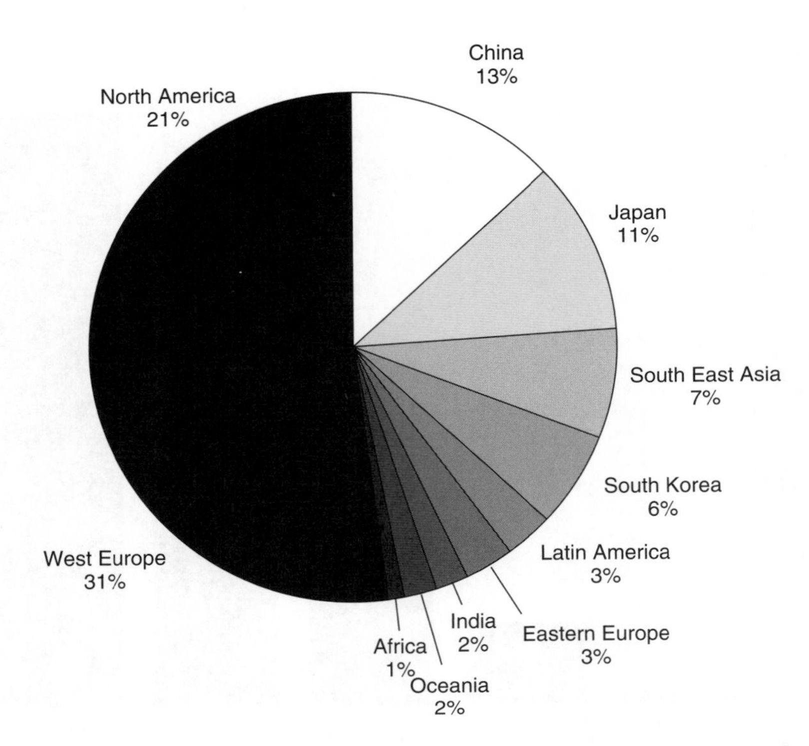

Figure 1.3 Total 3G femtocell deployment market in 2012 (Data from www.fwdconcepts.com/femtocell)

and receiving signals. This is caused by a number of factors: first, the efficiency of the amplifiers is very low (typically 10–15%) as they work at the linear rather than the saturation region as the sophisticated modulation techniques used in 3G and beyond systems require linear amplification; second, a base station requires an air-conditioning system in order to keep running at atmospheric temperature; third, a backup system is also needed to account for loss of power supply. The base station power consumption problem leads to a high demand on the so-called 'green communications systems' or 'green radio'.

- Femtocell provides an ideal solution for FMC (Fixed Mobile Convergence).
- Femtocell plays an important role in mobile broadband and ubiquitous communications.
- Femtocell represents a major paradigm shift. Users will pay to install femtocells. Hence, the first phase of the rollout of high data rate networks such as LTE can start from indoor where high data rates are needed most. As future terminals will support GSM, WCDMA (or other 3G technologies) and LTE, the rollouts of LTE can be very different from the rollouts of GSM and UMTS. This is really an important paradigm shift as far as future mobile communications network rollouts are concerned.

1.4 Deployment of Femtocells

Femtocells can bring a lot of advantages for both operators and subscribers.

1.4.1 Operator's Perspective

As a large amount of traffic (up to 70–80%) can be offloaded from macrocells, which means that fewer outdoor macrocells will be needed. The reduction of macrocell sites will result in a huge CAPEX saving for operators in their radio access networks. The reduction of traffic from macrocell sites will also result in significant saving in the backhauling. This will also lead to associated saving on the OPEX.

The reduction of macrocell sites will simplify the site survey and planning process; it also means less rent will be paid for the usage of base station sites. In the rollout of 3G/4G networks, site acquisition is a big challenge for operators, in particular, in urban areas. It has become increasingly difficult for operators to find base station sites.

Femtocells can help operators cost-effectively to build out network capacity and achieve a more cost-effective evolution plan with reduced risks and financial burdens. This is due to the facts that first, femtocells are low cost solutions for indoor coverage compared with other approaches; second, users will at least share a substantial amount of the installation cost of FAPs and the operation of FAPs will be largely financed by users (operators will also carry out remote maintenance, etc.). In particular, operators can encourage open access for femtocells and further reduce the demand of outdoor macrocells.

Femtocells will improve service quality; hence it will increase customer loyalty and reduce churn, which is a major issue and can cost operators millions of dollars a day. Surveys show that poor service quality is the most important factor for a subscriber leaving a mobile operator.

Femtocells will help mobile operators to drive data usage and provide richer services (for example, through home zone plans and bundled services), which will boost ARPU (Average Revenue Per Unit). Voice alone is no longer enough to future-proof revenue generation.

Compared with picocells and other indoor technologies, femtocells are a low cost solution to increasing indoor coverage and improving service quality.

Femtocells will help operators to deliver a seamless user experience across outdoor and indoor environments, at work, on the move or at home, and provide a basis for next-generation converged services that combine voice, video, and data services to a mobile device.

Even in areas that can be served by macrocells, femtocells can still bring a lot of benefits to operators as they will remove the need to deliver indoor services from macrocells and decrease the overhead incurred by delivering signals indoors.

The reduced demands on macrocells may allow operators to share the outdoor LTE network macrocells.

Femtocells can offer new mobile operators new alternative approaches to network roll-out. For example, new mobile operators can provide indoor solutions in hotspots using picocells and femtocells, provide femtocells solutions for home users, build macrocell networks where there is real need and reach roaming agreements with some established operators.

So far, we have only discussed the benefits that femtocells can bring to operators. The deployment of femtocells will potentially also cause some problems to operators. One of the drawbacks of femtocells for operators is that interference becomes more random and harder to control. This is particularly problematic for CDMA based networks such as UMTS. In order to improve overall network capacity, it is beneficial for operators for both the macro and femto layers to use the same frequency band to operate. Thus, the randomness of interference from femtocells may cause problems on macrocell operation, for example, causing coverage holes. As CDMA networks are interference limited, macrocell capacity can be affected if the interference from femtocells is not controlled well. Operators will not be able to access subscribers' premises, thus femtocell self-configuration is very important. Remote monitoring and maintenance will also be important for operators.

It needs to be pointed out that UTRAN was developed under the assumption of coordinated network deployment whereas femtocells are typically associated with uncoordinated and large scale deployment [4].

1.4.2 Subscriber's Perspective

For those who experience no or poor indoor coverage at home, femtocells can enable subscribers to use their mobiles at home. With femtocells, in addition to voice service, multimedia, video and high speed data services will also become available. As the indoor performance of the network can be much improved, so can the user experience for both voice and data services. Femtocells will offer users a single address book and one billing account for land line phone, broadband and mobile phone. Users can benefit from home zone plans and bundled services that will be more cost effective than using services from more than one provider. Femtocells can act as the focal point to connect all domestic devices to a home server and act as the gateway for all domestic devices to the Internet. Femtocells will deliver converged services (voice, video and data services) at home and enable users a seamless user experience across both outdoor and indoor environments with personalized converged services for UEs. Femtocells will save UE power. As the distance between the UE and the FAP is much shorter than that between the UE and the macrocell site, transmitting power on the uplink can be much reduced, which will result in power saving on the UE. The battery is one of the biggest bottlenecks in providing high speed data services to mobile devices. As the transmitting power of UE can be greatly reduced, health concerns on using mobile devices can be reduced. It needs to be pointed out that if there are any health concerns arising from using mobile communications, they would mainly come from uplink, as the UE is very close to the users (in particular, the head).

1.5 Important Facts to Attract More Customers

1.5.1 Access Control

There are two possible access methods for femtocells, both of them having some advantages and drawbacks.

Public Access Femtocells

In femtocell networks, an outdoor user could receive a stronger signal from a nearby femtocell than from a distant macrocell. With public access a connection is possible to this femtocell. This method benefits outdoor users, who are able to make use of nearby femtocells, thus reducing the overall use of system resources (power/frequency) and therefore interference. Moreover the situation is identical between neighbouring femtocells. It is possible that in some situations (for example in dense population areas or multi-floored buildings) the signal power of neighbouring femtocells will be higher than the femtocell of the customer. With public access connection to other femtocells would be possible.

Private Access Femtocells

In private access, only a list of registered users can access a femtocell. How the users will enter the list of authorized users has to be defined. Moreover such an approach will increase the interference. For example passing users, if the signal coming from the macrocell is low, will have to increase their power, thus producing more interference with the neighbouring femtocells.

Choice Depending on the Scenarios

The femtocell customers pay for the femtocell themselves, but also for the broadband Internet connection being used for backhaul. That is why access control methods are an important concern for them: should all users passing close to the customer's building have the right to use the femtocells if they do not pay for them? A recent survey [5] shows that customers would prefer femtocells in a private access mode where only a few users are allowed to connect to their femtocells. But in other scenarios like enterprise FAP, many femtocells will have to be used to cover a large area, and also many different passing users can go from offices to offices, which is why the public access will be preferred in nonhome scenarios. Finally there are some concerns to be discussed concerning emergency calls. A FAP is required to provide emergency services, as is the case for VoIP phone providers. That is why in home scenarios, even if private access is preferred, some resources could be released in a public access mode, so that emergency call services can be ensured by everybody.

1.5.2 Standardization

Before the market of femtocells can reach a massive success, a standardization is necessary. It is one of the aims of the Femto Forum to promote this standardization [2]. As a result of the joint work of the Femto Forum, 3GPP and Broadband Forum, which are the three main standards-related organizations for Femto technology, a series of the Femtocell standards has been officially published by 3GPP. The new standard forms part of 3GPP Release 8.

The new Femtocell standard covers four main areas: Network Architecture, Radio and Interference aspects, Femtocell Management/Provisioning and Security. In terms of network architecture, it re-uses existing 3GPP UMTS protocols and extends them to support the needs of high-volume femtocell deployments, detailed in TS25.469. In Rel-8, 3GPP has specified the basic functionalities for the support of HNB and HeNB. The requirements for these basic functionalities were captured in TS 22.011. From Rel-9 onward, it has been agreed to consolidate all the requirements from Rel-8 and further requirements for HNB and HeNB in the TS22.220. TR23.832 describes an IMS capable HNB SubSystem (the HNB and the HNB Gateway) as an optional capability of HNB that, for example, allows an operator to offload CS traffic to the IMS.

From Rel-8, another important feature called Closed Subscriber Group (CSG) is introduced. A HNB may provide restricted access to only UEs belonging to a Closed Subscriber Group (CSG). One or more of such cells, known as CSG cells, are identified by a unique numeric identifier called CSG Identity. The related description can be found in TS25.367.

The following is a summary.

- TS22.220 : Service requirements for HNB and HeNB
- TR23.830 : Architecture aspects of HNB and HeNB
- TR23.832 : IMS aspects of architecture for HNB
- TS25.467 : UTRAN architecture for 3G HNB
- TS25.367 : Mobility procedures for Home Node B (HNB)
- TS25.469 : UTRAN Iuh interface HNB Application Part (HNBAP) signalling
- TR25.820 : 3G Home Node B (HNB) study item
- TR25.967 : FDD Home Node B (HNB) RF Requirements
- TS32.581 : HNB OAM&P, Concepts and requirements for Type 1 interface HNB to HNB Management System
- TS32.582 : HNB OAM&P, Information model for Type 1 interface HNB to HNB Management System
- TS32.583 : HNB OAM&P, Procedure flows for Type 1 interface HNB to HNB Management System
- TR32.821 : Study of Self-Organizing Networks (SON) related OAM Interfaces for Home HNB
- TR33.820 : Security of HNB/HeNB
- TS25468 : UTRAN Iuh Interface RANAP User Adaptation (RUA) signalling

According to the Femto Forum [2], the new standard has adopted the Broadband Forum TR-069 management protocol which has been extended to incorporate a new data model for Femtocells developed collaboratively by Femto Forum and Broadband Forum members and published by the Broadband Forum as Technical Report 196 (TR-196). TR-069 is already widely used in fixed broadband networks and in set-top boxes, and will allow mobile operators to simplify deployment and enable automated remote provisioning, diagnostics checking and software updates.

Work has already been done to incorporate further Femtocell technology in the 3GPP Release 9 standard, which will address LTE Femtocells and also support more advanced functionality for 3G Femtocells. Femtocell standards are also being developed for additional air interface technologies by other industry bodies.

1.5.3 Business Models

Femto Forum published in February 2009 the research [2] conducted by a US-based wireless telecommunications consultancy – Signals Research Group (SRG). The company used data that had been provided by a group of mobile operators and vendors, and they found that femtocells can generate attractive returns for operators by significantly increasing the expected lifetime value of a subscriber across a range of user scenarios. Subscribers in turn could realize cost savings and other benefits from femtocells. Offloading voice and data traffic from the macro network will become a more important factor in the business case as mobile data traffic continues to grow rapidly around the globe. Value-added services that are made possible by the presence of the femtocell in the home will strengthen the business case. Operators can use femtocells to provide deep in-building mobile broadband coverage in a very cost effective manner:

- to provide 2.5 Mbps broadband at home, it costs 320 dollars using the femto solution.
- to provide 2.5 Mbps broadband at home, it costs 900 dollars using the macro solution.

The study implicitly highlights many benefits of femtocells to the consumer:

- fewer dropped calls,
- better voice quality,
- higher data rates,
- potentially attractive tariffs or voice and data bundles.

However, the business case with femtocells is not simple because it often requires two different entities: the mobile network operators that provide the femtocells and the Internet provider who delivers the broadband backhaul connection.

1.5.4 Applications

The first application of femtocells is phone calls, which will be free or at a low price. Moreover, thanks to femtocells, some new applications can be proposed by the operators. Because the indoor coverage will be maximum, some new data intensive services will be experienced by the customers. As described in [6], the new services can be divided into two kinds of application: Femtozone Services and Connected Home Services.

Femtozone Service

These services are web/voice services that are activated when the phone comes to the range of the FAP. Some examples include:

- Receive an SMS to indicate when someone enters or leaves the home.
- A virtual number to reach all the people currently in the home.

Connected Home Services

In these kinds of service, the phone accesses the LAN via the femtocell, in order to control a range of networked services, for example:

- upload some musics from the mobile phone to the PC,
- use the mobile phone to control other devices (TV, HiFi).

Table 1.1 UMA/FAP air interfaces

	At home	Outdoor
Unlicensed Mobile Access (UMA)	WiFi	GSM/CDMA
FAP	GSM/CDMA	GSM/CDMA

The development of new applications will be very important in the future, if femtocells are to succeed and attract more customers.

1.5.5 Femtocells vs Unlicensed Mobile Access (UMA)

Unlicensed Mobile Access extends voice and data applications over IP access networks. The most common approach is the successful dual-mode handset, where the customer can roam and handover between the GSM/UMTS and their WiFi network (see Table 1.1). UMA and femtocells are both efficient solutions to providing Fixed Mobile Convergence. However it is important to note that femtocells have the following advantages:

- Femtocells do not require the use of special dual mode handsets. Every mobile phone can use femtocells.
- Femtocells save the battery compared with dual mode handsets where GSM/UMTS and WiFi interfaces have to coexist, largely increasing the battery consumption.
- With femtocells, thanks to the handover with the outdoor network, users can smoothly use their mobile when they enter or leave their house.
- As detailed before, femtocells are a more interesting solution for operators, because they increase at a low cost their indoor radio coverage, allowing them to provide new services and get more revenues.

In reality, UMA and femtocells are not competitors, but are more complementary, which is why operators are interested by both approaches, which will have to coexist.

1.6 The Structure of the Book

This book is organized as follows.

In Chapter 2 an overview of the different indoor coverage techniques is given. The evolution from macrocell to femtocells is presented and the different methods are compared. Advantages and drawbacks of the different techniques, like Distributed Antenna System (DAS), repeaters, and picocells are given and the main challenges related to femtocells are introduced.

Different possible approaches have been proposed to include a femtocell network within a mobile operator network. Therefore the different femtocell architectures will be presented in Chapter 3, where the different options for ensuring the connectivity of the femtocell to the core network will be described.

The different possible air interfaces for femtocells are presented in Chapter 4: first Global System for Mobile communication (GSM), which is still the main wireless interface in many locations, will be presented. The first deployments will be more focused on Universal Mobile Telecommunication System (UMTS) and High Speed Downlink Packet Access (HSDPA), and in the future, Orthogonal Frequency Division Multiplexing (OFDM) based femtocells like Wireless Interoperability for Microwave Access (WiMAX) and LTE are expected to be produced.

In Chapter 5, the problem of how to simulate femtocells is detailed. The development of a femtocell simulation tool is presented, from radio coverage level, to system level. This chapter considers the different technologies (Code Division Multiple Access (CDMA) or Orthogonal Frequency Division Multiple Access (OFDMA)). This tool will be used in the next chapters to provide some interesting results.

Chapter 6 analyses the important problem of interference due to femtocells. The two main kinds of interference are described and discussed: interference between femtocell and macrocell called *cross-layer*, and interference between neighbouring femtocells called *co-layer*. Some interference cancellation and avoidance techniques will be presented.

In femtocell networks, it is important to manage correctly the mobility of the users, so that they can hand over from the macro to the femto network. The high density of femtocells will require the operators carefully to take into consideration the problem of mobility, such as, for example, how to manage the neighbouring cell lists. That is why the topic of mobility management will be investigated in Chapter 7.

To reduce the negative impact of the femtocells on the network due to interference, the femtocells should be able to configure their main parameters. It is important that the configuration is done by the femtocell device itself, and so in Chapter 8 self-organization is presented.

In Chapter 9, some other important challenges that have to be solved will be presented, such as the solutions to ensuring the timing accuracy, security and the location of the femtocell devices. Finally some other more commercial issues related to access methods, applications and health will be investigated.

In this book, the term FAP is used to denote the femtocell device, while the term *femtocell* makes reference to the coverage area of an FAP. Similarly, Macrocell Base Station (MBS) refers to the hardware that creates a macrocell, while *macrocell* denotes the area of coverage created by the MBS.

References

[1] J. Cullen, 'Radioframe presentation,' in *Femtocell Europe 2008*, London, UK, June 2008.

[2] 'Femtoforum,' http://www.femtoforum.org.

[3] S. Carlaw, 'Ipr and the potential effect on femtocell markets,' in *FemtoCells Europe*. ABIresearch, 2008.

[4] 3GPP, '3G Home NodeB Study Item Technical Report,' 3rd Generation Partnership Project – Technical Specification Group Radio Access Networks, Valbonne (France), Technical Report 8.2.0, Sep. 2008.

[5] M. Latham, 'Consumer attitudes to femtocell enabled in-home services – insights from a European survey,' in *Femtocells Europe 2008*, London, UK, June 2008.

[6] ip.access, 'Oyster 3g: The access point,' http://www.ipaccess.com/femtocells/oyster3G.php, 2007.

2

Indoor Coverage Techniques

Guillaume de la Roche and Jie Zhang

2.1 Improvement of Indoor Coverage

As explained in the previous chapter, improving indoor radio coverage has recently become more and more important. This is the reason why different solutions have been proposed. In the past, indoor coverage was only provided by an outdoor antenna, as presented in Figure 2.1. In this approach, the only way to increase indoor coverage was to increase the power or to add more cells. This led to the creation of more small outdoor cells (microcells) providing more capacity for the network. Unfortunately, this approach is expensive for operators because they have to install more sites, which dramatically increases the maintenance costs. Moreover, this solution also creates more problems concerning interference, as more cells will overlap each other. Finally, improving indoor radio coverage by adding outdoor cells is not optimal because it does not directly optimize indoor coverage and thus the efficiency of such a method is not optimal.

To overcome the limitation of outdoor cells, different approaches have been proposed in order to increase the indoor signal directly. The solution consists of adding antennas directly inside the buildings. First, Distributed Antenna System (DAS) has been developed. Different antennas are distributed inside the building to create a homogeneous coverage. These antennas are connected to a common source (the Base Station).

Another solution, initially to cover tunnel and also long distance corridors, called radiating cable, has been proposed. The radiating cable will replace the antenna to make the signal propagate along it.

Finally, the more recent solution is the installation of small indoor base stations like picocells or femtocells. All these techniques will be detailed in the following sections.

2.2 Outdoor Cells

In the past, with the deployment of GSM systems, the network coverage has always been provided by base stations installed in rural areas and with cell radius of a few kilometres,

Femtocells: Technologies and Deployment Jie Zhang and Guillaume de la Roche

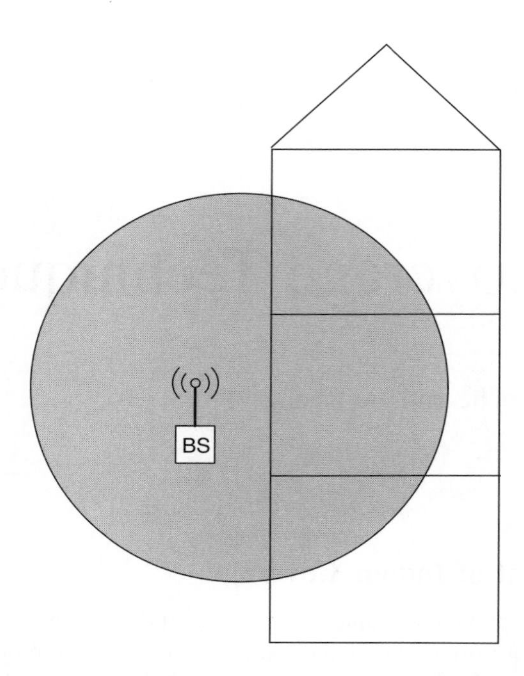

Figure 2.1 Outdoor cell radio coverage of a three-floor building

or in urban areas with a cell radius of a few hundred metres. The main part of the network provides voice service, thus a very high data rate was not really useful. That is why most of the indoor coverage was only provided by outdoor emitters. This fact still applies today, because specific indoor solutions (that will be described in the next section) only started to be deployed over the last ten years.

In Figure 2.1, a schematic view represents how indoor coverage is provided. Of course, the walls of the buildings, depending on the properties of its material, will attenuate more or less the signal. This is why operators had to find some solutions for increasing indoor coverage, which include adding new base stations, or modifying the existing ones. Due to health regulations, a solution that consists of increasing the radiating power of the base stations was not possible. Two main kinds of environment can usually be found and each requires different approaches: rural areas where macrocells cover large distances, and urban areas where the high population density requires a higher number of small cells.

2.2.1 In Rural Areas

In rural areas, high base stations are installed to cover longer distances. These are called macrocells. The power of the macrocell base stations is high, in order to maximize the covered distance. In this kind of environment, due to the low number of customers and the high price of powerful base stations, the approach for operators has always been to try to ensure minimal coverage so that voice calls can be performed outside. The deployment of such a network was very often done by combining wireless network planning tools and real measurements.

The disadvantage of a macrocell network is that the network is deployed by only taking into account the outdoor coverage, which is why in many rural areas it is still necessary to go outside of the building to be able to make a call. Then, in order to optimize indoor coverage, the only approach for operators was to add more macrocells. Since such equipment is expensive, not only in terms of buying cost, but also in terms of maintenance, the operators always had to deal with an economic compromise: add macrocells in areas where the number of customers is big enough to make the installation profitable, and leave the other areas either without or with a minimal coverage.

2.2.2 In Urban Areas

In urban areas, the problem is similar to that with rural environments, except that the high density of possible customers makes operators seek some efficient solutions to increase the coverage. The common approach to increasing the capacity of a cellular network is to add more cells, each of them covering smaller areas. Hence, in an urban environment, the operators had to install more base stations with lower power. They also had to face the same dilemma as in a rural environment where the coverage inside buildings is ensured only by improving the outdoor signal quality.

In this kind of environment, where multiple reflections on walls and diffractions on roofs occurs, it was very often not sufficient to use free space or empirical radio propagation models. Deterministic radio propagation models, such as those based on a ray tracing approach, can be used to compute efficiently the diffractions and reflections of the signal to compute accurate coverage maps. Even if this kind of tool helped the operators, they mainly helped to optimize the street level coverage but not the coverage inside the building. This is due to the fact that operators can have access to the building data of cities thanks to aerial pictures like Google maps for example, but it is quite impossible to have the whole data related to the content of the buildings of the city. Moreover the complexity of such a tool would be drastically high.

Outdoor Microcells

With the recent development of UMTS or HSPA, new data services appear, requiring higher coverage. To increase the capacity, operators started to install smaller outdoor base stations called microcells. These are deployed in specific areas in which extra capacity is known to be needed, for example near a train station or in a city centre. They are also often temporarily deployed during special occasions like sporting events, for example. Adding microcells in the urban environment allows the operator to subdivide the cells, leading to an optimization of the use of the spectrum and ensuring a better capacity.

All the previously described methods (macrocells, microcells) have been used for a long time, but they only indirectly optimize the indoor coverage by increasing the outdoor coverage. The need to optimize directly the indoor coverage led to the development of more specific technologies, that will be described in the following sections.

2.3 Repeaters

Due to the attenuation of the walls of the buildings, the idea of using a component that amplifies the outdoor signal and sends it inside a building has been proposed in order to increase the radio coverage. As represented in Figure 2.2, a repeater can be used so that it retransmits the outdoor signal inside the building.

Two kinds of repeaters have been proposed:

- Passive repeaters, which amplify the signal in a certain frequency band, regardless of its nature.
- Active repeaters, which are capable of modifying the signal before retransmitting.

2.3.1 Indoor Passive Repeaters

Indoor repeaters work at a certain frequency band. They are usually made of three components: an external antenna to receive the outdoor low level signal, an amplifier, and another antenna to retransmit the amplified signal inside the building.

External Antenna

Usually, the external antenna is directional and oriented in the direction of the closest outdoor antenna sector. The use of high gain antennas allows provision of better signal

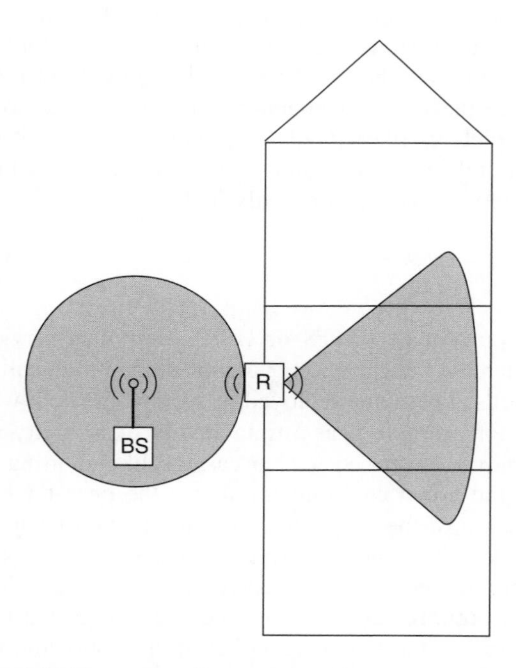

Figure 2.2 Repeater radio coverage of a three-floor building

quality than with small gain antennas. In some cases where smaller repeaters are used, the gain is not so high, which is why it is important to compensate by ensuring that the directional antenna is perfectly oriented in the correct direction. In-building planning engineers use signal strength monitors for this task.

Amplifier

If P_0 is the power of the signal received at the external antenna, and P the power of the signal emitted by the indoor antenna, the power gain G (in dB) is computed with the following equation:

$$G = 10 \times \log\left(\frac{P}{P_0}\right) \tag{2.1}$$

Usually the gain value of the amplifiers can vary from 30 to 50 dB.

Indoor Antenna

The indoor antenna rebroadcasts the amplified signal. Depending on the installation omni-directional or directionnal antennas can be used.

2.3.2 Active Repeaters

Active repeaters are more advanced repeaters, because they are capable of decoding the signal and reshaping it before retransmitting it. With such a system the noise can be removed. Moreover, more advanced functionalities are proposed, like, for example, receiving on one channel and retransmitting on another. That is why active repeaters, if correctly installed, offer better performance than passive repeaters. Not only do they extend the coverage, but also they increase the data rate by decreasing the errors. Because active systems are expensive compared with passive, an idea proposed in [1] is to use active repeaters only where it is necessary (very confined areas) and combine them with cheap passive repeaters in the rest of the environment.

2.3.3 Development of In-Building Repeaters

In most cases, indoor repeaters are used to amplify the outdoor signal and retransmit it inside the building, and thus extend the outdoor coverage inside the building. They can also be used indoors only, to transmit the signal from one part of the building to another, or to make the network cross large concrete walls. In [2], many repeaters are used to increase the indoor coverage. These multiple repeaters are combined with phase shifters. This combination is to ensure that signals in the desired areas are combined constructively by tuning the phase. This is similar to the case of forming an antenna array, and as a result, the radiation pattern is changed. By tuning correctly the phases of the different repeaters, it is possible to make the best server area fit as much as possible with the areas to cover. This approach is also interesting because more repeaters create more multi-path and more uncorrelated signals, which helps in producing more diversity, and improves the performance of Multiple Input Multiple Output (MIMO) systems for example.

Interference Cancellation

In a repeater system, both antennas have to receive and transmit signal. It is very important that these antennas are sufficiently isolated to avoid oscillating effects. For example, in Figure 2.3, if the isolation between the antennas were not sufficient, it would be possible for the external antenna to receive not only the outdoor signal, but also a part of the indoor signal from the indoor antenna. This received signal would then be amplified, thereby adding noise to the initial signal, and thus greatly degrading the performance of the system. To avoid oscillation it is important to maintain a minimum isolation value. Typically a minimum of 15 dB is recommended. This is why it is very important, when installing indoor repeaters, to check that the attenuation due to the obstacle between the two antennas (usually the separating wall) is sufficient. Moreover, it is also important to choose correctly adequate antennas to avoid an overlap (but choosing directional antennas, for example). Finally, increasing the distance between the antennas will also reduce interference. Some more advanced repeaters as in [3], equipped with interference cancellation systems, have been proposed. In this approach, interpolated filters are used as input filters for frequency band selection, and an output filter for spectrum mask control. The system can estimate the feedback system and filter it.

Gain Control

As explained before, with repeaters, the noise is amplified, causing high degradations in performance. The choice of the gain of the amplifier is not an easy task, because there needs to be a compromise between a high gain offering theoretically a better coverage but with more noise, and a low gain reducing the amplification of the noise, but also the size of the covered area. Repeaters with Automatic Gain Control (AGC) have been developed [4]. AGC repeaters adapt automatically, the gain depending on the capacity, thus allowing the size of the service area to remain constant.

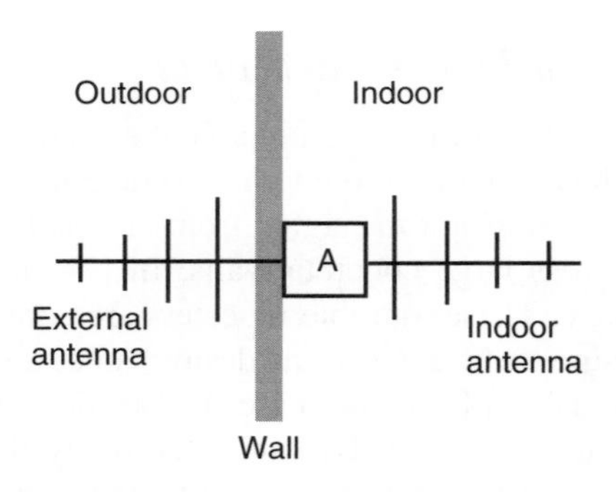

Figure 2.3 Through-wall repeater

2.3.4 Conclusion

Passive repeaters are cheaper than active repeaters, but their performance is lower because of the noise that is also amplified. Active repeaters can reshape the signal and improve the performance of the systems. Moreover, the use of multiple repeaters can help in improving the performance of the system. However, this idea has not been investigated much, because of the success of the development of the DAS (Distributed Antenna Systems), which also resides in the idea of using multiple antennas inside the building. This technology will be described in the next section.

2.4 Distributed Antenna Systems (DAS)

The idea of DAS is to split the transmitted power between separated antenna elements. For example, these antennas can be located on different floors of a building to provide homogeneous coverage. The idea of such a system can be found at the beginning of the 1980s [5] and the first paper about indoor distributed antenna systems was proposed by Saleh in 1987 [6]. He proposed replacing an antenna radiating at a high power, with multiple small antennas using low power to cover the same area. DAS will much improve the efficiency of the network, if both the overlap between the coverage areas of the different antennas is reduced, and the coverage areas of the antennas fit as much as possible to the shape of the building. The task of the Indoor Planning Engineer is to try to make coverage as homogeneous as possible. Different DAS systems have been proposed on the market.

Passive DAS use passive elements to make the output signal of the base station go to different antennas. Later, more advanced DAS systems have been developed, based on active components making the performances of the system better, and making the installation easier [7].

2.4.1 Passive Distributed Antenna Systems

Passive Components

In passive systems, different components are used to split the signal power between the antennas. These components are passive, which means they do not need external power supply.

- *Coaxial cables:* Coaxial cables are used to split the signal and form the link between the different elements of the DAS. Their main disadvantage is high signal loss depending on the distance. The engineers, in order to ensure they have the right radiated power at each antenna, must take into account the length of the cables to compute the global loss. For example, with a 0.5 inch coaxial cable on 1800 MHz, the loss is about 0.1 dB per meter.
- *Splitters:* This component equally splits the input signal into N output signals. It is used as an interconnection to split the signal between the different antennas.
- *Taps:* Taps are similar to splitters, but are able to divide the input signal into two output signals with different ratios. They are used to adjust the power to allocate to different floors, for example.

- *Attenuators:* These simply attenuate the signal with the value of the attenuator. They are used to bring the signal to a lower level.
- *Filters:* These are used to separate frequency bands, for example, a triplexer can separate the incoming signal into three output signals corresponding to the frequencies 900 MHz, 1800 MHz and 2100 MHz.
- *Other components:* Some other components can also be used to design indoor networks, for example, terminators are used to end a line, circulators to protect a port against reverse reflections due to a disconnected cable in the system, and couplers are used to combine signals from different incoming sources.

Deployment of Passive DAS

With passive DAS, the signal is distributed between the antennas using the previously described passive elements. A typical example of installation is illustrated in Figure 2.4. The number of antennas to use and the output power are important parameters that must be planned carefully and will be very dependent on the kind of environment. As shown in [8], both in single cell and multi-cell environments, DAS using an efficient power control increase capacity and reduce interference.

It is interesting to compare the performance of DAS with a system using a single antenna to evaluate the interest of such an approach, see Figure 2.5. In [9] some interesting relationships are discussed. The Path Loss (*PL*) at point *r* due to one antenna represents

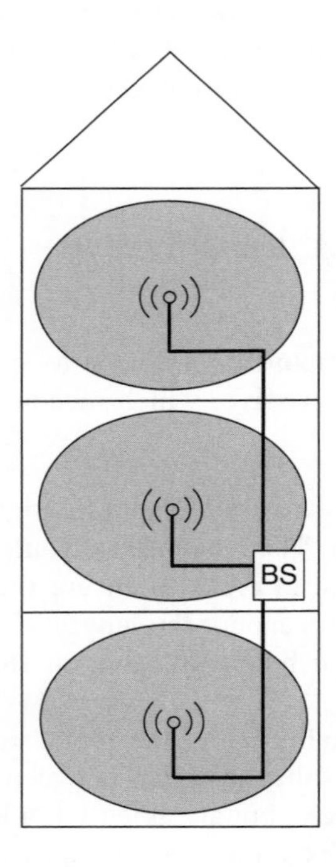

Figure 2.4 DAS radio coverage of a three-floor building

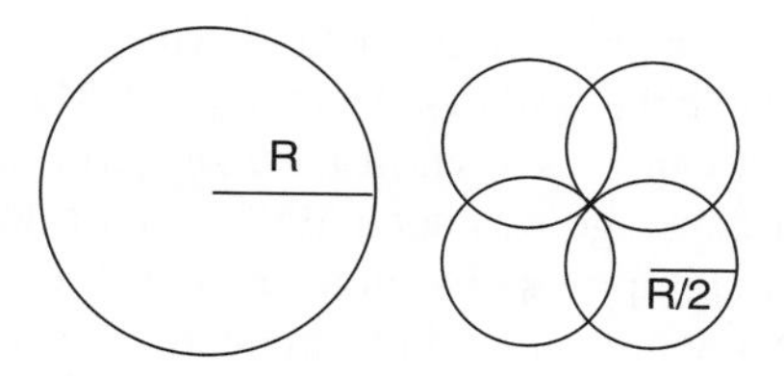

Figure 2.5 Areas covered by a single antenna, and four distributed antennas

the signal loss from distance dr to reference distance d. It can also be written as:

$$PL = \frac{P_{dr}}{P_d} \tag{2.2}$$

If the loss is computed from $d = 1m$ the PL can also be modelled as:

$$PL = Cd^{\alpha} \tag{2.3}$$

with d the distance from the source, C a constant, and α the Path Loss exponent. Typically, with the 2 slopes Path Loss model, α is between 1 and 3 near the antenna, and between 3 and 7 in far distance.

If the coverage of an antenna is assumed to be a circle, the area A of such a coverage is:

$$A = \pi d_c^2 \tag{2.4}$$

with d_c the distance to the cell boundary. From the previous equations it is possible to write:

$$PL = C\left(\frac{A}{\pi}\right)^{\frac{\alpha}{2}} \tag{2.5}$$

Maximizing Coverage

For a given radiated power, an N antenna system, compared to a single one, will produce a new coverage area A_N so that:

$$A_N = N^{1-\frac{2}{\alpha}} A \tag{2.6}$$

For example, with four antennas, and supposing $\alpha = 5$, the improvement factor is 2.29.

Minimizing the Radiated Power

For a given coverage area, the radiated power P_N compared with the one of a single antenna P, is reduced so that:

$$P_N = N^{1-\frac{\alpha}{2}} P \tag{2.7}$$

With the same parameters as previously, the power reduction is 9 dB.

DAS can be connected to repeaters in order to distribute inside the building the signal coming from outside. In [10] the authors deployed a UMTS system made of a DAS connected to the repeater. It can be shown in their scenario that with such a system the improvement of the indoor coverage was about 35% gain of downlink capacity. With this deployment the indoor SIR is improved by on average 3.4 dB, leading also to an increase of capacity in HSDPA. This approach lead to a reduction of the transmit power, thus the neighbour cells suffer less interference.

DAS can also be used to cover long distance environments like a tunnel, for example. In [7], a DAS is successfully used to ensure indoor coverage inside a tunnel.

The DAS as a Space Diversity Solution

In MIMO systems, an efficient solution for increasing the performance of a system is by combining different antennas so that, by allocating the correct power to each antenna, a global antenna pattern with a particular shape is created. The shape of the resulting antenna pattern can be computed, so that different beams are oriented in the direction of the different users. This process, also called beam forming produces space diversity, so that, for example, two users allocated to two different beams could use the same channel or code without interfering. This access method is also called SDMA (Spatial Division Multiple Access).

In a similar manner, with distributed antenna systems, it is also possible to exploit the space diversity by ensuring that the best coverage areas of the different antennas overlap as little as possible [11]. The number of antennas, and where to deploy them is the main task of in-building engineers. For example, a common approach in multi-floored building is the use of one antenna per floor, or to equally split the number of antennas between the floors. For example, a GSM measurement campaign has been performed in [12] using an office environment and eight antennas. To cover this building made of two floors, four antennas per floor were used to give the homogeneous coverage. This results in a big increase in the capacity of the network due to better channel reuse between the different floors. In [13], a CDMA DAS in an office environment is also presented. It results, on average, in a decrease in power of more than 10 dB for the distributed antenna system (three nodes in this case). In general, if the power is optimally split between the antennas, and if their position is well calculated it is possible to allocate the antennas to produce the space diversity, and thus increase the capacity of the system.

2.4.2 *Drawback of Passive Systems*

Passive elements were successfully used for more than 15 years for GSM, but the disadvantage of passive systems is that they are made of passive elements. Unfortunately, when using higher frequencies, the degradation of the signal can greatly affect the quality of the transmission. When designing indoor solutions for 3G, this degradation of the signal in the components is a main issue.

Another point is that, when the size of the building is huge, installing coaxial cables is not always possible, because of the signal loss along such a distance. Very often the only solution is to increase the emitted power to ensure an adequate quality of signal.

Finally, if passive components are very cheap compared with those of an active system, the installation of coaxial cable is expensive. These cables are heavy and rigid, and installing them is not an easy and cheap task. It is a reason why, with passive systems, due to installation problems, the possibilities of positions of antennas are often limited. All these drawbacks lead to the development of active systems, easier to install and to control, and trying to compensate for the attenuation of the coaxial cables.

2.4.3 *Active Distributed Antenna Systems*

Active Components

Unlike passive systems which do not require the use of electronic components, active DAS use different active elements as described below.

- The master unit: The master unit (MU) can be connected to the base station or the repeater. It distributes the signal via the optical fibre to the different expansion units. The master unit is the intelligent part of the distributed antenna system that controls all the signals to deliver and adjust the signal levels thanks to internal amplifiers and converters.
- Remote unit: The RU is installed near the antenna to minimize the losses and is connected to the antenna. The RU converts the signal from the RU into downlink radio signal, and converts the uplink radio signal into signal to the EU.
- Cable: First active systems where deployed using standard connections like coaxial. In this case the problem of the losses between cables is still important. However, the installation is made easier because the active remote unit can compensate for the loss depending on the distance. Later, with the development of cheap optical fibre, some systems using such a technology have been proposed in order to transport the signal over longer distances. These methods will be presented in the following paragraph.

Deployment of Active DAS

The passive components are not needed any more, which is why the installation of an active system is easier. Indeed, the losses are automatically compensated by the master unit and the remote unit: there is no need to choose splitters and attenuators to adjust the losses, and no need to measure accurately the length of the cables.

Basic Active DAS for Small/Medium Size Buildings

In the case of shorter distances, the use of standard cables is possible. The RU is installed near the antenna, so that all the losses in the cable are compensated for and the only loss will be those in the coaxial cable between the RU and the antenna. This system is easier to install than a passive system because there is no need to install components like splitter taps or attenuators, or take into account the length of the cables, to adjust the losses in the cable. Such active systems can cover distances from a few hundred metres.

Active Fibre DAS (Radio over Fibre)

Active fibre DAS is the most efficient in term of performance. Optical fibres are used to make the link between the MU and the RU. They can cover very long distances (up to 6 km) and support multiple radio services. With such a system the RU directly converts the optical signal into radio signal and vice versa. The other advantage is that optical fibre is very cheap and easy to install. Radio over fibre is now the most common technique used for indoor radio coverage [14, 15]. As detailed in [16], radio over fibre is today the optimal solution to extending indoor coverage, because it provides scalability, flexibility, easy expandability, and also because the signal degradation is very low compared with DAS using standard connections.

2.4.4 Choice between Passive and Basic Systems

Passive vs Basic Active DAS vs Active Fibre DAS

In Table 2.1 a comparison between passive and active system is given.

Hybrid DAS

Some other solutions combining passive DAS and active DAS have been installed. The idea is to connect the remote units via fibre optic, but to use passive coaxial cabling to link these remote units to the antennas. The combined method has the advantages of covering a long distance thanks to the fibre optic connection, and a cheaper price due to the passive components. Hybrid DAS can also combine different systems with different frequency bands. Due to the simplicity of radio over fibre installation, there are many possible solutions, such as combining the distributed antenna system with repeaters for example.

Best Solution

In terms of buying costs, passive systems are cheaper but suffer from high installation price due to the coaxial cabling. Active systems offer better performance and easier installation but are more expensive. With passive systems, no electronic systems have to be installed or power supplied. Hybrid DAS is sometimes a good compromise but

Table 2.1 Comparison between the different DAS technologies

	Passive DAS	Basic active DAS	Active fibre DAS
Covered distance	Up to 400 m	Up to 400 m	Up to 6 km
Equipment price	Cheap	Standard	Standard
Installation	Difficult	Easy	Very easy
Multi-standard	No	No	Yes
Input (base station or repeater)	High power	Low power	Low power

still requires installation of coaxial cables. Active systems are easier to manage because automatic diagnostics and alarms are integrated into the remote units, making the problems of system failures easier to solve.

2.5 Radiating or Leaky Cable

The radiating cable, also called the leaky feeder, is a metallic wire that acts as a long antenna. The electromagnetic energy can be received or transmitted all along the cable, which is why it is well adapted to long narrow environments like corridors, elevators or tunnels. For example, in London, a radiating cable system is used in the underground for their internal communication network. In general, as represented in Figure 2.6, the radiating cable is directly connected to the base station.

2.5.1 Principle of Operation

A radiating cable is similar to a standard coaxial cable, however some tuned slots are positioned on the surface of the outer conductor. A schematic representation of the transversal cut of such a cable is represented in Figure 2.7. The slots are tuned to the specific RF wavelength of operation or tuned to a specific radio frequency band. They will leak a part of the electromagnetic energy propagating in the cable in the form of electromagnetic waves. The antenna pattern is quasi-omni-directional in the transversal plan of the cable. To reach a better efficiency, it is advised to leave a space between the cable and the walls. Moreover, metallic fixings and parts are not recommended because they affect the antenna pattern.

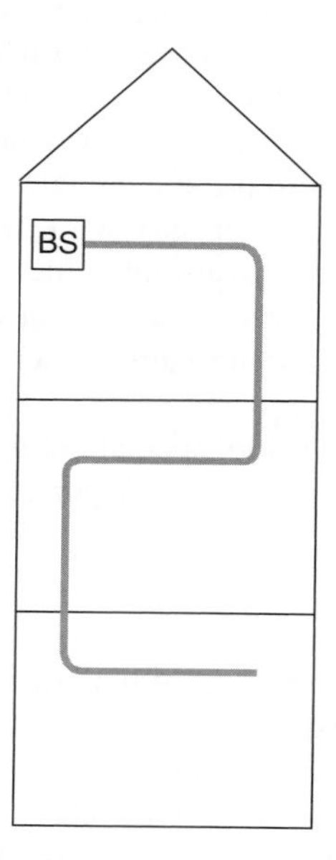

Figure 2.6 Radiating cable radio coverage of a three-floor building

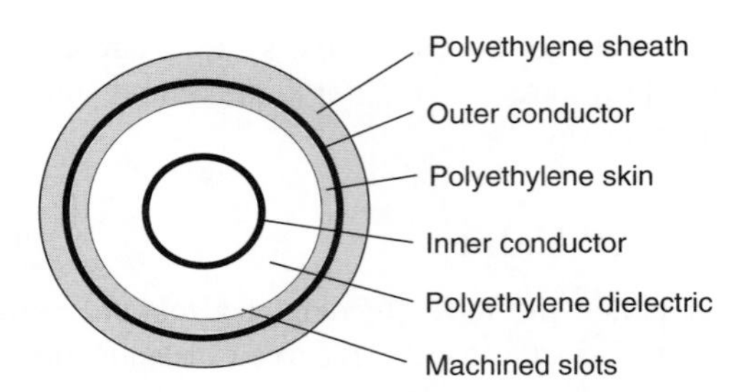

Figure 2.7 Transversal cut of a leaky cable

If the cable is uniformly homogeneous, the ratio between the radiated energy and the energy propagating in the cable is constant all along the cable. The higher the losses in the cable, the more the ability to radiate energy is important. Some recent solutions have been proposed involving an adjustable coupling loss, allowing the user to adapt the radiated energy in complex environments [17].

This is why such a system has a limited range, especially in the high frequency range, where the losses are more important.

2.5.2 Deployment

As explained before, this technology is ideal for covering long narrow spaces. The main advantage of the radiating cable is that the energy is well distributed. For example, in a corridor, it provides homogeneous coverage all along the cable compared with the use of numerous base stations along the corridor, where the energy is distributed around the base stations. With a radiating cable, a single base station may be able to provide coverage over a large area, reducing the cost of system implementation.

A disadvantage is the difficult and expensive installation. The installation is time consuming and it is not always easy to find the available space to install it. The cable must be aligned perfectly so that the slots can leak with minimum loss. Moreover the cable must not be installed directly against a wall but some space must be left, thanks to some special adaptors. Finally, especially inside tunnels (for example with trains making dust), the dirt degrades the performance of the cable. Therefore the cable must be regularly cleaned.

Due to the attenuation inside the cable, special configurations like cascaded BDAs and T-feed have been proposed in order to cover longer distances.

Cascaded BDAs

This solution to overcoming the problem of signal attenuation uses Bi-Directional Amplifiers (BDAs) when the maximal distance of cable to obtain a minimum quality is reached. For long distances, BDAs are installed at certain intervals (see Figure 2.8), and their gains are configured so that the signal level is maintained at a certain level. In practice no more than three or four BDAs can be used because of the noise levels that are also amplified in the BDAs.

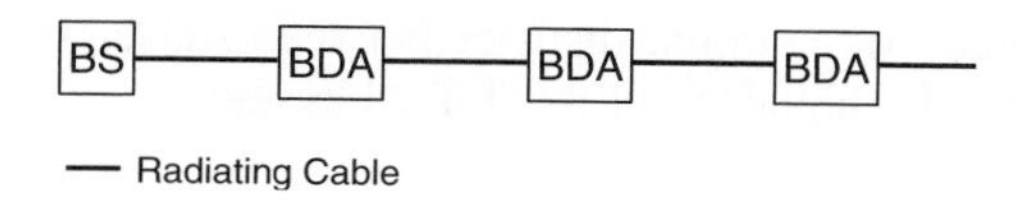

Figure 2.8 Radiating cable fed by cascaded BDAs

T-Feed

The T-feed uses an optical converter that converts the BS signal and distributes it via optical fibre, as represented in Figure 2.9. Each BDA has an optical interface to convert the optical signal and send it in both directions of the cable. This system can cover longer distances and has a better control of noise, because the different BDAs do not feed each other as in the cascaded BDAs configuration.

Comparison

As said before, the T-feed system is preferred because it reduces noise so can be used to cover longer distances. Moreover, because with T-feed the signal is sent in both directions of the BDAs, the resulting signal reaches higher levels and is more homogeneous. To illustrate this idea, in Figure 2.10 the attenuation along the cable with the two previous

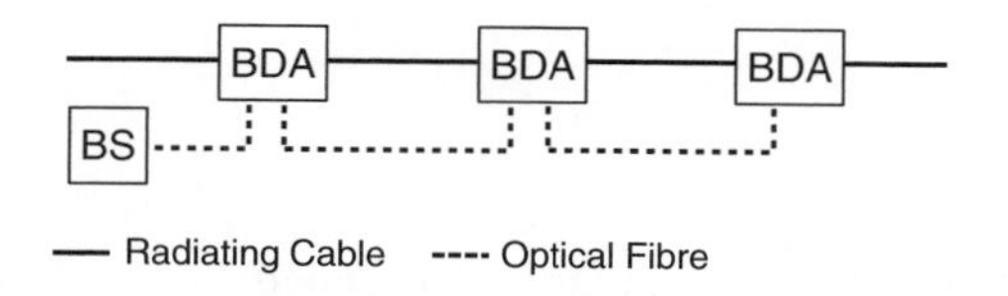

Figure 2.9 T-fed radiating cable system

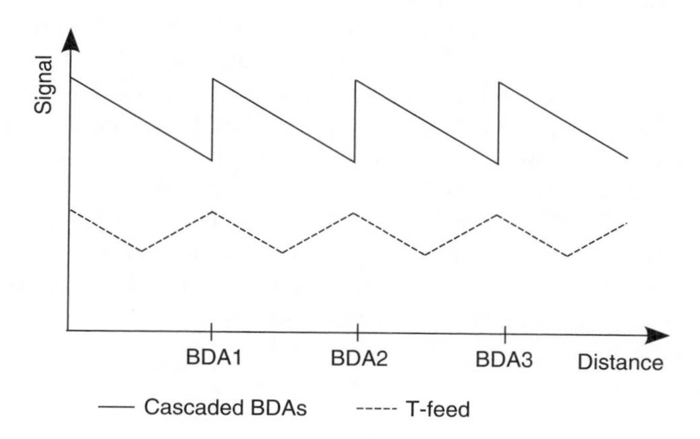

Figure 2.10 Attenuation along a cable for the cascaded BDAs and the T-feed radiating cable systems

approaches is represented: for a similar distance between BDAs, the signal level is higher so the signal quality will be higher with the T-feed system.

2.5.3 Alternative to Radiating Cables

Radiating cable is an efficient solution for covering long distances, which is why it is widely used. However, some alternative solutions are possible, such as using a DAS along the environment. In [18], it is explained that radiating cable solutions outperform DAS because they contain coverage much better, but in practice, since the installation of this is not always straightforward for certain buildings, as well as installation costs and interference issues arising from the interaction with surrounding objects, DAS is still the preferred option in most installations.

2.6 Indoor Base Stations

In the previously described methods for increasing radio coverage, the first ones to be developed aimed at increasing the indoor signal level by using the signal emitted by outdoor base stations (like macrocell, microcells and repeaters), and the later ones aimed at extending the coverage of the base stations inside a building by using radiating cable or distributed antenna systems. A new proposal for increasing both the coverage and the capacity, is to deploy small base stations directly inside the buildings as represented in Figure 2.11. Two main approaches have thus been proposed: the picocells and the femtocells.

2.6.1 Picocells

With the success of IEEE 208.11 standard, also called WiFi (Wireless Fidelity), which allows people to access to their broadband Internet connection via air interface, operators started to think about extending this concept to their mobile networks. In WiFi, the user connects to a device called an Access Point (or AP), which integrates an antenna to make the link between the user, and the connection to the Internet. A picocell is a small base station very similar to an access point. It is usually small (typically A4 paper size, and a few centimetres thick), and integrates an antenna that radiates a low power signal. Indeed, a picocell is a simplified base station, with low power and lower capacity than microcell or macrocell base stations. It connects to the Base Station Controller (BSC) of the operator. As with standard base stations, the BSC manages the transmission of data between the picocell and the network, and performs the hand-overs between the cells and the allocation of the resources to the different users. The picocell is connected to the core network via standard in-building wiring, fibre optic or Ethernet connection. Usually an omni-directionnal antenna is integrated into the picocell [19].

Advantages

The main advantage of picocells is that they are cheaper than standard base stations, and the installation cost is also lower. They effectively increase indoor coverage because they

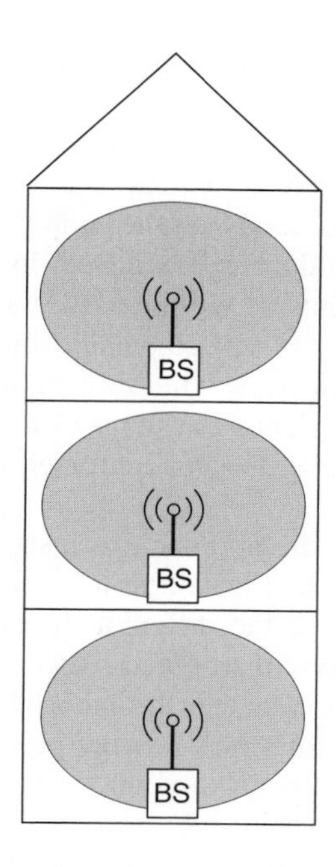

Figure 2.11 Indoor base station radio coverage of a three-floor building

are installed indoors. The coverage area of the cells is small compared with outdoor base stations, because the radiated power is lower, and also because of the numerous reflections and diffractions due to the walls and other obstacles inside the building. Thus, covering the inside of a building requires the use of many picocells compared with outdoor cells. This allows the operator to have more cells, and thus increase the capacity of the network inside the building. That is why picocells are deployed in indoor areas containing a high density of users.

Moreover, by installing a picocell inside a building, the operator can have more capacity in the outdoor network because the outdoor cells that are used to cover the building become available just to outdoor users. In many situations, it also gives an opportunity for the operator to reduce the radiated power of the outdoor cells used to cover the building. That is why picocells not only increase the capacity inside the building, but also increase the outdoor capacity, and reduce the outdoor interference because the overlap between outdoor cells can be reduced. Picocells, because they are small cells, can also be used in scenarios where localization is important [20]. Indeed, Indoor localization using outdoor cells and triangulation methods, is not accurate enough due to the reflections from the obstacles. However, with small cells it can be easily ascertained which building the user is in, and if there are many picocells a more accurate position of the user is known.

Deployment

Like DAS or repeaters, deployment of picocell networks needs special attention. Where the picocells should be located and what are the best parameters is the main challenge. For the deployment of picocells, an optimal solution would be that where the number of picocells is sufficient to ensure the coverage and capacity requirements, but also a solution where the number of picocells is not too high in order to avoid interference. Picocells are mainly deployed in large areas like commercial centres or airports, thus such an installation can be challenging. When installing picocells another important effect to be minimized is the interference with neighbouring outdoor cells. Indeed, in buildings close to outdoor cells, a part of the signal coming from outside will interfere with the indoor signal. Moreover, a part of the signal emitted by picocells will go outside and interfere with the outdoor cells. These phenomena mainly occur through windows, because the signal reflects less from glass than from stone or concrete. Thus the main challenge when installing picocells is taking into account the outdoor cells. Hence, combined indoor/outdoor network planning is an important issue.

In typical multifloor buildings, the approach to installing picocells is quite similar to that used when installing DAS. When installing a DAS, antennas connected to the same base station are distributed between the different floors, whereas in a picocell network installation, base stations are directly distributed between the floors. The main challenges with picocells are the interference between the picocells that are at the edge of outdoor cells [21], and managing the handoffs between the picocells [22]. In [23], some simulations for overlapping GSM picocells are described, and it is shown that to improve the performance a good solution is to take into account the picocells positions accurately. This can be done by entering in the Base Station Sub-system (BSS) database the relative positions between the picocells.

Applications

As detailed before, the advantage of picocells is that they efficiently increase both the indoor radio coverage and the capacity. They are a very good way of fulfilling two requirements:

- Filling the macro network coverage holes where the signal level is too low.
- Offloading traffic from the macro network in dense urban areas.

Some examples of applications include business environments and shopping centres. They are also useful in high-rise buildings. Indeed in many cities high-rise buildings are more and more common, but the macrocell signal strength tends to get weaker the higher you go. This is due to the fact that very often, in densely populated areas, operators have to add more outdoor cells. To increase the capacity, the size of these cells has to be reduced, which is why operators tend to minimize the antenna tilt. The immediate effect of this approach is poor coverage at the high levels of many buildings. Picocells placed at the high level of office environments can efficiently solve this issue.

Picocells are also useful in difficult buildings such as historical buildings (huge walls made of stone absorbing most of the signal coming from macrocells) or buildings made of complex shapes and materials (like metal structures or special glass windows for thermal efficiency).

Finally, picocells are also used in vehicular applications where the backhaul connection is ensured by satellite for example. Thus picocells can be used to provide phones for passengers in aircrafts or cruising ships.

Proposition of Small Picocells

With the success of picocells for multi-user indoor environments, in 2002 a group of engineers at Motorola started to develop the smallest UMTS base station. The main idea was to propose a WiFi-like solution but for mobile phone networks to deploy in the home environment. A few years later, the concept of a residential base station appeared, which aimed at a low power indoor solution for the home market. As represented in Figure 2.12, the idea of such small picocells is to cover only one house, which is why the low power should be adapted so that the cell size is between 20 and 30 metres maximum. This figure illustrates well how the cell sizes evolved over time, by reducing the size of the cells to fulfill the networks requirements, which are always more and more capacity demanding.

Such very small cells were called femtocells and will be presented below.

2.6.2 Femtocells

To extend the idea of picocells to home networks only, with an approach more similar to WiFi access points, femtocell base stations have been proposed. The femtocell is a simplified picocell directly installed by the customer in their home. It combines, in the same device, all the functionalities of a picocell and a BSC. Thus, instead of being connected to the operator's BSC (like a picocell), the femtocell is connected directly to the Internet as represented in Figure 2.13. With femtocells, all the communications go to the operator's network through the Internet, and there is no need for BSC/MSC infrastructure. Femtocells, because they typically cover a smaller area and have fewer users than picocells, and because they have to be cheap, are limited in output power and capacity (between 10 and 20 dBm, between four and six users). Within femtocell networks, outdoor users connect to the macrocells and when they enter their home they

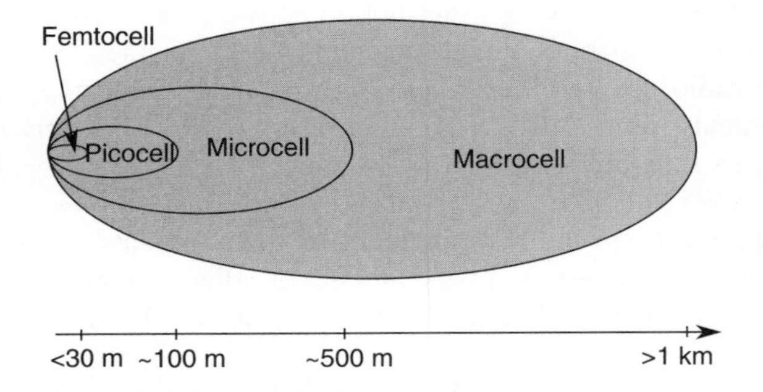

Figure 2.12 Comparison of cell sizes for different technologies

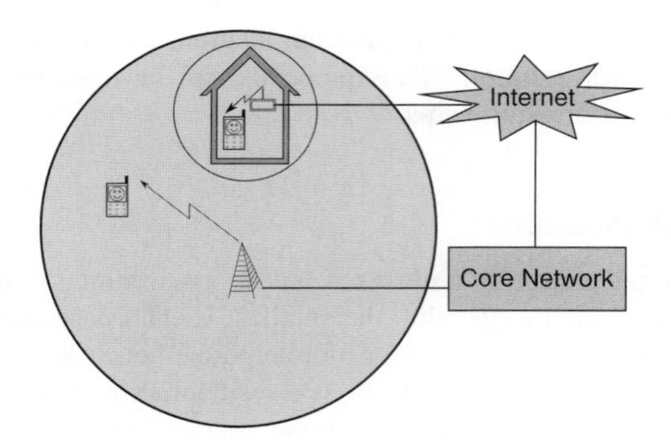

Figure 2.13 Typical femtocell and macrocell scenario

handoff and connect to their femtocell. This ensures a smooth communication for the user and a maximal coverage is obtained inside the home [24].

Advantages

Femtocells are installed by users inside their homes, so this solution will ensure good coverage for subscribers. This will not only increase the coverage but also the number of cells, and thus the capacity of the network. For the operators, femtocells are not only an efficient solution to increasing the indoor coverage, but also a cheap solution because femtocells are paid for by the customers. The alternative consisting in increasing the indoor coverage by adding more outdoor cells would be a lot more expensive for operators.

Deployment

Femtocells could be deployed in different kinds of scenario. The first market is in the home, and the requirement is access to a broadband Internet connection. In rural environments, without sufficient macrocell coverage, femtocells can be a suitable approach to allowing new customers to access the mobile network and have a maximal coverage inside their home. Thus, there is a high potential market in the USA where huge areas are still not well radio covered, but where high data rate Internet is largely deployed. In urban environments, femtocell is a good approach to covering dense buildings areas, where the customers, apartments are rarely in direct view of a macrocell, and where the losses due to the other buildings cause poor indoor coverage.

The second market for femtocells would be office environments, where the low capacity of femtocells would be sufficient to cover a number of offices. In this case, buildings would be covered by a set of femtocells that should be carefully planned, in a similar way as with picocells.

In the home environment, femtocells will probably be deployed in a private access mode. This means that only the subscriber of a femtocell will define the list of allowed users of a femtocell. This is mainly due to the fact that it is very unlikely that subscribers, who pay for a femtocell, will allow everybody free access to their femtocell. In the office environment, because of moving users, the femtocells would have to be deployed in public access. In this mode all the users will have the right to use the femtocells.

Finally, a future idea would be to deploy femtocells outdoors in small hot spot areas where customers would enjoy a maximal coverage. In this scenario the cheap price of femtocells, and their auto-configurability, would allow the creation of many very small cells with high coverage. An example could be to install femtocells in the bus stops of a city. Of course there are still many challenges to overcome before femtocells can be installed outside. The main challenge is that femtocells will have to configure themselves in order to avoid interfering with macrocells or other femtocells.

2.6.3 Differences between Picocells and Femtocells

Taking into account the previous discussions, a comparison between picocells and femtocells is presented in Table 2.2. To summarize, femtocells are small picocells where the properties have been simplified to reduce the cost and simplify the installation. Contrary to picocells, Femtocell Access Point (FAP) have the particular following characteristics.

Connection

FAP, unlike picocells, are connected to the operator's network by the broadband connection. The femtocell is a self-contained base station and is linked to the core network using IP. The femtocell is self-configurable and the interface between the femtocell and the core network has to be simple to avoid any action from the operator. Moreover, it is important to standardize this interface because if the operators want to use different femtocells from different manufacturer they have to be compatible.

Installation

The femtocells are installed by the customers inside their home. This is why such installation must be as simple as possible (Plug and Play). A newly installed femtocell has to configure its parameters automatically depending on the surrounding environment, to avoid a negative impact (interference) with the neighbouring macrocells and femtocells. Ideally, a user should only have to plug in the power supply and connect the femtocell

Table 2.2 Comparison between picocells and femtocells

Parameter	Picocells	Femtocells
Installation	By the operator	By the user
Connection to the core network	Coaxial or fibre optic	ADSL, cable
Price	Cheap	Very cheap
Capacity	10–50 users	3–5 users
Covering range	<100 m	<30 m

to the broadband connection. This self-configuration is very important, because if many femtocells are deployed, operators cannot afford to optimize the parameters of all the femtocells in order to reduce the interference with their macrocells. However, in an open access scenario, where numerous femtocells are required to ensure radio coverage, femtocells can be deployed by the operators themselves in an approach similar to picocells.

2.7 Comparison of Indoor Coverage Techniques

In this chapter, an overview of the different indoor radio coverage techniques has been presented. First solutions like macrocells, microcells and repeaters have been proposed. They increase indoor coverage by extending the area of the outdoor coverage inside a building. Thus, the indoor coverage can be extended with different methods, such as increasing the outdoor signal power by adding more cells, or retransmitting the outdoor signal inside the building by using repeaters. However, these solutions are not optimal because they do not really optimize the indoor coverage, but only extend the effects of outdoor coverage inside the buildings.

Second, some more efficient solutions have been proposed. They consist in installing base stations directly inside the buildings. On the one hand, the radiating cable is a good approach for long narrow environments like tunnels, but its installation in large

Table 2.3 Comparison between the different indoor coverage techniques

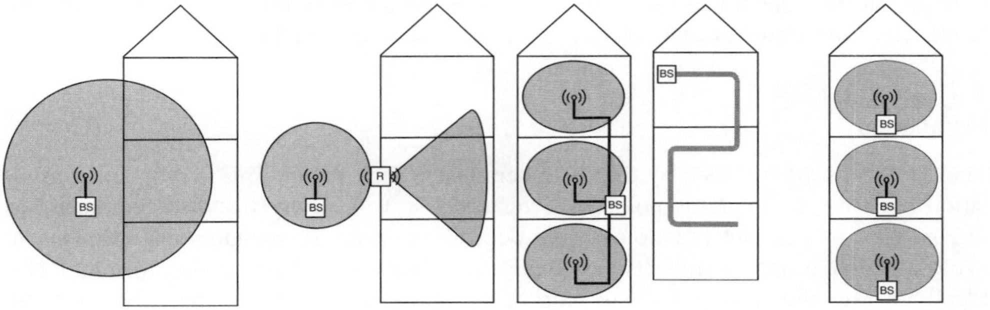

Macro/microcell	Repeater	DAS	Radiating cable	Pico/femtocell
Expensive price	Convenient price	Convenient price	Convenient price	Cheap
Expensive installation	Difficult installation	Easy installation	Difficult installation	Very easy installation
High power	Low power	Low power	Low power	Very low power
Bad indoor coverage	Acceptable coverage	Good coverage	Good coverage	Good coverage

environments like buildings is too challenging and expensive to be implemented. On the other hand, picocells and femtocells are efficient solutions because, if their positions are judiciously chosen, the indoor signal power can be efficiently improved, and the number of cells will increase the potential capacity of the system.

Picocells are installed by the operators, that is why they are well adapted to commercial centers or office buildings for example. However, in home environments, femtocells seem to be the optimal approach for both customers and operators.

In Table 2.3 a summary of the main ideas of the previously described indoor coverage techniques is presented.

In-building radio coverage is still insufficient in many rural or urban places. Moreover, new emergent mobile applications like video conference require higher data rates. Because most of the time the mobiles are used indoors, increasing the indoor radio coverage is the main objective of most operators. Increasing indoor coverage by optimizing the outdoor network was, until recently the only solution. It could be done by adding more cells or increasing the output power of the existing cells, or using repeaters to redirect the signal inside buildings. However, the maximum limits have been reached and now operators have to look for new approaches. In the last few years, DAS, repeaters and radiating cables have started to be the common approach to overcoming indoor coverage problems. These technologies have been shown successfully to increase the coverage in large public places or companies, but there are still many places where the indoor coverage is still insufficient, and where the previous technologies are too complex to cover them. These areas, such as personal homes, represent a huge potential market for operators. In order to cover inside homes, operators need a solution that will fulfill the following requirements:

- a solution that covers small areas,
- a low capacity solution,
- a solution installed and maintained at a low price.

It seems femtocells will be a good option in future years.

In this chapter the evolution of indoor coverage techniques has been presented. The femtocells have been introduced and their advantages presented. The next chapter will present the possible network architectures used in femtocell networks.

References

[1] H. Hristov, R. Feick, and W. Grote, ‘Improving indoor signal coverage by use of through-wall passive repeaters,’ in *IEEE Antennas and Propagation Society International Symposium*, vol. 2, July 2001, pp. 158–161.

[2] Y. Huang, N. Yi, and X. Zhu, ‘Investigation of using passive repeaters for indoor radio coverage improvement,’ in *Antennas and Propagation Society International Symposium, 2004. IEEE*, vol. 2, June 2004, pp. 1623–1626.

[3] M. Lee, B. Keum, Y. Son, K. Joo-Wan, and H. S. Lee, ‘A new low-complex interference cancellation scheme for wcdma indoor repeaters,’ in *IEEE Region 8 International Conference on Computational Technologies in Electrical and Electronics Engineering, SIBIRCON*, July 2008, pp. 457–462.

[4] Y. Ito, S. Fujimoto, and M. Kijima, ‘Novel repeater with automatic gain-control for indoor area,’ in *Asia-Pacific Microwave Conference Proceedings, APMC*, vol. 5, Dec 2005, p. 3.

[5] F. Coperich and C. Turner, 'Fcc requirements for type accepting radio signal boosters and licensing distributed antenna systems,' in *32nd IEEE Vehicular Technology Conference*, vol. 32, May 1982, pp. 247–254.
[6] A. A. M. Saleh, A. J. Rustako Jr, and R. S. Roman, 'Distributed antennas for indoor radio communications,' *IEEE Transactions on Communications*, vol. 35, no. 12, pp. 1245–1251, 1987.
[7] C. Briso-Rodriguez, J. Cruz, and J. Alonso, 'Measurements and modeling of distributed antenna systems in railway tunnels,' *IEEE Transactions on Vehicular Technology*, vol. 56, no. 5, pp. 2870–2879, Sept 2007.
[8] H. Yanikomeroglu and E. Sousa, 'Power control and number of antenna elements in cdma distributed antenna systems,' in *IEEE International Conference on Communications, ICC 98*, vol. 2, June 1998, pp. 1040–1045.
[9] P. Chow, A. Karim, V. Fung, and C. Dietrich, 'Performance advantages of distributed antennas in indoor wireless communication systems,' in *IEEE 44th Vehicular Technology Conference*, vol. 3, June 1994, pp. 1522–1526.
[10] J. Borkowski, J. Niemela, T. Isotalo, P. Lahdekorpi, and J. Lempiainen, 'Utilization of an indoor das for repeater deployment in wcdma,' in *IEEE 63rd Vehicular Technology Conference, VTC 2006-Spring*., vol. 3, 2006, pp. 1112–1116.
[11] J. Yang, 'Analysis and simulation of a cdma pcs indoor system with distributed antennae,' in *Sixth IEEE International Symposium on Personal, Indoor and Mobile Radio Communications PIMRC'95*., vol. 3, Sept 1995, p. 1123.
[12] T. Sorensen and P. Mogensen, 'Radio channel measurements on an eight-branch indoor office distributed antenna system,' in *IEEE VTS 53rd Vehicular Technology Conference, VTC 2001 Spring*, vol. 1, May 2001, pp. 328–332.
[13] H. Xia, A. Herrera, S. Kim, and F. Rico, 'A cdma-distributed antenna system for in-building personal communications services,' *IEEE Journal on Selected Areas in Communications*, vol. 14, no. 4, pp. 644–650, May 1996.
[14] K. Utsumi, H. Sasai, T. Niiho, M. Nakaso, and H. Yamamoto, 'Multiband wireless lan distributed antenna system using radio-over-fiber,' in *International Topical Meeting on Microwave Photonics, MWP 2003 Proceedings*, Sept 2003, pp. 363–366.
[15] Z. Uykan and K. Hugl, 'Hsdpa system performance of optical fiber distributed antenna systems in an office environment,' in *IEEE 16th International Symposium on Personal, Indoor and Mobile Radio Communications, PIMRC*, vol. 4, Sept 2005, pp. 2376–2380.
[16] M. Fabbri and P. Faccin, 'Radio over fiber technologies and systems: New opportunities,' in *9th International Conference on Transparent Optical Networks. ICTON '07*, vol. 3, July 2007, pp. 230–233.
[17] J. H. Wang, 'Leaky coaxial cable with adjustable coupling loss for mobile communications in complex environments,' *IEEE Microwave and Wireless Components Letters*, vol. 11, no. 8, pp. 346–348, Aug 2001.
[18] I. Stamopoulos, A. Aragon, and S. Saunders, 'Performance comparison of distributed antenna and radiating cable systems for cellular indoor environments in the dcs band,' in *Twelfth International Conference on Antennas and Propagation (ICAP 2003)*, vol. 2, March-April 2003, pp. 771–774.
[19] M. Smith, A. Bush, P. Gwynn, and S. Amos, 'Microcell and picocell base station internal antennas,' in *IEEE Wireless Communications and Networking Conference, WCNC*, vol. 2, Sept 1999, pp. 708–711.
[20] E. Kudoh, A. Shibuya, T. Ogawa, D. Uchida, M. Nakatsugawa, H. Suda, and S. Kubota, 'Picocell network for local positioning and information system,' in *26th Annual Confjerence of the IEEE Industrial Electronics Society*, vol. 2, Oct 2000, pp. 1165–1170.
[21] C. Johnson and J. Khalab, 'Load based radio resource management for umts picocells,' in *4th International Conference on 3G Mobile Communication Technologies*, June 2003, pp. 88–92.
[22] Y.-U. Chung and D.-H. Cho, 'Performance analysis of handoff algorithm in fiber-optic microcell/picocell radio system,' in *IEEE 51st Vehicular Technology Conference Proceedings, VTC 2000-Spring*, vol. 3, May 2000, pp. 2408–2412.
[23] D. Molkdar, 'Simulation results of a typical gsm picocellular system,' in *52nd IEEE Vehicular Technology Conference, VTC-Fall*, vol. 4, Sept 2000, pp. 1590–1596.
[24] V. Chandrasekhar and J. G. Andrews, 'Femtocell networks: A survey,' *IEEE Communication Magazine*, vol. 46, no. 9, pp. 59–67, September 2008.

3

Access Network Architecture

Enjie Liu and Guillaume De La Roche

3.1 Overview

Femtocells promise improved indoor coverage and increased throughput for mobile data services while off-loading traffic from expensive macro radio access networks onto the low cost public Internet. While the mobile industry holds high hopes for femtocells, a number of key technical challenges must first be addressed before the femtocell market can see significant commercial success. One such challenge is to define and standardize an approach for integrating femtocells back into mobile core service networks, i.e. device-to-core network connectivity. The Radio Access Network (RAN) in use today comprises hundreds of base stations connected to a single Radio Network or Base Station Controller (RNC/BSC). The interface is Iub running the Asynchronous Transfer Mode (ATM) protocol over dedicated leased lines. Unlike macro 3G RAN, femtocell access networks require operators to integrate hundreds of thousands of low-capacity home base stations that can be moved, added, and changed by end users at any time, all connected over the unsecured and untrusted public Internet. This raises a number of important issues:

1. Is it scalable?
2. Is it secure?
3. Is it standardized?

Because the latest technological progress allows powerful processing capabilities to be applied to low-cost home base stations, the network protocol stacks can now be substantially collapsed. In addition, the standard Internet Protocol (IP) has rapidly replaced hierarchic telecom-specific transmission protocols. The combination of the collapsed protocol stacks and IP transport enables femtocells to utilize flat networks – such as the Internet – as a backhaul transport to operator core networks, as illustrated in Figure 3.1.

Femtocells: Technologies and Deployment Jie Zhang and Guillaume de la Roche

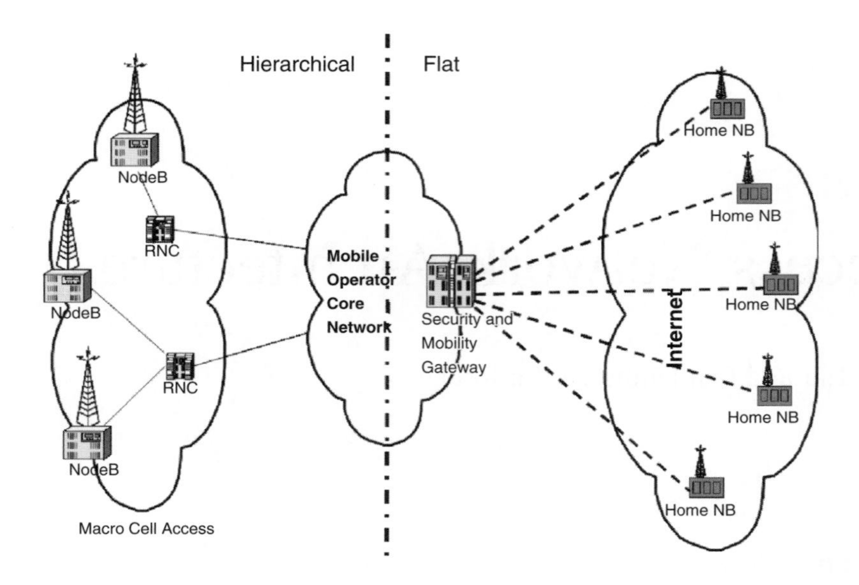

Figure 3.1 Flat architecture for femtocells integrating into mobile networks

There has been no commonly understood description of the fundamental architecture of a femtocell access network and its interconnection to an existing Universal Mobile Telecommunication System (UMTS) system. Four different 3G femtocell architectures have been proposed.

3.1.1 Legacy Iub over IP

This is the earliest femtocell access network architecture generally referred to as 'Iub over IP', as illustrated in Figure 3.2. These solutions looked to exert leverage on the existing 3GPP defined Iub interface that exists between 3G Radio Network Controllers (RNCs) and 3G base stations (NodeBs). Primarily proposed by RNC vendors, these approaches allowed operators to influence the same RNC to support Home NodeBs in addition to macro network NodeBs. Each femtocell is connected to the RNC over the standard 3GPP Iub interface (TS 25.434) [1]. The Iub protocol stack is encapsulated within the IP signalling, also called a tunnelling Iub. Network security is handled by the Internet Protocol Security (IPsec) protocol.

As Iub over IP solutions enable operators to operate their existing core networks through standard interfaces (Iu-CS and Iu-PS), they meet the operator requirement for full service transparency, as well as the requirement for low initial deployment cost and network disruption.

The main concern with this approach is the ability of the RNC to scale up to serving hundreds of thousands of Home NodeBs (HNBs). The challenge with scaling this approach is in the basic design of RNCs, which are typically optimized to support a relatively low number of very high-capacity macro NodeBs. The fact that (despite it being a standard

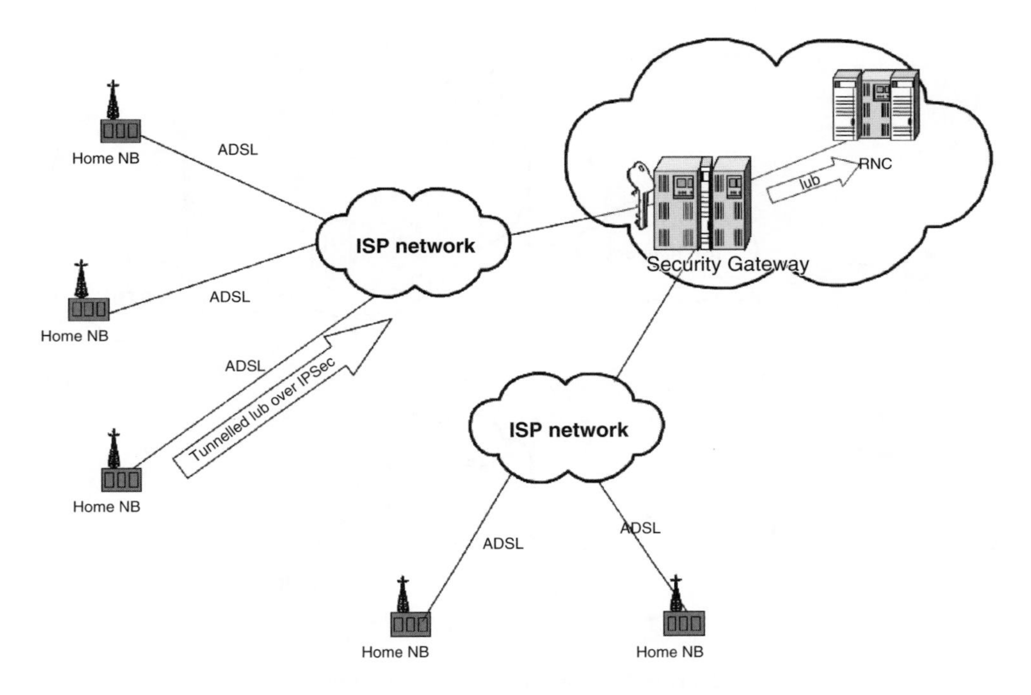

Figure 3.2 Tunnelling Iub with modified RNC

interface) the Iub typically has vendor-specific features makes this approach only suitable for equipment manufacturers with an installed RNC base.

The Iub over IP architecture for integrating the femtocell access network into the operator's macro layer network was only recently studied and attempted. The RNC's lack of scalability in accommodating a large number of HNBs resulted in this alternative architecture no longer being considered after an initial feasibility study carried out in 3GPP standardization. Therefore, it will not be described in detail in the subsequent sections.

3.1.2 Concentrator

To overcome the concerns with conforming to proprietary Iub interfaces and getting current RNCs to handle thousands of Home NodeBs, an alternative architecture that uses a proprietary Concentrator/RNC that can handle thousands of Home NodeBs has been presented as shown in Figure 3.3. This approach allows functions to be partitioned differently between the Home NodeB and RNC, thus enabling the concentrator to handle a large number of Home NodeBs. This approach fits seamlessly into a mobile network operator's RAN by replacing their current RNCs with this proprietary concentrator to serve thousands of femtocells.

The proprietary technology based concentrator architecture did not go any further after 3GPP started standardizing architecture for femtocell access networks. Therefore it will not be described in detail in the subsequent sections.

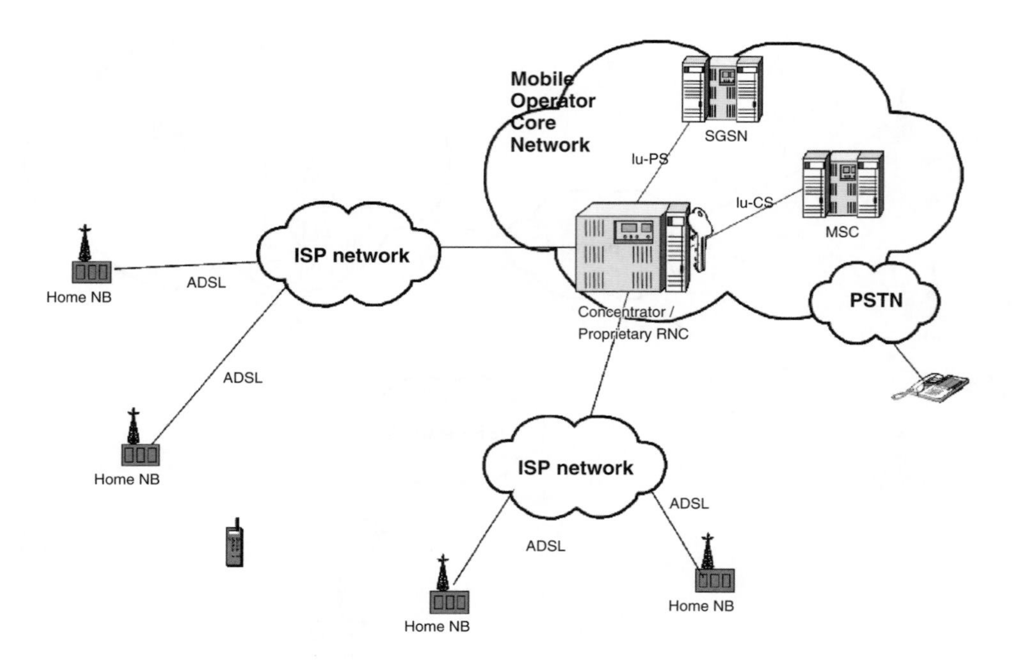

Figure 3.3 Proprietary concentrator/RNC

3.1.3 Generic Access Network (GAN)-Based RAN Gateway

The most recent proposals for femtocells integrating to core network are generally referred to as RAN Gateway solutions. The RAN Gateway approach is based on a new, purpose-built, network controller (RAN Gateway) that resides between an operator's existing core network and the IP access network, akin to an RNC. On its Internet side, the RAN Gateway aggregates traffic from a large number of femtocells over the new Iu-over-IP interface. The RAN Gateway then integrates the traffic into the existing mobile core network through standard Iu-CS and Iu-PS interfaces on the core network side. As the RAN Gateway solutions influence an operator's existing core network through standard interfaces, they allow for full-service continuity as well as a low initial cost of deployment. The RAN Gateway approach employs a 'Flat IP' architecture, in which a number of the functions of a standard RNC are moved to the femtocell itself, and the scaling issues associated with the Iub over IP approach are avoided. Since this architecture removes the RNC, the functionality associated with this is moved into the Home NodeBs (RLC/RRC tasks to support radio channel set up, etc.); as such, the Home NodeB is now more intelligent or autonomous and is often renamed as an 'Access Point' or 'Femto Access Point'. These tasks are significantly simpler than those required in a traditional RNC; for example, given the constrained environment of a femtocell, support for mobility is simpler, there is no need for soft-handoff, etc. This architecture is often referred to as 'flattened', 'collapsed stack' or 'Base Station Router'.

GAN-based Home NodeB architecture, as shown in Figure 3.4, is being considered in 3GPP, and the defined standard femtocell interface between HNB and HNB Gateway,

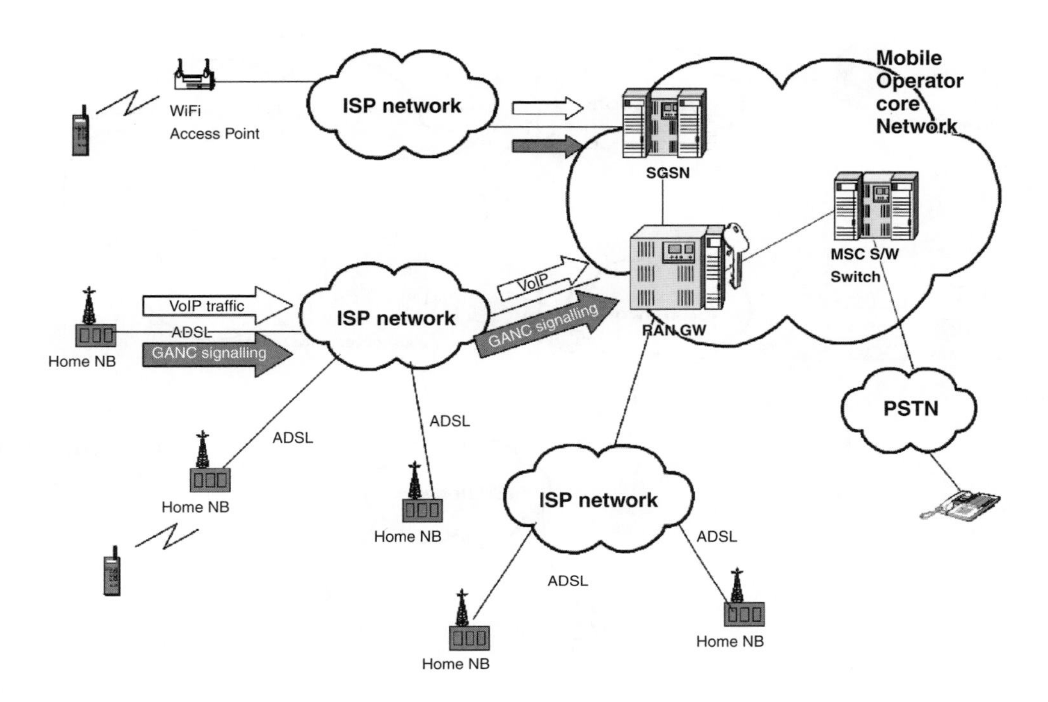

Figure 3.4 GAN-based femtocell access network architecture

Iuh, is likely to evolve from GAN-based Home NodeB architecture. This architecture alternative will be described in detail in the subsequent sections.

3.1.4 IMS and SIP

One alternative approach to femtocells integrated into core network connectivity is to use a new SIP-based protocol between the mobile core network and the Home NodeB. Figure 3.5 shows the approach that breaks from the existing network architecture and embraces the protocols of an all-IP network as envisaged by the 3GPP IP Multimedia Subsystem (IMS). These include Voice-over IP (VoIP) using the Session Initiated Protocol (SIP), with the RNC function now fully integrated into the femtocell AP. Operators would deploy a new SIP-based core network that operates in parallel with their existing circuit and packet-based core network. When a handset is connected to a femtocell, it receives all of its services from the new SIP core network. Parenthetically, this architecture is more aligned with the Wireless Interoperability for Microwave Access (WiMAX) architecture, which is IP-based from the start.

Many operators believe that they will eventually transition their core networks toward an IMS and SIP-based infrastructure and these solutions are also viewed positively. SIP-based approaches also hold the promise of cost-effective support for large-scale deployments.

As handsets are served by a different core network when connected to femtocells as compared with when they are connected to the macrocell network, service continuity

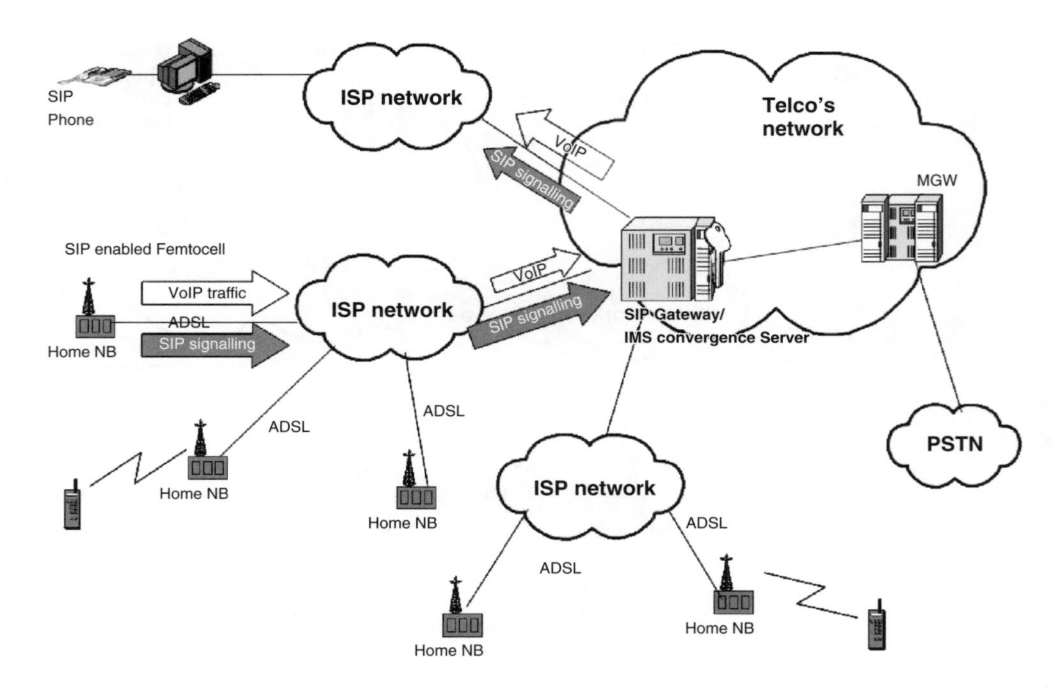

Figure 3.5 SIP/IMS enabled femtocell

between the indoor and outdoor base stations becomes potentially more complex due to the very different technologies involved. As the SIP-based approach requires operators to acquire and integrate a new core service network, the initial deployment costs are much higher than with other approaches.

From Release-8 onwards, 3GPP started standardizing integration of femtocell access network into IMS infrastructure. More details are addressed in the subsequent sections.

3.2 GAN-Based Femtocell-to-Core Network Connectivity

Generic Access Network as defined in 3GPP TS 43.318 [2] and TS 44.318 [3] is a current 3GPP standard that may be used to support Home NodeB. The following sections describe the application of GAN to Home NodeB.

3.2.1 GAN Variant of Iu-Based Home NodeB Architecture

Figure 3.6 illustrates HNB functional architecture utilizing GAN Iu mode [3] as a basis for that architecture. The key consideration for the architecture is the functional splitting of the traditional RNC role between the HNB and Home NodeB Gateway (HNBGW). In this architecture, HNB is responsible for the radio aspects and the HNB GW is responsible for Core Network (CN) connectivity.

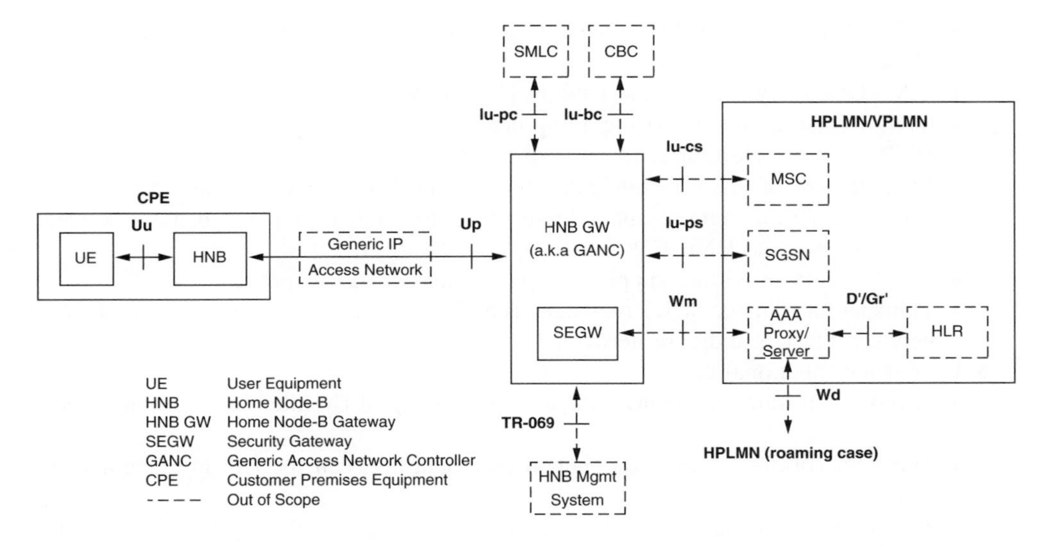

Figure 3.6 HNB functional architecture based on GAN Iu mode

The main features of the GAN-based HNB architecture are:

- Coexistence with the UMTS Terrestrial Radio Access Network (UTRAN) and interconnection with the CN via the standardized interfaces defined for UTRAN:
  - Iu-CS interface for circuit switched services as overviewed in 3GPP TS 25.410 [4].
  - Iu-PS interface for packet switched services as overviewed in 3GPP TS 25.410 [4].
  - Iu-PC interface for supporting location-based services as described in 3GPP TS 25.450 [5].
  - Iu-BC interface for supporting cell broadcast services as described in 3GPP TS 25.419 [6].
- User Equipment (UE): A standard 3G handset device as defined in 3GPP TS 23.101 [7].
- Home Node-B. The HNB is Customer premises Equipment (CPE), which offers a standard radio interface (Uu) for UE connectivity. The HNB provides the radio access network connectivity to the UE and uses the GAN Iu mode Up interface as defined in 3GPP TS 43.318 [2] and the extensions to connect to HNB-GW.
- Home Node-B Gateway. The HNB-GW entity is the same as the Generic Access Network Controller (GANC) defined for GAN Iu mode. The functionality of the GANC defined for GAN Iu mode in TS 43.318 is modified so as to allow a different CPE device type (i.e., the HNB as opposed to the WiFi AP, which requires a dual mode handset) to be connected over the generic IP access network. The HNB-GW entity works between the Iu interfaces and the GAN Iu mode Up interface using the following functionality:
  - Control plane functionality:
    - Security Gateway (SeGW) for the set-up of a secure IPSec tunnel to the HNB for mutual authentication, encryption and data integrity.

   - SeGW Encapsulating Security Payload (ESP) processing of Up interface control plane packets.
   - GAN Discovery support and Default HNB-GW assignment.
   - GAN Registration support including provision of GAN system information to the HNB and possible redirection to a different serving HNB-GW.
   - Management of GAN bearer paths for CS and PS services, including the establishment, administration, and release of control and user plane bearers between the HNB and the HNB-GW.
   - Functionality providing support for paging and handover procedures.
   - Transparent transfer of L3 messages (i.e., Non-Access Stratum (NAS) protocols) between the UE and core network.
 - User plane functionality:
   - SeGW Encapsulating Security Payload processing of Up interface user plane packets.
   - The interworking of circuit switched user data between the Up interface and the Iu-CS interface.
   - The interworking of packet switched user data between the Up interface and the Iu-PS interface.
- A generic IP access network provides connectivity between the HNB and the HNB-GW. The IP transport connection extends from the HNB-GW to the HNB. A single interface, the Up interface, is defined between the HNB-GW and the HNB.
- Transaction control (e.g. CC, SM) and user services are provided by the core network (e.g. Mobile Switching Center (MSC)/Visitor Location Register (VLR) and the Serving GPRS Support Node (SGSN)/Gateway GPRS Support Node (GGSN).
- Use of Authentication, Authorization and Accounting (AAA) server over the Wm interface as defined by 3GPP TS 29.234 [8]. The AAA server is used to authenticate the HNB when it sets up a secure tunnel.
- HNB management system entity is introduced to manage the configuration of HNB in a scalable manner. The HNB management system utilizes standard CPE device management interface as described in Digital Subscriber Line (DSL) Forum technical specifications TR-069 [9]. It should be noted that the TR-069 interface, although shown to be extending from HNB-GW, is between the HNB and HNB Mgmt system channeled via the Up interface's secure tunnel.

3.2.2 Component Description

Femto Access Point (FAP)

The FAP is a 'zero touch' plug-and-play consumer device, which is installed at the subscriber premises and is connected to the operator's core network over the subscriber's broadband connection. The FAP provides localized 3G coverage and dedicated capacity in a home, enhancing the end-user experience through improved quality of service. The FAP incorporates adaptive and distributed radio management, without the need for central Radio Frequency (RF) management and with the ability to obtain optimal local coverage with minimal macro network interference. The FAP also incorporates a periodic monitoring of the RF environment where it is located to ensure that it remains aware

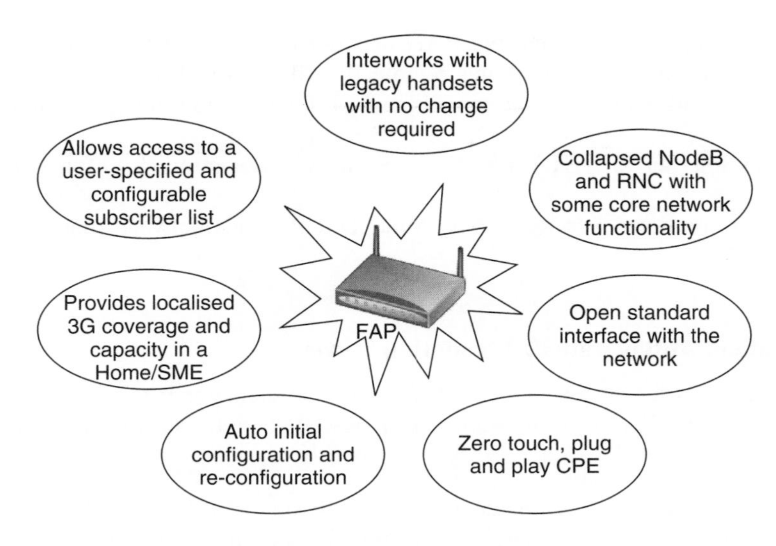

Figure 3.7 FAP key features

of 'macro and femtocell' network and dynamically adapts to any possible change. The FAP is responsible for managing the connection to the operator's core network, mediating all CS call and PS session functions between the network and the User Equipment. The FAP supports the 'sticky coverage' concept, which is auto-configured at initial start-up to encourage the end-users in the home zone for as long as possible, maximizing the reach of the home zone services and minimizing unwanted ping-pong mobility effects with the macro network. The FAP interacts with the legacy 3G handsets using the 3GPP Uu interface with no change needed in the handset. The key features of FAP are illustrated in Figure 3.7 above.

Security Gateway (SeGW)

Security Gateway (SeGW) is a highly scaleable, 3GPP standards-based product.

- It provides secure access over the RAN GW to the core network, authenticating and terminating IPsec tunnels that originate from the FAP.
- It interfaces with the AAA server via the 3GPP standard Wm interface, securing access control through the execution of authentication, authorization and accounting procedures.
- It interfaces with multiple FAP over the 3GPP standard Up interface for Radio Resource Connection (RRC)-equivalent signalling and keying material exchange.
- It provides a secured access for the GPRS Tunnelling Protocol (GTP) tunnels that terminate on the Core Network Serving GPRS Support Node over the 3GPP standard Iu-PS interface. IP is used for the transport of the GTP tunnels.
- It provides high-capacity IPsec tunnel termination services for each femtocell in the access network. One IPsec tunnel per FAP is required.
- Its functionality is compliant with the following 3GPP standards: [2, 3, 8, 10–14].

- A single SeGW can serve multiple RAN network controller and media gateways.
- It manages, allocates and distributes remote IP addresses to FAPs.
- It establishes and manages IPsec tunnels to each FAP for both integrity and encryption purposes.
- It manages Internet Key Exchange (IKE) Security Association (SA) for authentication purposes.
- It maps Differentiated Services Code Point (DSCP) fields between inner and outer IP packet headers.

Authentication, Authorization, Accounting Server

The AAA server is interfaced by the SeGW using the diameter Wm interface and by the RAN network controller using the standard S1 (RADIUS) interface enhanced with vendor specific attributes.

- It improves the level of secure access that can be provided by the RAN GW.
- It supports an SS7 MAP-D interface to enable the security access of each International Mobile Subscriber Identity (IMSI) that is attempting to secure an IP access register with the RAN GW.
- It provides EAP-SIM/EAP-AKA authentication services between the FAP and the Home Location Register (HLR)/Home Subscriber Server (HSS) as per standard 3GPP security architecture.
- The AAA is also the platform for operator-defined, scripted, external business logic that can provide additional Service Access Controls, UE session parameters and logging of UE registration events (for legal purposes).
- A single AAA can serve multiple SeGW and RAN-GW platform requests.

RAN Network Controller

RAN Network Controller is the centralised component of the RAN GW that provides a conduit between the Core Network and the RAN GW.

- It provides a transparent transfer of the 3GPP standards nonaccess stratum protocols between the FAP and the core network MSC and core network SGSN. It provides relay of layer 3 NAS messages from each UE to the core network.
- It supports generic access network discovery and default RAN-GW assignment, based on the evolved procedures [2].
- It provides management of the RAN GW bearer paths for circuit switched and packet switched services between the FAP (and each UE camped on the FAP) and the RAN GW.
- It supports RAN GW registration and possible redirection to another serving RAN-GW based on the evolved procedures [2]. It provides UMTS transaction layer support for services such as paging and handover.
- It provides soft-switch functionality for the control of circuit switched media, interfacing with the Media Gateway (MGW) via H.248 interface. It provides MGW and circuit switched/packet switched bearer path management.

- It interfaces with the SeGW via RANAP carried via Signaling Transport (SIGTRAN) M3UA if the Mobile Network Operator (MNO) core network does not support SIGTRAN, otherwise there is a direct association with the MNO core network SIGTRAN.
- It interfaces with the core network SGSN over the 3GPP standard Iu-PS interface, for the signalling control of packet switched services. Iu-PS uses IP for the transport of signalling messages.
- It provides Up session management to every attached FAP and UE (UE management is by FAP proxy). It provides separate contexts maintained for each FAP and UE served.
- It provides RAN GW registration and authorization services for each FAP and UE.

Media Gateway Controller (MGW)

Media gateway controller handles the circuit switched media streams and interaction with the MSC or Rel-4 MGW in the MNO core network.

- It interfaces with the core network MSC over the 3GPP standard Iu-CS interface, for the support of circuit switched services. ATM is used as the transport of the media between the MGW and the MSC.
- It supports H.248 MGW control protocol.
- It provides the SIGTRAN to Iu-CS/PS control plane (ATM/AAL5) conversion for interface to an ATM-based MSC.
- It can connect to multiple MSCs.
- It can be controlled from multiple RAN network controllers to maximize capacity and prevent stranded resources.

Signalling Gateway

Signalling Gateway (SG) is required when the MNO core signalling network does not support SIGTRAN and interaction is required to transport the RANAP messages to and from the MSC.

Access Point Management System

Access Point Management System (AP-MS) provides Operational, Administration, Maintenance and Provisioning (OAM&P) functions for the FAP that are distributed at the end-user's location. Entities constituting the AP-MS system are AP-MS Server, AP-MS workstation and performance management system, see Figure 3.8.

The key features of AP-MS are illustrated in Figure 3.9.

- It manages the FAP using the procedures and methods described in the DSL Forum TR-069 specifications.
- It is responsible for the provisioning of the FAP during the installation process.
- It monitors for faults reported by the managed FAPs.

- It provides a means for the operator to manage the configuration of each FAP.
- It provides user interface with security to restrict the functions to which the user has access.
- It interfaces with the FAP over the Zz interface using a secure IP connection.
- It provides the means to manage the upgrade of the software for the FAPs.
- It collects the performance metrics reported by the FAPs.
- It interfaces with customer care systems.

3.2.3 Functional Split between HNB and NHB-GW

Tables 3.1 and 3.2 are an extension to Table 5.1 in TS 25.410 [4], which defines the functional split between the core network and the UTRAN. This is used to capture the functional split between the HNB and HNB GW. As seen from Tables 3.1 and 3.2, the radio management functions are delegated to the HNB, whereas the core network connectivity functionality is maintained in the HNB-GW. Additionally, certain functions require coordination between the HNB and HNB-GW and as such these functions are expected to be managed by both HNB and HNB-GW. 'Paging' is an example of such functionality where coordination between HNB and HNB-GW is necessary. Paging from CN must be processed by the HNB-GW in order to determine the specific HNB, which must be targeted for the paging due primarily to the uncoordinated nature of the HNB deployment.

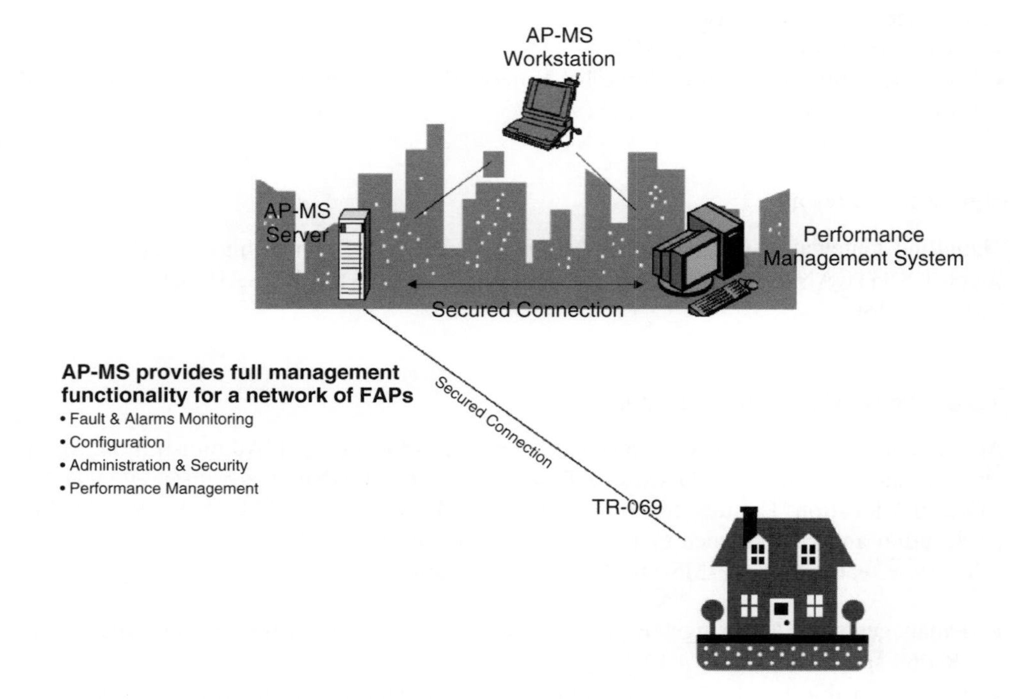

Figure 3.8 AP-MS architecture

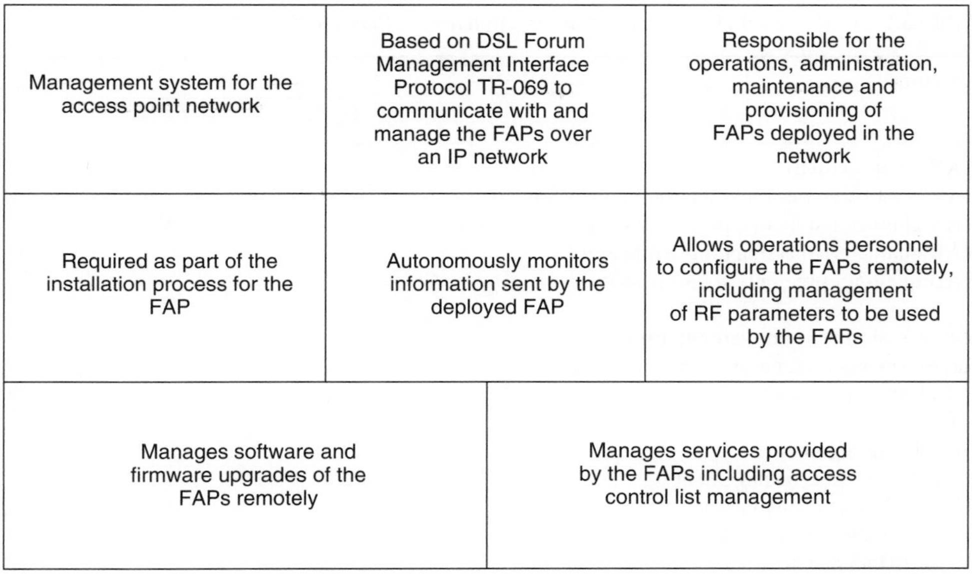

Management system for the access point network	Based on DSL Forum Management Interface Protocol TR-069 to communicate with and manage the FAPs over an IP network	Responsible for the operations, administration, maintenance and provisioning of FAPs deployed in the network
Required as part of the installation process for the FAP	Autonomously monitors information sent by the deployed FAP	Allows operations personnel to configure the FAPs remotely, including management of RF parameters to be used by the FAPs
Manages software and firmware upgrades of the FAPs remotely	Manages services provided by the FAPs including access control list management	

Figure 3.9 AP-MS key features

3.2.4 Internal and External Interfaces (Standard Conformance)

Interfaces between RAN GW and Core Network

The RAN GW supports the following internal and external standard interfaces between the RAN GW and the Core Network. *D′ Interface* D′ Interface is a standard 3GPP interface used to support authentication services between the access network and the HLR. The AAA server queries the HLR during the femtocell authentication process with the RAN GW. The embedded Subscriber Identity Module (SIM) within the FAP provides the unique identity for the device, which is then authenticated using standard MAP procedures with the HLR. *Iu-PS Control and User Plane* Iu-PS interface is a standard 3GPP interface. Both the Iu-PS bearer and control plane from the RAN GW towards the core network transport over an ATM (STM-1) or IP/Ethernet transport. *Iu-CS Control and User Plane* Iu-CS interface is a standard 3GPP interface. The Iu-CS traffic transports over ATM towards core network. In the future, a backhaul transport mechanism will be supported where Iu-CS user plane traffic is transported over IP from RAN GW towards core network.

Interfaces within RAN GW

The femtocell network supports the following standard interfaces within the RAN GW. *H.248* H.248 is the MGW control protocol to enable the RAN network controller to manage the MGW bearer paths. *SIGTRAN* SIGTRAN is the Iu-CS/PS control plane over a standard SIGTRAN transport between the RAN network controller and the signalling gateway. The signalling gateway is embedded within the MGW and performs

Table 3.1 GAN based HNB architecture functional split Part 1 [15] [16]

Function	UTRAN		CN
	HNB	HNB GW	
RAB management:			
RAB establishment, modification and release	X		X
RAB characteristics mapping Iu transmission bearers		X	
RAB characteristics mapping Uu bearers	X		
RAB queuing, preemption and priority	X		X
Radio resource management functions:			
Radio Resource admission control	X		
Broadcast information	X		X
Iu link management:			
Iu signalling link management		X	X
ATM VC management		X	X
AAL2 establish and release		X	X
AAL5 management		X	X
GTP-U tunnels management		X	X
TCP management		X	X
Buffer management	X	X	
Iu U-plane (RNL) management:			
Iu U-plane frame protocol management			X
Iu U-plane frame protocol initialization		X	
Mobility management:			
Location information reporting	X	X	X
Handover and relocation			
Inter RNC hard HO, Iur not used or not available	X	X	X
Serving RNS relocation (intra/inter MSC)	X	X	X
Inter system hard HO (UMTS-GSM)	X	X	X
Inter system change (UMTS-GSM)	X	X	X
Paging triggering			X
GERAN system information retrieval	X	X	X
Security:			
Data confidentiality			
Radio interface ciphering	X		
Ciphering key management			X
User identity confidentiality	X		X
Data integrity			
Integrity checking	X		
Integrity key management			X

Table 3.2 GAN based HNB architecture functional split Part 2 [15] [16]

Function	UTRAN		CN
	HNB	HNB GW	
Service and network access:			
CN signalling data	X	X	X
Data volume reporting	X		
UE tracing	X	X	X
Location reporting	X	X	X
Iu coordination:			
Paging coordination	X	X	X
NAS node selection function		X	
MOCN re-routing function		X	X
Multicast Broadcast Multimedia Service (MBMS):	X	X	X
MBMS Radio Access Bearer (RAB) management	X	X	X
MBMS UE linking function	X	X	X
MBMS registration control function	X	X	X
MBMS enquiry function	X	X	X

the protocol translation between the RAN GW and the core network for the Iu control plane. *Simple Network Management Protocol (SNMP)* is the protocol for Enhanced Messaging Service (EMS). *Wm* is the protocol for Extensible Authentication Protocol (EAP)-SIM/Authentication and Key Agreement (AKA) authentication between the SeGW and the AAA server. *Remote Authentication Dial-In user Services (RADIUS)* is the protocol for access controls and authorization between the RAN network controller and the AAA server.

Interfaces between FAP and RAN GW

The femtocell network supports the following standard interfaces between the FAP and individual network elements in the RAN GW. *Up/Iu-h* is now known as Iu-h. Iu-h interface is being standardized in 3GPP. Up is the standard Iu mode protocol for the transport of 3G UMTS protocols and services over the IP access network. *Real Time Transport (RTP)* RTP is the protocol for circuit switched bearer traffic over the public Internet between the FAP and the MGW. *GTP-U* is the protocol for packet switched bearer traffic between the FAP and the SGSN. *IPsec* is the protocol for integrity and encryption of all traffic between the FAP and the RAN GW. *TR-069* is the management protocol for managing the FAP community from the RAN GW.

3.2.5 Protocol Architecture

In Figure 3.10, all unshaded boxes represent standard protocols, as defined for the respective interfaces. Shaded boxes represent modifications to the standard protocols, which are

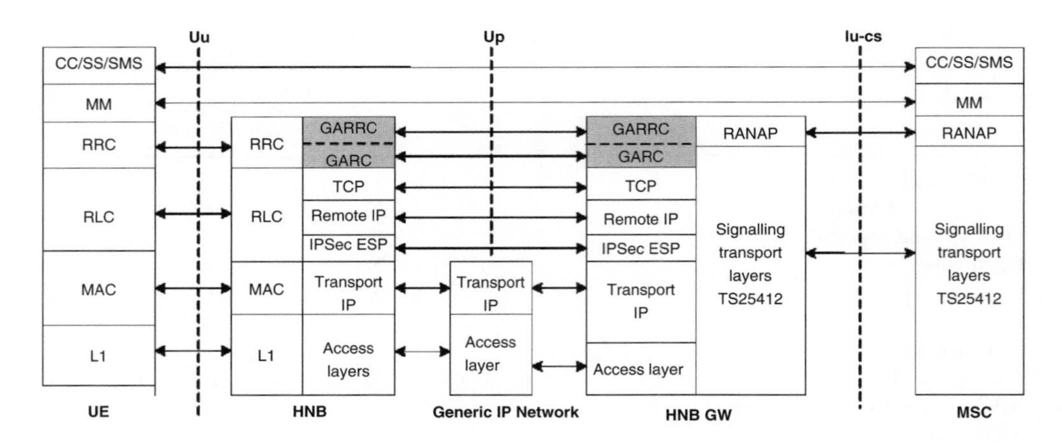

Figure 3.10 CS domain-control plane architecture

explained in the next section. The descriptive text supporting the figure is functionally equivalent to that of 3GPP TS 43.318 and uses HNB terminology to explain how GAN Iu-mode is applied to the HNB architecture.

CS Domain-Control Plane

The HNB architecture in support of the CS domain-control plane is illustrated in Figure 3.10.

The main features of the HNB CS domain-control plane architecture are as follows:

- The underlying access layers and transport IP layer provides the generic connectivity between the HNB and the HNB-GW.
- The IPsec layer provides encryption and data integrity between the HNB and HNB-GW.
- The remote IP layer is the ‘inner’ IP layer for IPSec tunnel mode and is used by the HNB to be addressed by the HNB-GW. The remote IP layer is configured during the IPsec connection establishment.
- A TCP connection is used to provide reliable transport for both the GA-RC and GA-RRC signalling (described below) between the HNB and HNB-GW. The TCP connection is managed by GA-RC and is transported using the remote IP layer.
- NAS protocols, such as MM and above, are carried transparently between the UE and MSC.
- The Generic Access Resource Control (GA-RC) protocol manages the Up session, including the GAN discovery and registration procedures.
- The Generic Access Radio Resource Control (GA-RRC) protocol performs functionality equivalent to the UTRAN RRC protocol, using the underlying Up session managed by the GA-RC. Note that GA-RRC includes both CS service and PS service-related signalling messages.
- The HNB-GW terminates the CS-related GA-RRC protocol and interlinks it with the RANAP protocol over the Iu-CS interface.

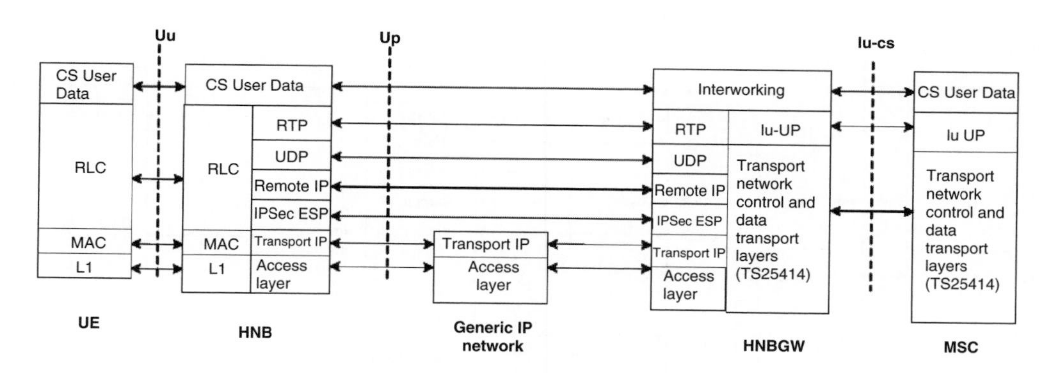

Figure 3.11 CS domain-user plane architecture

- The Iu-CS signalling transport layer options (both ATM and IP-based) are defined in 3GPP TS 25.412 [17].

CS Domain-User Plane

The HNB architecture in support of the CS domain-user plane is illustrated in Figure 3.11 above.

The main features of the HNB CS domain-user plane architecture are as follows:

- The underlying access layers and transport IP layer provides the generic connectivity between the HNB and the HNB-GW.
- The IPSec layer provides encryption and data integrity.
- The CS user-plane data transport over the Up interface does not change from that described in 3GPP TS 43.318 [2].
- The HNB-GW provides interaction between RTP/UDP and the circuit-switched bearers over the Iu-CS interface.
- The HNB-GW supports the Iu User Plane (Iu UP) protocol. Each Iu UP protocol instance may operate in either transparent or support modes, as described in 3GPP TS25.415 [18]; the mode choice is indicated to the HNB-GW by the MSC using RANAP protocol.
- The Iu-CS data transport layers (both ATM and IP-based) and associated transport network control options are defined in 3GPP TS25.414 [19].

PS Domain-Control Plane

The HNB architecture in support of the PS domain-control plane is illustrated in Figure 3.12.

The main features of the HNB PS domain control plane architecture are as follows:

- The underlying access layers and transport IP layer provides the generic connectivity between the HNB and the HNB-GW.

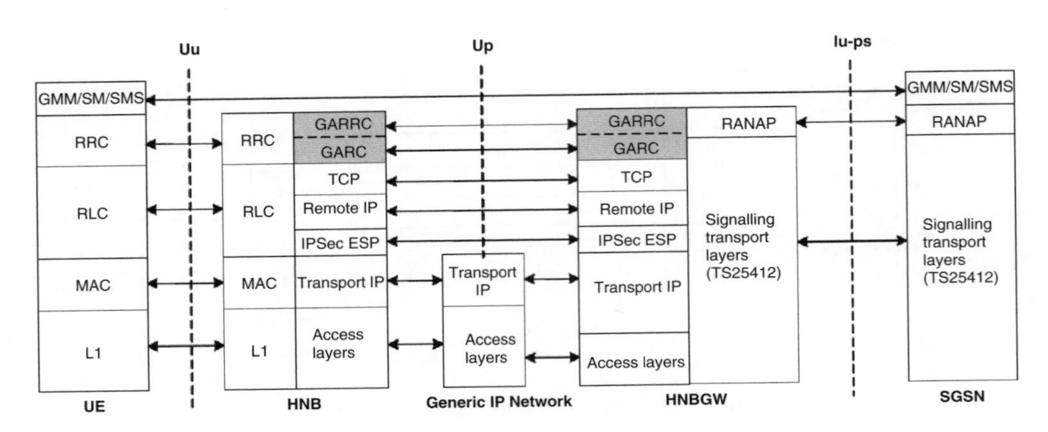

Figure 3.12 PS domain-control plane architecture

- The IPSec layer provides encryption and data integrity between the HNB and HNB-GW.
- TCP provides reliable transport for the GA-RRC between HNB and HNB-GW.
- The GA-RC manages the IP connection, including the GAN registration procedures.
- The Generic Access Radio Resource Control (GA-RRC) protocol performs functionality equivalent to the UTRAN RRC protocol, using the underlying Up session managed by the GA-RC. Note that GA-RRC includes both CS service and PS service-related signalling messages.
- The HNB-GW terminates the GA-RRC protocol and interconnects it with the RANAP protocol over the Iu-PS interface.
- NAS protocols, such as for GMM, SM and SMS, are carried transparently between the UE and SGSN.
- The Iu-PS signalling transport layer options (both ATM and IP-based) are defined in 3GPP TS25.412 [17].

PS Domain-User Plane

The HNB architecture in support of the PS domain-user plane is illustrated in Figure 3.13. The main features of the HNB PS domain-user plane architecture are as follows:

- The underlying access layers and transport IP layer provides the generic connectivity between the HNB and the HNB-GW.
- The IPSec layer provides encryption and data integrity.
- The GA-RRC protocol operates between the HNB and the HNB-GW transporting the upper layer payload (i.e. user plane data) across the Up interface. The GA-RRC protocol for the PS domain user plane uses the GTP-U G-PDU message format, fully compatible with the GTP-U G-PDU message format used over the Iu-PS and Gn interfaces.
- PS user data is carried transparently between the UE and CN.

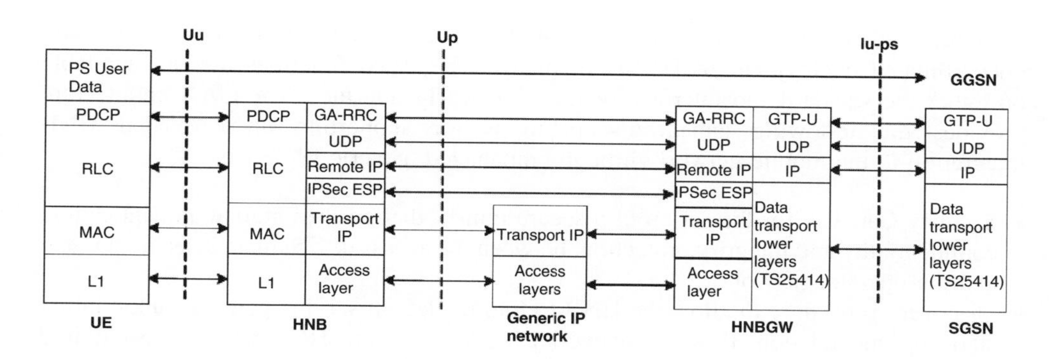

Figure 3.13 PS domain-user plane architecture

- The HNB-GW terminates the GA-RRC protocol and interconnects it with the Iu-PS interface using GTP-U.

3.2.6 *GAN Specification Extensions for HNB Support*

The GAN specification needs to be extended primarily to relay radio attributes between the HNB and the HNB gateway. These extensions are limited to information elements added to existing GAN procedures. No new procedures are required to support the HNB application on GAN. The following lists the key extensions to the GAN specifications [3] for HNB support:

- Extend GA-RC REGISTER REQUEST message with an additional IE to include HNB identity (e.g. IMSI).
- Update GAN Classmark IE with additional device types for HNB/HNB-UE and also an Emergency Call request flag (for unauthorized UE emergency call registration).
- Extend RAB Configuration attribute in GA-RRC ACTIVATE CHANNEL and GA-RRC ACTIVATE CHANNEL ACK message to transparently relay radio attributes between HNB and CN via the HNB-GW.
- Extend GA-RRC RELOCATION INFORMATION message to relay radio attributes between HNB and HNB-GW.
- Extend GA-RRC SECURITY MODE COMMAND to include CK, IK so that the HNB can protect the air interface.
- Use of a single IPsec tunnel between HNB and HNB-GW for multiplexing separate UE sessions.

3.2.7 *Advantages of GAN HNB Architecture*

GAN is a standard design that provides large scale/uncoordinated dual-mode-handsets access to the PLMN via generic IP networks. Therefore, it defines functions and capabilities that readily address many of the issues arising from large scale and uncoordinated deployment of HNBs. In many cases, such functions are not available on either the standard Iub interface or the standard Iu interface. The GAN architecture provides the following enhanced capabilities for supporting uncoordinated HNB integration to the mobile network through unmanaged, generic IP networks:

- Security Gateway for the set-up of a secure tunnel, that ensures mutual authentication, confidentiality and integrity protection, between the access device in an insecure domain and the operator network.
- Discovery procedure to allow the HNB to find its default serving gateway, upon initial start-up. This function allows the network to scale to as many HNB gateways as required knowing that HNBs will be able to find the one HNB gateway best suited to serving that HNB.
- Registration procedures to allow the access device to register for service, obtain/update system information for operation and, when needed, be redirected to a different serving gateway. Registration enforces access controls for the HNB and for each UE since only Registration Accepted devices are served by the HNB Gateway.
- QoS enhancements such as RTP redundancy for the preservation of VoIP audio quality across unmanaged IP networks (i.e. Internet) using standard RFC 3267 features.
- QoS enhancements for the detection of degraded Uplink VoIP quality across the unmanaged IP network and the ability to initiate handover from the HNB GW to the macro UTRAN to preserve the service quality for the affected UE.

GAN is designed for access to core network services over generic IP networks, and while the initial application is for access by WiFi-enabled GSM/UMTS handsets, there is no inherent limitation that prevents it from being used for access by HNB. Rather, the support for the above functions indicates that GAN is actually very well suited to the HNB architecture.

3.3 3GPP Iuh (Iu-Home) for Home NodeB

3.3.1 Iub and Iuh for HNB

Figure 3.14 shows two ways to provide very small area coverage cells in UMTS. One is with a traditional NodeB scaled down to 'femto' size to support a very small area cell and reporting to a conventional RNC. The other approach is a more Home NodeB tailored approach, which has elements of an RNC collapsed into it and connected to the CN with many other HNBs via a gateway tailored for this purpose [20].

The introduction of an HNB gateway approach for collapsed or partially collapsed architecture HNB solutions enables options for better mobility and OAM for a Home NodeB Access Network than with a conventional NodeB and RNC approach. The resulting HNB Access Network has the following entities:

- **HNB Access Network:** The full access network would therefore comprise of N × HNB and M × HNB-GW and is envisaged to provide the following:
  - same NAS messaging procedures between the UE and CN as for established UMTS macro system,

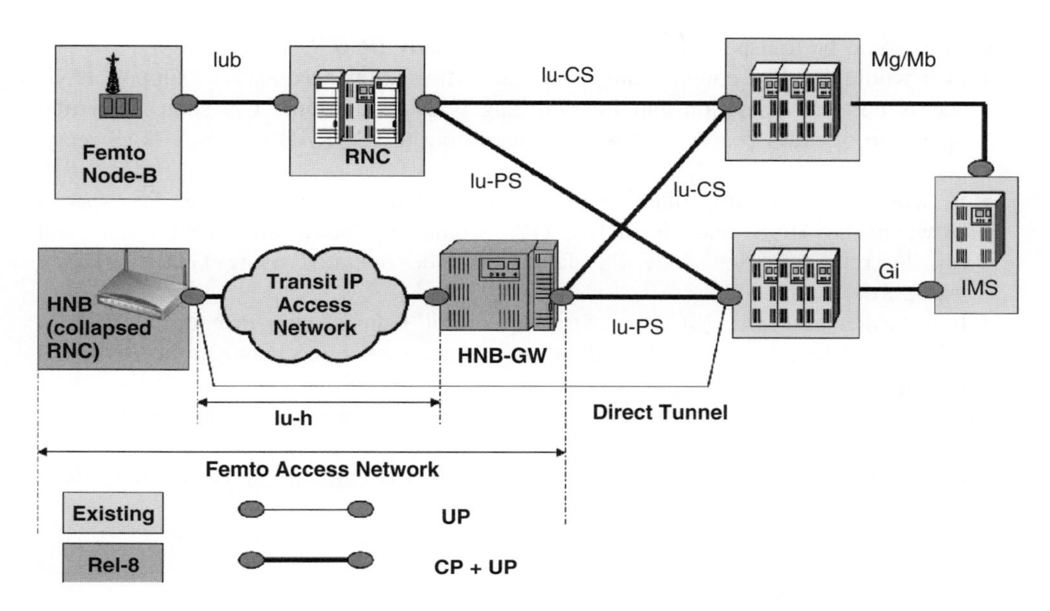

Figure 3.14 HNB access network architecture

- same types of services as established macro network,
- same established UMTS security techniques UEA, UIA. UE authentication for HNB access network as for established UMTS macro system,
- support of direct tunnel between HNB and GGSN.

- **HNB:** The HNB has RF parameters and performance as per the HNB (TR of RAN-4 currently under construction) and contains part or all of the functionality normally associated with an RNC. The HNB is envisaged as operating a standard Uu towards existing UMTS UEs and an interface towards the HNB gateway designated, say Iu-h.
- **HNB Gateway:** This is envisaged as a Network Element that connects with many HNBs over an Iu-h and presents a standard Iu towards an existing UMTS and supports the following basic functionality:
  - Provides a mechanism to support other enhanced features for the HNB access network such as coordinated clock sync distribution from the UMTS CN and/or other connected transit network towards the HNB including for example: the use of assisting IP based synchronization techniques such as IEEE1588 and/or IETF Network Time Protocol (NTP) in standard or enhanced form.
  - Operates two-way security authentication/certification brokering between HNB and HNB-GW, potentially implementing 3GPP GAA principles for NE-NE authentication.
- **Iu-h:** It is proposed that this reference point be introduced between the HNB and an HNB gateway in order to standardize common features for an HNB access network and allow a common agreed transport for proprietary enhancements to allow for product differentiation. The kinds of basic principle being considered for standardization are:

- Iu-h would be transported over IPv4 and optionally IPv6.
- Iu-h would need to enable transport of all information necessary to support HNB access network operation and interworking with a traditional CN such as normal operation of radio bearers at the HNB according to CN RAB requests, NAS relay, etc.
- It would be assumed that relay of user plane for CS and/or PS speech between the HNB and the HNB GW would be based on RTP/UDP/IP and for the relay of user plane PS data would be operated using GTP-U/UDP/IP transport.
- It is further assumed that either Stream Control Transmission Protocol (SCTP) or TCP would be adopted for signalling and control messaging between the HNB and the gateway
- Specific encryption protection should be provided for the Iu-h (e.g. IPSec)
- Iu-h would need to support integrity checking between the HNB and the HNB-GW

3.3.2 Iu-h for HNB

3GPP RAN Working Group 3 is currently at draft stage for Stage 2 UTRAN architecture for 3G Home NodeB. The complete specification will be TS 25.467 [16]. The reference model shown in Figure 3.15 contains the network elements that make up the HNB access network. There is a one-to-many relationship between HNB-GW and HNBs. The support of the 3G Home NodeBs is ensured with enhancements to the Iu interface architecture to cater for scalability issues. Whereas the Iu interface is specified at the boundary between the core network and UTRAN, the Iuh interface is specified between the HNB GW and the HNBs. The HNB and HNB-GW in combination supports all of the UTRAN functions. The legacy UTRAN functions in the HNB are supported by RANAP, whereas the functions HNB Registration, UE Registration and HNB-GW discovery are supported by the new protocol Home NodeB Application Protocol (HNBAP) between the HNB and the HNB-GW. The HNB-GW provides concentration function for the control plane and may provide concentration function for the user plane.

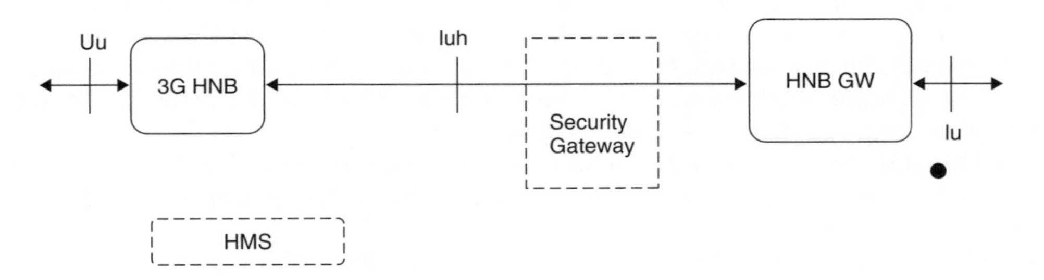

Figure 3.15 Iuh reference model

The HNB-GW serves the purpose of an RNC presenting itself to the CN as a concentrator of HNB connections. The Iu interface between the CN and the HNB-GW serves the same purpose as the interface between the CN and an RNC. The security gateway is a logically separate entity and may be implemented either as a separate physical element or integrated into, for example, an HNB-GW. The HNB access network includes the following key functional entities.

- HNB Management System (HMS)
  - is based on TR-069 family of standards,
  - facilitates HNB-GW discovery,
  - provisions configuration data to the HNB,
  - performs location verification of HNB and assigns appropriate serving elements (HMS, security gateway and HNB-GW),
  - has security gateway (SeGW),
  - terminates secure tunnelling for TR-069 as well as Iuh,
  - provides authentication of HNB,
  - provides access to HMS and HNB-GW.
- HNB Gateway (HNB-GW)
  - terminates Iuh from HNB. Appears as an RNC to the existing Core Network using existing Iu interface,
  - provides service to the HNB.
- HNB
  - customer premise equipment that offers the Uu interface to the UE,
  - provides the RAN connectivity using the Iuh interface,
  - supports RNC alike functions,
  - supports HNB registration and UE registration over Iuh.

3.4 Evolution to IMS/HSPA+/LTE

Standardization work in 3GPP focusing on femtocells/Home NodeBs in Long Term Evolution (LTE) networks is well under way (3GPP TR R3.020, Rel-8). Several evolution paths from an initial Radio Access Network (RAN) centric solution to the LTE solution are possible, catering to different operators' circumstances. Figure 3.16 shows the ultimate evolution of femtocell access network architecture towards LTE.

As for the X2, from a logical standpoint, the X2 is a point-to-point interface between two eNBs within the E-UTRAN. Refer to TS 36 series, such as [21] for more details.

The natural first step from the current stage (Figure 3.17) on the road to LTE/Enhanced Packet Core (EPC) is the addition of serving GPRS support node functionality to the radio access network gateway as shown in Figure 3.18. This has two main advantages: (i) It collapses the packet switched domain architecture reducing costs and latency; (ii) The Gn interface is future proof as it allows the RAN GW to connect directly to the Mobility Management Entity (MME) and System Architecture Evolution Gateway (SAE GW) of the upcoming EPC, as shown in Figure 3.19.

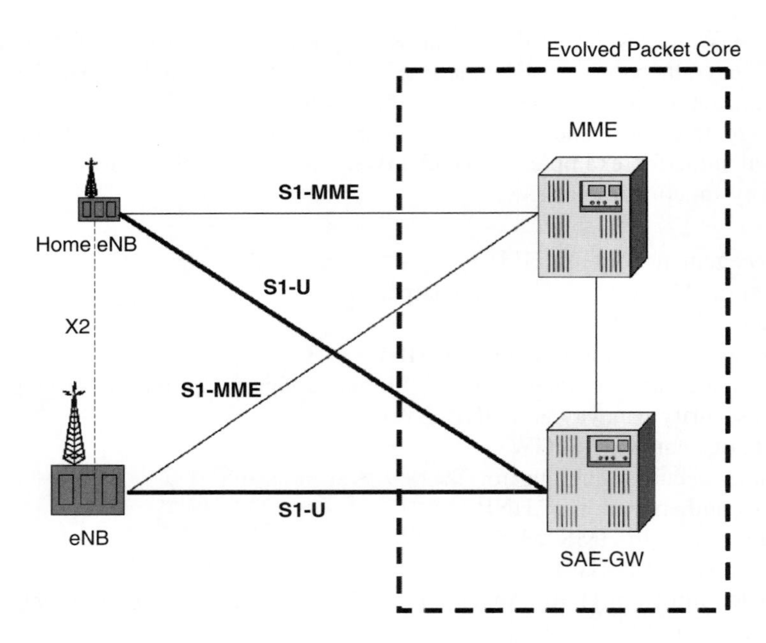

Figure 3.16 Evolution to LTE

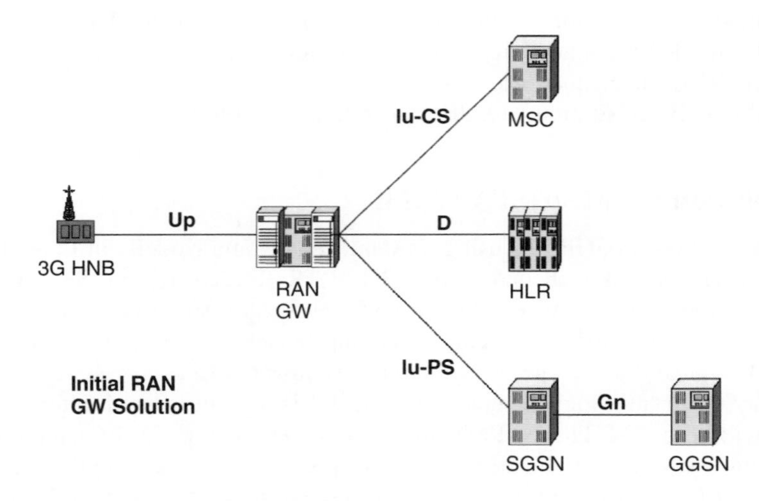

Figure 3.17 Current RAN GW solution

In the first two steps the circuit switched services are still provided via the legacy circuit switched core network as it is expected that IMS based voice-over-IP capable UEs will not be widely developed yet. When that time comes the circuit switched core network can then be discontinued and the IMS core network will provide VoIP services to VoIP capable UEs as shown in Figure 3.20. Throughout all these steps the interface

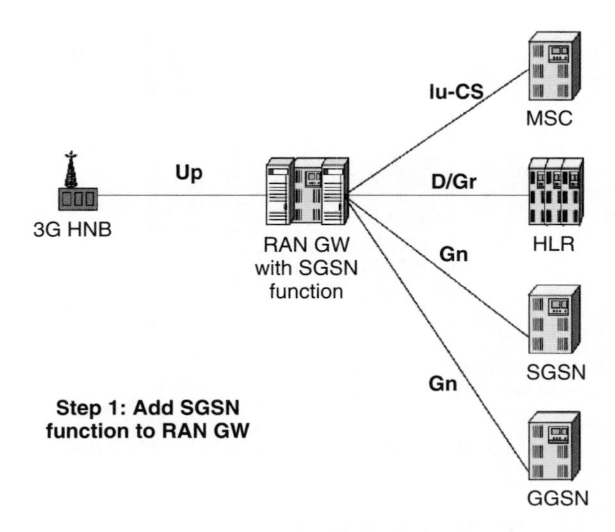

Figure 3.18 Step 1: RAN GW solution evolution

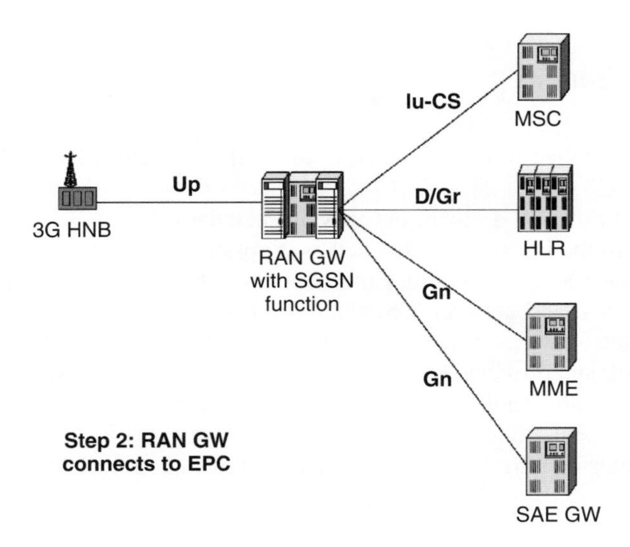

Figure 3.19 Step 2: RAN GW solution evolution

between the 3G home node B and the RAN GW is little changed. However the 3G air interface is improved with Rel-7 enhancements like continuous connectivity for packet data users, MIMO, downlink higher order modulation using 64 QAM for High Speed Downlink Packet Access (HSDPA), uplink higher order modulation using 16 Quadrature Amplitude Modulation (QAM) for High Speed Uplink Packet Access (HSUPA), etc.

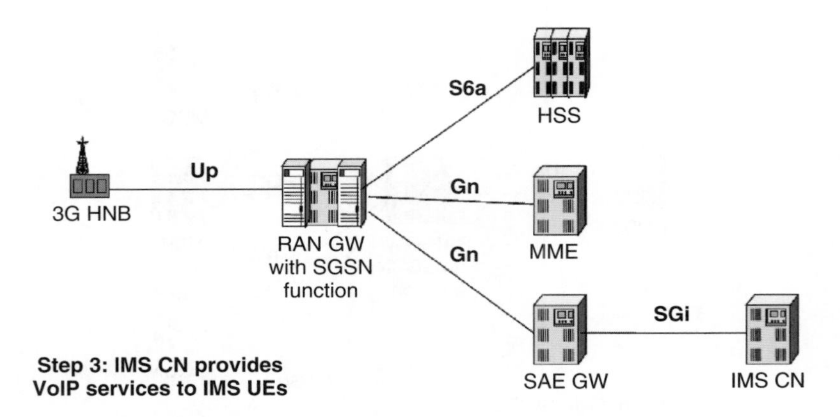

Figure 3.20 Step 3: RAN GW solution evolution

3.5 Architecture with IMS Support

At the time when this book was being written, 3GPP had just finished release 9. The following contents are mainly referred to [22].

3.5.1 Added Features

A non-IMS capable HNB subsystem is defined in [23], and based on it, the following main features were added to the architecture of an IMS capable HNB subsystem:

- support access to the CS domain, but it is not mandatory,
- support access to the PS domain: this makes it possible for a non-IMS-capable HNB to support the same PS access mobility mechanisms between different Closed Subscriber Group (CSG) cells, and between CSG and non-CSG cells,
- enable originated services requested by UEs with CS-specific NAS signalling to be interworked with and provided by the IP multimedia core network subsystem. Similarly, the architecture shall enable terminated services in IMS to be delivered to UEs with CS-specific NAS signalling.
- The CS/IMS interworking functionality shall be transparent to UE.

3.5.2 Alternative Architectures

In the standard, there are the following alternatives.

Option 1

IMS-capable HNB subsystem provides IMS-based services to CS UEs and IMS UEs. A reference architecture for the NAS control plane is shown in Figure 3.21. IM-Interworking Function (IWF) in the figure stands for IMS interworking function.

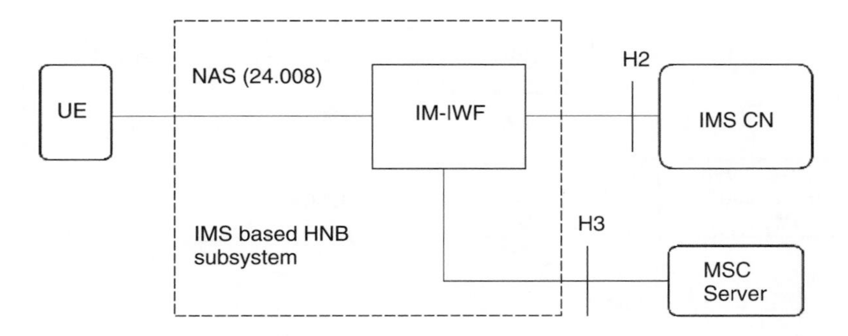

Figure 3.21 NAS control plane

Option 2: IMS-Capable HNB Subsystem Using IMS Centralized Services IWF

This option enhances the HNB Subsystem with IMS functionality by reusing the IMS centralized service (ICS) approach. ICS was defined in Rel-8 [24].

The IMS functionality (i.e. SIP UA) is provided by the IWF, which contains the functions equivalent to MSC server enhanced for ICS. Neither the UE, the HNB nor the HNB GW needs to be enhanced by IMS specific functions. The ICS IWF is connected via Iu-cs reference point to the HNB GW and reuses I2 and I3 from [24] for interworking to IMS. Figure 3.22 shows the general architecture.

Option 3: Interworking of IMS at HNB

The architecture reference model is illustrated in Figure 3.23. CS-to-IMS interworking is performed at the HNB in the architecture. Network elements and reference points that are introduced to support the HNB Subsystem, and the IMS Capable HNB Subsystem in particular, are shown by way of external boxes with heavy black lines. It consists of network elements of HNB, Home User Agent (HUA), HNB-GW, Enhanced MSC Server and defined reference points of HGm, Hi, Iuh, Iu-cs.

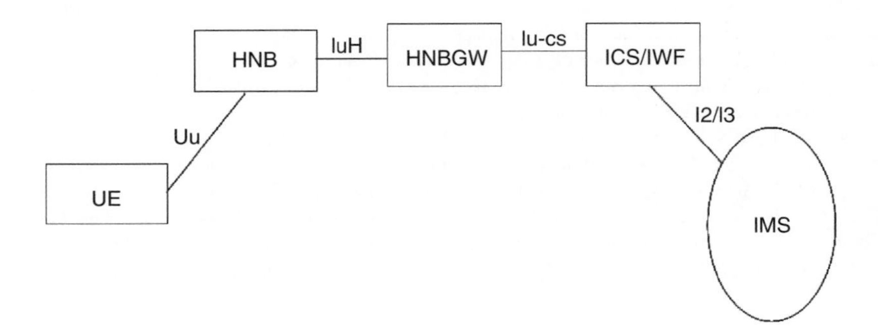

Figure 3.22 IMS capable HNB subsystem using ICS

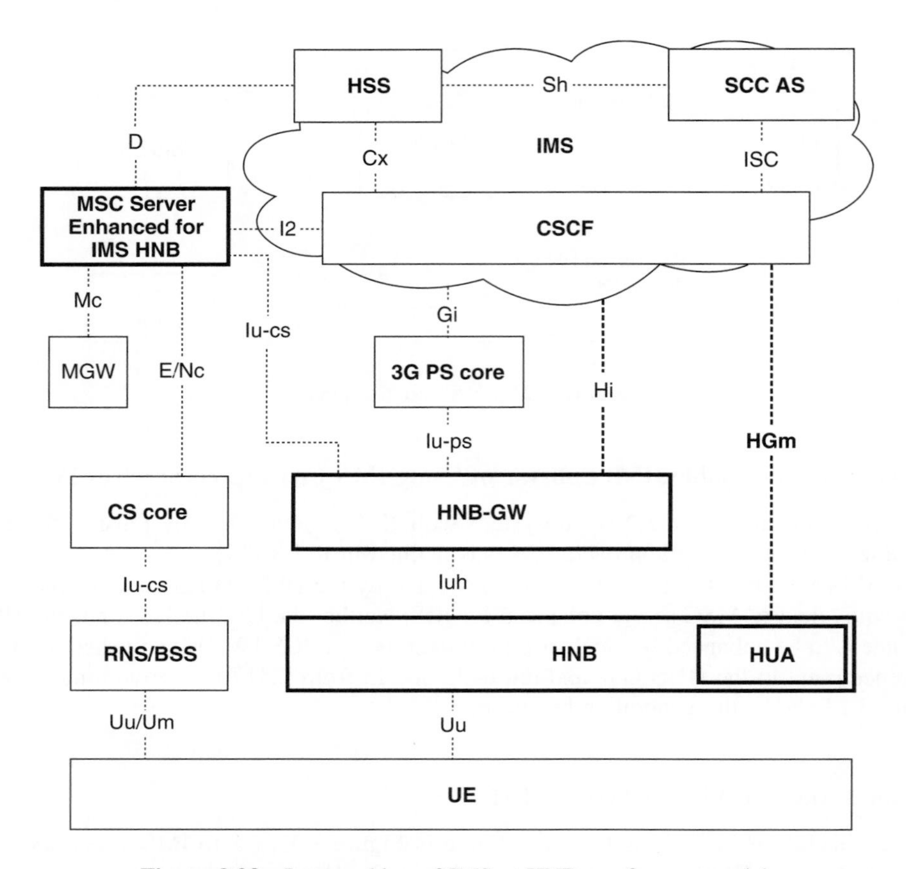

Figure 3.23 Interworking of IMS at HNB – reference model

References

[1] *TS 25.434 UTRAN Iub Interface Data Transport and Transport Signalling for Common Transport Channel Data Streams, Rel-7*.

[2] *TS 43.318, Generic Access Network (GAN) Stage 2, Rel-5*.

[3] *TS 44.318, Generic Access Network (GAN); Mobile GAN Interface Layer 3 Specification, Rel-5*.

[4] *TS 25.410, UTRAN Iu Interface: General Aspects and Principles, Rel-5*.

[5] *TS 25.450, UTRAN IuPC Interface General Aspects and Principles, Rel-5*.

[6] *TS 25.419, UTRAN IuBC Interface: Service Area Broadcast Protocol (SABP), Rel-5*.

[7] *TS 23.101, General UMTS Architecture, Rel-5*.

[8] *TS 29.234, 3GPP System to Wireless Local Area Network (WLAN) Interworking, Rel-5*.

[9] *DSL Forum TR-069, CPE WAN Management Protocol*.

[10] *TR 33.234, 3G Security; Wireless Local Area Network (WLAN) Interworking Security, Rel-5*.

[11] *TS 29.161, Interworking between the Public Land Mobile Network (PLMN) Supporting Packet Based Services with Wireless Local Area Network (WLAN) Access and Packet Data Networks (PDNs), Rel-5*.

[12] *TS 24.234, 3GPP System to Wireless Local Area Network (WLAN) Interworking; WLAN User Equipment (WLAN UE) to Network Protocols; Stage 3, Rel-5*.

[13] *TS 23.234, 3GPP System to Wireless Local Area Network (WLAN) Interworking; System Description, Rel-5*.
[14] *TS 33.234, 3G Security; Wireless Local Area Network (WLAN) Interworking Security, Rel-5*.
[15] *R3-080105, Kineto, NEC and Motorola GAN Variant of Iu-based 3G HNB architecture*.
[16] *TS 25.467, UTRAN architecture for 3G Home NodeB, Stage2, Rel-8 (Draft)*.
[17] *TS 25.412, UTRAN Iu Interface Signalling Transport, Rel-5*.
[18] *TS 25.415, UTRAN Iu Interface User Plane Protocols, Rel-5*.
[19] *TS 25.414, YTRAN Iu Interface Data Transport & Transport Signalling, Rel-5*.
[20] *R3-072309, Nomenclature and Architecture Proposal for HNB Access Network*, Motorola Std.
[21] *TS36.420, X2 General Aspects and Principles*.
[22] *TS 23.832, IMS Aspects of Architecture for Home Node B (HNB)*.
[23] *TR 23.830, Architecture Aspects of Home NodeB and Home eNodeB*.
[24] *TS 23.292, IP Multimedia Subsystem (IMS) Centralized Services; Stage 2*.

4

Air-Interface Technologies

Alvaro Valcarce and Enjie Liu

4.1 Introduction

Today, several technologies coexist that are capable of providing wireless data access in an indoor environment. For instance, Unlicensed Mobile Access (UMA) [1] allows data connectivity by means of technologies based on an unlicensed spectrum (e.g. Bluetooth). Furthermore, IEEE 802.11 or WiFi is also a common alternative for indoor connections and is typically presented as a competing technology to femtocells. Since WiFi is a well established technology, femtocell-sceptics argue that it will not be easy to convince WiFi users to change to femtocells in order to enjoy Small Office/Home Office (SOHO) connectivity. Why would someone be interested in aquiring a new device to emit in a licensed spectral band, while other systems have worked well so far? To answer this question, it is crucial to understand the characteristics of the various available technologies that will be described in this chapter. The RF technologies on which current and future femtocells are based, are therefore presented here to let the readers create their own opinion about the suitability of the different types of home base stations.

Furthermore, there also exist some fundamental differences that must be taken into account when evaluating distinct systems. For instance, the Medium Access Control (MAC) layer of WiFi relies on collision avoidance, which compels different users to compete continuously for the resources of the access point. As a consequence, users with better signal quality (e.g. users closer to the access point) might obtain better resources and more frequently than other users. The consequence of this is that services requiring a certain Quality of Service (QoS) level (e.g. VoIP, online gaming, media streaming, remote surgery, ...) can not be guaranteed for more than a few users. On the other hand, the MAC layer of mobile technologies such as UMTS typically relies on scheduled approaches, allowing different User Equipments (UEs) to specify QoS requirements. This might lead one to conclude that UMTS is better suited to support this type of service than is WiFi.

Femtocells: Technologies and Deployment Jie Zhang and Guillaume de la Roche

It has been shown in the previous chapters that the femtocell concept can be based upon a wide variety of wireless technologies, so great care must be taken before choosing which is the most appropriate one for a given scenario. In order to serve as a quick reference, an overview of the main features of different technologies for the air interface is presented in the following sections. This is intended to serve as a coarse description of the diverse existing approaches that will help the reader grasp the fundamentals of femtocell air interfaces and to find the appropriate reference for a more in-depth study.

4.2 2G Femtocells: GSM

Although most of the FAP manufacturers have concentrated on producing 3G femtocells, there are also some reasons for building Global System for Mobile communication (GSM) air interfaces into femtocells. GSM is an old system compared with UMTS and LTE. Nevertheless, it is well tested and still holds the biggest number of subscribers compared with newer networks. In 2009, in some countries like India, GSM macrocell networks were in rapid expansion, being UMTS barely considered for newer rollouts. This is mainly due to its dramatically lower cost compared with more modern networks. The fact that GSM femtocells are substantially cheaper to produce than 3G ones, makes them thus more competitive against other technologies such as voice over WiFi or UMA.

For instance, the Swedish telecom giant Ericsson was one of the first manufacturers to produce this type of femtocell. In February 2007, Ericsson launched its first GSM FAP model and in September 2008 they signed a contract for its deployment with the British supermarket company Tesco [2]. This agreement will have GSM femtocells from Ericsson installed in Tesco shops all around the United Kingdom, which will be used by their employees to roam onto the Orange network. Another example of investment on GSM femtocells is the Scottish network operator Hay Systems Ltd (HSL), who announced in January 2009 [3] the production of 2.75G femtocells with support for GSM, General Packet Radio Service (GPRS) and EDGE air interfaces.

There are however also several reasons argued by other members of the industry, for not producing GSM femtocells. For instance, the power control mechanism in GSM is not as flexible as in 3G and this might be an important source of interference to overlayed macrocells. Furthermore, GPRS's achievable throughputs are quite low compared to newer systems, meaning that 2G femtocells would not be able to provide much more than a high quality voice service. If that is the case, would an indoor user be interested in purchasing a femtocell to be used just for voice? Hence, the economic viability of this type of femtocell is still arguable.

In the following sections, the fundamentals of the GSM standard are presented and described in relationship to femtocell networks.

4.2.1 *The Network*

The fundamental structure of a GSM macrocell network is shown in Figure 4.1. Additions to this architecture have been made over the years, in particular the introduction of GPRS, which extended the network's use from voice to data transfer at throughputs close to 100 kbps.

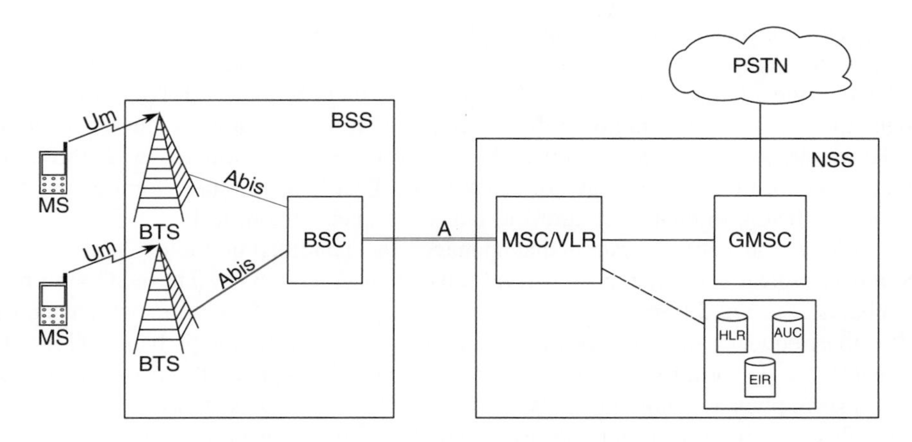

Figure 4.1 Main GSM network architecture

The basic GSM network is divided in two main parts:

- The Base Station Subsystem (BSS): This sets up and maintains radio connections between Mobile Stations (MS) and the core network. Although this part traditionally comprises different network elements, the functions carried out by the BSS must be fully integrated by a single Femtocell Access Point (FAP).
- The Network Switching Subsystem (NSS): This is the core network of the Mobile Network Operator (MNO) and it performs call routing as well as subscriber-related tasks.

Base Station Subsystem (BSS)

The BSS is the point of entrance of mobile users to the GSM network and has two well defined elements: The Base Transceiver Station (BTS) and the Base Station Controller (BSC). Due to the difficulties of controlling the radio parameters of millions of femtocells within the network, the BSC functionalities must be carried out by the femtocell itself in a self-configuring manner. Therefore and although the tasks assigned to the BTS and to the BSC are different in the macrocell GSM network, these are carried out at the FAP in a GSM femtocell network.

The BTS contains the necessary radio equipment to send and receive data to and from the mobile users. This includes mainly transceivers and antennas, which are responsible for creating the coverage cell and the handling of wireless links. In the GSM standard, a total of eight Time Division Multiple Access (TDMA) time slots are defined in both downlink and uplink, which allow the BTS to serve a total of seven users. This is because one of those TDMA slots is used by the BTS for the broadcasting of signalling and system information (the Broadcast Control Channel (BCCH)). However, it is possible for a BTS with several transceivers at different frequencies to support more users. This is the case for instance, of sectorized BTS, which uses directional antennas to cover different regions from the same station. FAPs with sectorized antennas have been proposed [4] for interference avoidance, so this feature might be used in GSM femtocells to increase the achievable throughput. Among other tasks, the BTS is also responsible for the ciphering and deciphering of data exchanges with the MSs. In general, the BTS is the network element that provides the radio resources required for the establishment of a wireless link. However, such resources need to be controlled and coordinated with those from BTS at other locations in the case of macrocells. That is why there is another network element (the BSC), in charge of carrying out these management tasks.

The BSC is the control system that manages the handling of radio resources at the BTS and is typically in charge of several tens of macrocell BTS. The BSC assigns the radio channels to the different BTS, it controls the power levels within each channel and handles the frequency hopping for the mobile user when a cell change happens. It is thus responsible for performing the majority of handovers, except for the case when a BTS change also implies a change in the BSC. The BSC can also be seen as a concentrator of the data fluxes arriving from mobile users, which are then transmitted over to the Mobile Switching Center (MSC). BTSs communicate with the MSC via an Abis* interface, which is generally implemented by means of T1 lines. In the case of femtocells, this connection is all-ip and hence, the information and signalling is sent over to the MSC through the Internet. In some remote rural areas where it is hard to tend lines from each BTS, it is also possible to relay the information between close BTS by means of cheaper microwave links [5].

Network Switching Subsystem (NSS)

The Network Switching Subsystem (NSS) is the core of the GSM network. It connects mobile stations to the Public Switched Telephone Network (PSTN) and also between each other. Furthermore, it also performs tasks related to the billing of subscribers.

The main part within the NSS is the Mobile Switching Center (MSC), which can be considered as an enormous call switching block, routing connections from hundreds of BTS. The MSC arranges end-to-end communications by processing connection requests that arrive directly from the MS. When such a connection is successful, the MS will be registered in the Visitor Location Register (VLR), which is a database with temporary information about the mobile users. This database is typically integrated into the MSC itself although it can also be installed in another machine. The VLR contains information only about *visitors* to the MSC to which the VLR is associated. This means that information about MSs located in the neighbourhood of a given MSC will be stored in such VLR. Furthermore, the location of the mobile terminal† is stored in the Home Location Register (HLR), which is the main database of the mobile operator. Thus, all information regarding subscriptions, user profile, hired services and usage statistics is stored in the HLR. This implies that any user that purchases a GSM subscription from a network operator will be registered with the HLR right before performing the first call.

When a handover requires a change of BSC, the process is handled at the MSC. This does not happen very often in macrocell networks. However, a user leaving his home while

* Abis refers to the fact that this is the second A interface, being the first A interface the one connecting the BSC to the MSC.

† The precision of the location information is known only until the VLR where the MS is registered.

on a GSM connection, needs to be handed over from the femtocell to the macrocell. Since there is no common BSC between the femto and the macrocell, it is the responsability of the core network to perform the handover properly. This becomes even more complex in the case of open-access femtocells, where a using walking down the street might be continuously handed over between tens of femtocells.

The Gateway Mobile Switching Center (GMSC) is another network element built into the MSC, which is used to connect the PSTN network with the mobile network. When a call to a MS originates in the PSTN, it will first arrive at the mobile network via a GMSC, which will decide to which MSC the requested user is connected. Then, an appropriate route for the call will be set.

The Authentication Centre (AUC) is the unit providing authentication of mobile users. When a user requests access to the network, the AUC will send a random number which the MS will cipher using the key within his/her Subscriber Identity Module (SIM) card. The encrypted number is then sent back to the AUC where it will be deciphered and compared with the original one. If they match, the user is then authenticated into the network.

The Equipment Identity Register (EIR) is another database that keeps track of valid mobile devices. This allows the operator to block calls from stolen, broken or simply unauthorized devices. Mobile terminals are identified in this database by their International Mobile Equipment Identity (IMEI), which is a unique identifier for each mobile terminal.

The GSM network was initially designed for voice communications. Therefore, in order to provide support to data transmissions, the GPRS was introduced into the GSM network. This service requires a new network element called the Serving GPRS Support Node (SGSN), which connects to the BSC and serves as the entry point to the IP-based GPRS network. Finally, another element called the Gateway GPRS Support Node (GGSN) is introduced to communicate the SGSN to outside networks such as the Internet.

A main difference between the traditional GSM network and one containing femtocells is the number of cells. The macrocell GSM network of an operator has several thousands of macrocells. However, in a femtocell network this number scales up to hundreds of thousands or even millions. It is thus necessary for the core network to have the ability to receive all of the data fluxes from individual femtocells. These independent flows are concentrated in a new network element called the Femto Gateway (FGW). There are several proposed architectures (see Figure 4.2) for the way the FGW is to be connected to the core network. These are discussed in depth in Chapter 3.

4.2.2 The Air Interface

For historical reasons the frequency bands allocated to GSM systems vary between countries. However, there is a high level of agreement and most of the networks deployed worldwide function in the 900, 1800 and 2100 MHz frequency bands. For example, most countries in Europe use the 900 and 1800 MHz bands, while north America works on the 850 and 1900 MHz bands. However, countries from eastern Europe and Russia use the 450 MHz band. If it is assumed that FAPs are portable and roam internationally, they will need to support a wide variety of transmission frequencies. The incorporation of sufficient transceivers on these devices can increase their cost, making them not easily affordable. However, femtocells are oriented to the home market and it is not expected that they will

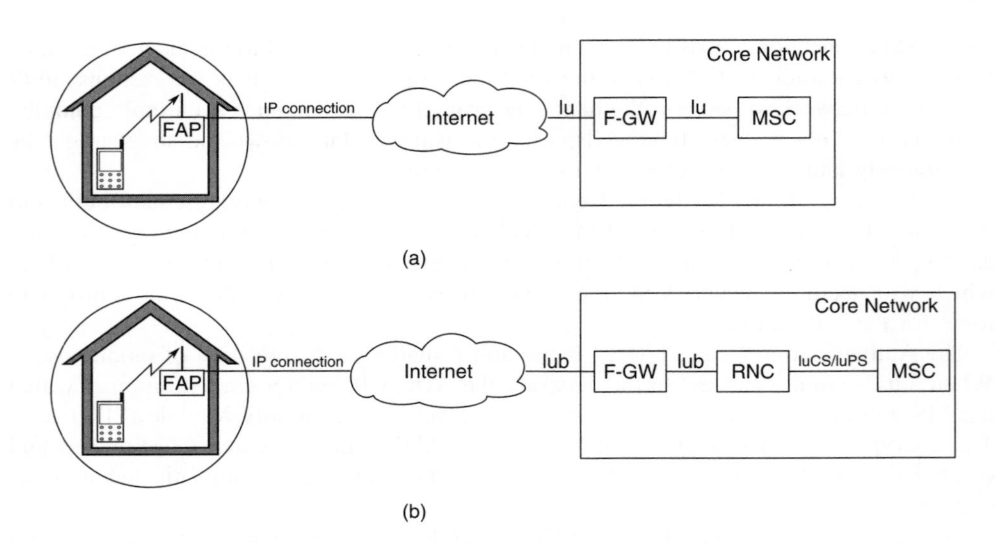

Figure 4.2 Generic femtocell network architectures. (a) Iu-over-IP. (b) Iub-over-IP

be relocated as often as mobile terminals. For the sake of economic viability, GSM FAPs will therefore be designed to include transceivers that work only in the licensed frequency band for a given country. Some proposals have even been made to detect the location of the FAP and block its transmissions when in a country different to the one it is intended to operate in.

Figure 4.3 shows an example of the frequency assignment in GSM systems using the 900 MHz band. As can be seen, the main duplexing technique for uplink and downlink transmissions in GSM networks is Frequency Division Duplexing (FDD), which assigns the lower frequency band to the MS and the upper band to the BTS. This design is based on the fact that transmitting on lower frequencies requires lower energy than it does at higher ones. It is therefore supposed to save the battery on mobile devices.

Of all the frequency channels available in the uplink and downlink bands, one is always left as a guard band. In GSM, transmission channels have a bandwidth of 200 kHz so it is easy to see that the number N_{ch} of available transmission carrier frequencies in each direction (downlink and uplink) is:

$$N_{ch} = \frac{25\,\text{MHz}}{200\,\text{KHz}} - 1 = 124 \tag{4.1}$$

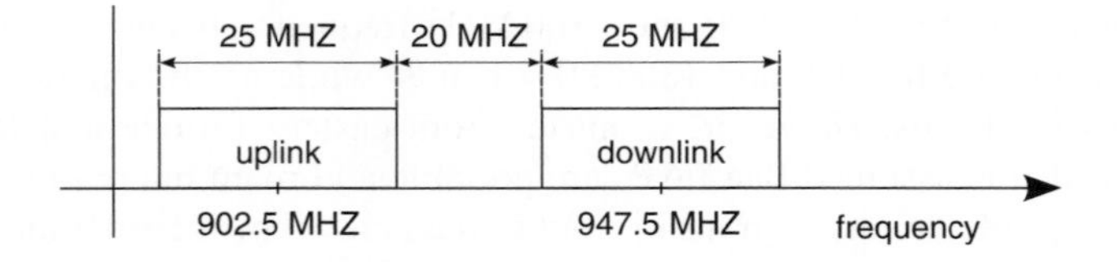

Figure 4.3 Spectral bands of GSM-900

Each country has followed its own procedures for the assignment of licenses in those channels, typically by auctioning emission permits to different operators. Therefore, each operator uses a subgroup of those N_{ch} channels containing more or fewer channels depending on the price paid for the license. It must however be highlighted that the channels are assigned in uplink/downlink pairs, i.e. it is not possible to pay a license for a downlink or uplink channel independently. Since there is a 20 MHz separation between the uplink and downlink bands, the total frequency separation between an uplink and downlink transmission channel results in 25 MHz + 20 MHz = 45 MHz, being the channel pair handled by the same transceiver within the radio equipment.

The price of a license for the GSM transmission channels raises up to several million Euros and there is thus a great interest by some of the operators to maximize the efficiency of spectral usage. The deployment of femtocells using the same frequency channels as the overlaid macrocells increases the overall network throughput without requiring additional GSM channels. Therefore, network operators could obtain higher profits by maximizing the Area Spectral Efficiency (ASE).

The GSM TDMA Frame

GSM manages the transmissions from different users using Time Division Multiple Access (TDMA) in a round-robin fashion. This means that each frequency channel is shared over time by several users, each being allowed to transmit in one of the 200 kHz channels for a given period of time. This is done by subdividing the frequency band into eight consecutive slots over the time domain. A single time slot is called a *burst period* and it has a duration of $15/26 \approx 0.577$ ms (see Figure 4.4). The set of eight consecutive time slots starting with slot 0 is called a *TDMA frame* and it has a duration of $120/26 \approx 4.615$ ms during which information bits are transmitted. For instance, a data burst has a time duration equal to the time it takes to transmit 156.25 bits, although not all of them contain useful information. As seen in Figure 4.4, a normal data burst carries the following 148 bits:

Tail bits: The 3 bits at the beginning and at the end of every data burst are always set to zero. These are used to clear the Viterbi equalizer in a GSM receiver by setting the equalizer to an initial state and leaving it ready to receive the next data burst.

Data bits: These are the containers of the useful voice data from and to the users.

Training bits: The content of this field is previously agreed by transmitter and receiver. Then, the receiver can use the received value for channel equalization.

Stealing flags: The single bits before and after the training bits indicate whether or not the current burst contains user data or control information.

Guard interval: This is a period of time during which nothing is transmitted. This avoids adjacent bursts interfering with each other because of propagation delays in a multipath environment.

Time slot 0 of a downlink TDMA frame is typically used to allocate the Broadcast Control Channel (BCCH), which is used by the BTS continuously to transmit the cell ID, Mobile Network Code (MNC), Location Area Code (LAC), etc. This leaves only seven time slots to be used for user traffic. However, the GSM standard allows several

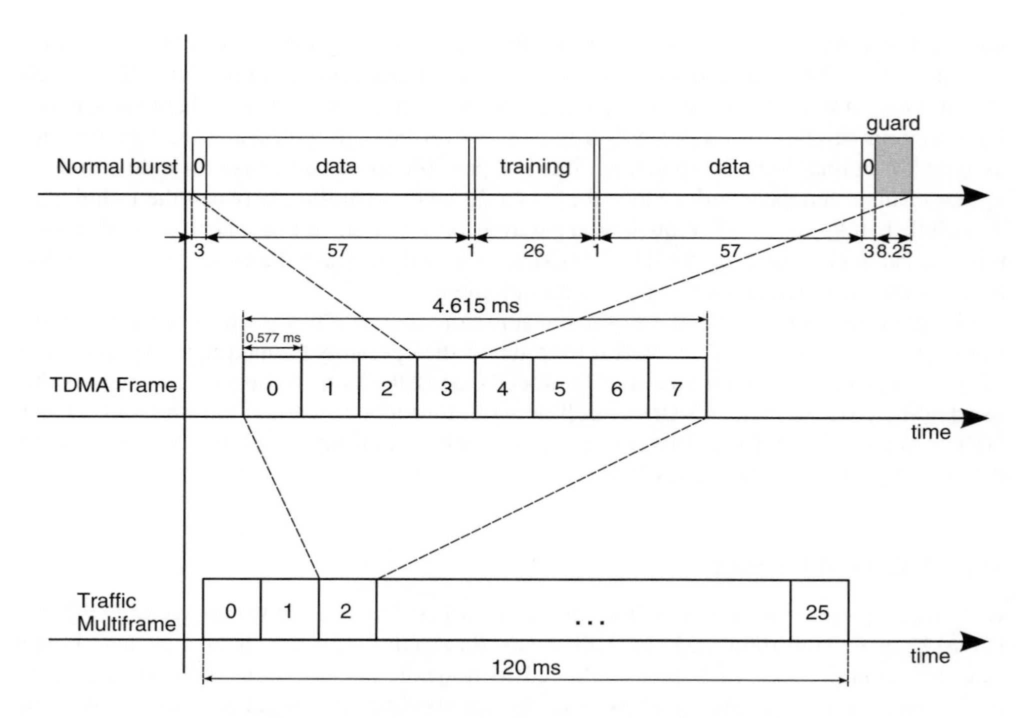

Figure 4.4 GSM traffic multiframe, frame and data burst

frequency channels to share a common BCCH, thus having all eight time slots available at other channels. Nevertheless, this must be carefully balanced by the FAP manufacturer depending on the traffic requirements of the services offered by the femtocell. This is because there might be some services for which having one additional traffic channel is crucial. However, having only one BCCH for several traffic channels might be unfeasible in terms of signalling. Hence, an appropriate dimensioning is necessary.

In GSM, a distinction is made between channels according to the type of information they carry. Thus there are traffic and control channels. Such channels are built up from several TDMA frames forming a much larger *multiframe*. The different types of multiframe are explained in the following.

The GSM Traffic Channel

The Traffic Channel (TCH) in GSM is a multiframe composed of 26 shorter TDMA frames (see Figure 4.4). However, only 24 of those frames are really used for user voice traffic, while the other two are reserved for transmitting the Slow Associated Control Channel (SACCH). The SACCH is used for the exchange of power control information between the MS and the BSC, which functionality is built into the femtocell access point. This is sent alternatively in frames 12 and 25 of the TCH multiframe and it carries in uplink power measurement reports from the MS, as well as power control commands in the downlink.

The GSM Control Channels

Figure 4.5 classifies the different control channels in GSM. The BCHs are used by the FAP to send network-related information to the MS and there are three of them:

- As explained above, the BCCH is used by the BTS to transmit continuously the cell ID and other network related information. The MS reads this channel to detect which cell it is currently in and to know the channel power.
- The Synchronization Channel (SCH) is used to help the MS synchronize to the frames sent by the BTS. This is done by sending a training sequence as well as the Base Station Identity Code (BSIC).
- The Frequency-Correlation Channel (FCCH) carries a sine tone. This is used by the MS to synchronize its frequency with that of the BTS.

The Common Control Channel (CCCH) is used for call establishment and is subdivided into three channels:

- The Paging Channel (PCH) is used to notify the MS that there is an incoming call.
- The Random Access Channel (RACH) is an uplink channel in which the MS requests access to the network. These requests follow the slotted Aloha protocol.
- The Access Grant Channel (AGCH) is used to notify the MS of the assigned slot after a request has been made through the RACH.

The Fast Associated Control Channel (FACCH) is an special channel used to transmit control information in cases where the signal quality drops quickly or during handovers. For example, if an obstacle suddenly appears between the MS and the FAP (someone passing by or an inner wall), the mobile transmitter might need to transmit with higher power to reach the FAP. Since this situation requires a quick intervention, the traffic channel is replaced temporarily by the FACCH to deal with this power increase request. The *stealing bits* or *flags* are transmitted before and after the training sequence in the burst of Figure 4.4. These indicate to the receiver that the current burst is a control message (e.g. the FACCH) so that the appropriate routines take control. The same applies to situations in which a user of an open-access femtocell is leaving his home and a handover to the macrocell outside takes place. As the power received from the femtocell drops quickly,

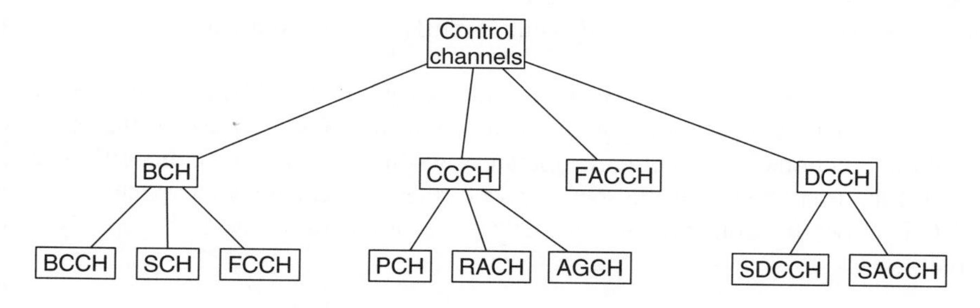

Figure 4.5 GSM logical control channels

the FACCH is used for a fast information exchange between the mobile user and the network so that the handover is done as rapidly as possible.

The Dedicated Control Channel (DCCH) is used for signalling purposes between the MS and the GSM network. It comprises the following channels:

- The Standalone Dedicated Control Channel (SDCCH) is used for signalling a call establishment.
- As explained above, the SACCH exchanges power control information between the MS and the FAP each multiframe.

4.3 3G Femtocells: UMTS and HSPA

As 3GPP specifications evolve, the network becomes more access independent, allowing connections to reach the core network by means other than just through the SGSN. This adds versatility to base stations, which are then capable of connecting through IP-based networks. These and other features make UMTS technology better prepared than GSM for the deployment of femtocells. To be consistent with the 3GPP terminology, the term Home NodeB (HNB) is used in this section to refer to the FAPs.

Thanks to interference averaging, WCDMA receivers are able of separating UMTS signals at very low levels of SINR. From the point of view of the air interface, UMTS is hence also better suited than GSM to cope with the high interference levels of two-layer networks. Besides, given the fact that UMTS delivers much higher data rates than GSM and that it is already a well tested technology, most manufacturers have concentrated on the development of UMTS-based HNBs.

The Universal Mobile Telecommunication System (UMTS) [6] is nothing other than the name given to a set of radio technologies specified by the 3rd Generation Partnership Project (3GPP) Radio Access Network (RAN) group [7]. The name given to the air interface of UMTS is UMTS Terrestrial Radio Access (UTRA), which is specified for functioning in FDD and Time Division Duplexing (TDD) modes. This is in opposition to GSM, which is only specified for FDD (see Figure 4.3). The main 3GPP Technical Specifications (TS) of the UMTS air interface are two (see Table 4.1): TS 25.101, which can be found in [8] specifies the minimum RF features that the FDD mode of UTRA must provide in the UE. Then, TS 25.102, can be found in [9] and it specifies the requirements of the TDD variant. Although the two options exist, most of the UMTS networks deployed worldwide use UMTS in FDD mode. This is mainly due to interference issues rising between adjacent NodeBs that transmit in the same frequency band.

In contrast to GSM, UMTS HNBs were standardized by 3GPP towards the end of 2008 in TS 22.220 [10]. This allows vendors to develop their FAPs in a way that guarantees a minimum functionality to network operators wishing to deploy femtocells. However, [10] is still the first version of the standard so further refinements are expected in the near future. Other specifications released and under development by the 3GPP are also shown in Table 4.1.

Table 4.1 UTRA specifications (selected)

Technology	Specification	Title	Release
W-CDMA HSDPA HSUPA	25.101	User Equipment (UE) radio transmission and reception (FDD)	R99 R4 R5 R6
W-CDMA HSDPA HSUPA	25.102	User Equipment (UE) radio transmission and reception (TDD)	R99 R4 R5 R6
3G	22.220	Service requirements for Home NodeBs and Home eNodeBs	R8
	25.469	UTRAN Iuh interface Home Node B (HNB) Application Part (HNBAP) signalling	R8
	25.820	3G Home NodeB Study Item Technical Report	R8
	25.967	FDD Home NodeB RF Requirements	R8

4.3.1 CDMA Fundamentals

Code Division Multiple Access (CDMA) is the medium access technology used in UTRA. However, the specific implementation of CDMA in UMTS is called Wideband Code Division Multiple Access (WCDMA), which distinguishes it from implementations used in other systems such as CDMA2000 or Evolution-Data Optimized (EVDO). The *wideband* clarification refers to the fact that the bandwidth used in UMTS systems is larger than that used in others. In contrast to other technologies, CDMA allows all users to simultaneously transmit over all the available bandwidth, being the different transmissions separated throughout the use of orthogonal codes.

The fundamental concept beneath CDMA is the *spread spectrum*, which consists of modulating data signals in such a way that the resulting signal has a much larger bandwidth than the original one. This is achieved by applying an XOR operator between the data and coding signals. Figure 4.6 illustrates this procedure, where s_b represents the original binary data signal and c is the spreading code or modulating signal. It can be proved that if the rate of c is larger than that of s_b, then the resulting signal x_c also has a larger bandwidth than the original s_b. The coding signal c is called the *spreading* signal because it modifies the spectral properties of the original signal by widening its spectral occupation. c has thus a much higher bit rate than the data signal and its elements are called *chips* to distinguish them from the binary data bits. In Figure 4.7 a data signal s_b

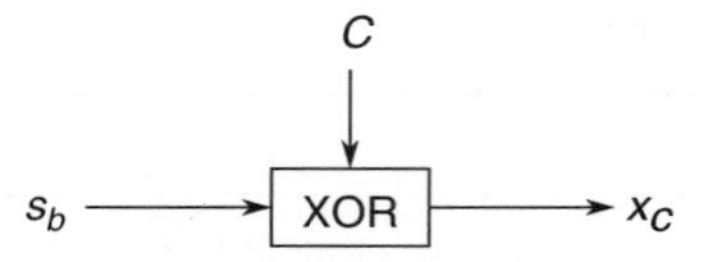

Figure 4.6 CDMA modulation process

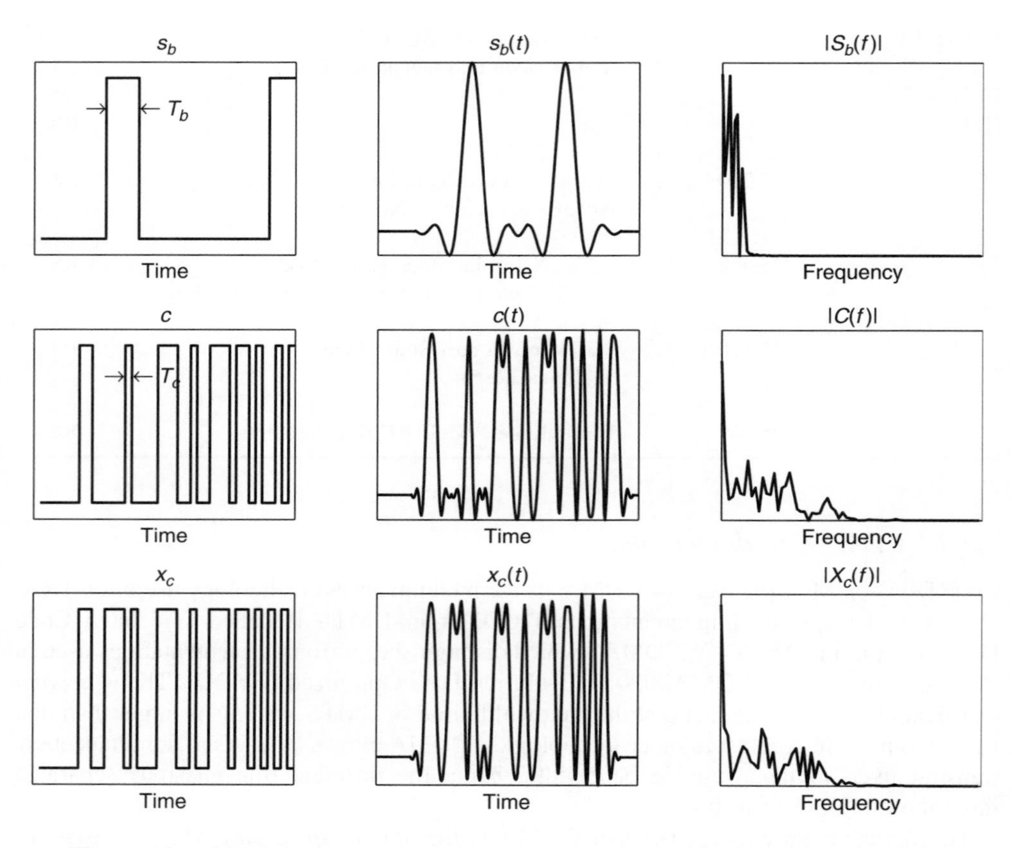

Figure 4.7 Time and frequency domain aspect of CDMA data and spreading signals

with period T_b is modulated by a spreading signal c with period T_c. The transmission filter in this example is a square root raised cosine filter and it can be seen that the resulting x_c has a larger bandwidth than s_b. The bit rate of s_b is $R_b = 1/T_b$, while the chip rate of c is $R_c = 1/T_c$. The Spreading Factor (SF) can be thus defined as:

$$SF = \frac{R_c}{R_b} = \frac{T_b}{T_c} \tag{4.2}$$

and it is an integer that gives the total number of chips per data bit.

The objective of the reception of CDMA signals is to separate the signals from different users. This is done by calculating the cross-correlation of the received signal and the user's spreading code. Only when the resulting cross-correlation reaches its maximum, the corresponding signal can then be extracted. The demodulation is performed by applying the inverse process of Figure 4.6, i.e. the received signal is XOR with the despreading code. This recovers the original data sequence by reducing the bandwidth of the received signal and increasing its power density. It is interesting to note that, if a narrowband interfering signal is present, the effect of this process is a spectral spread. Hence, the power density of a narrowband interfering signal is decreased while that of the data signal is increased. This property makes CDMA systems particularly resistant to narrowband interference. This does however not occur with broadband interference such as signals from other users, which remain as broadband interference even after the despreading process. The *processing gain* is thus defined as the increase in power density of the desired data signal and it is equal to the spreading factor *SF*.

The codes used to allocate several users in CDMA are ideally mutually orthogonal. However, this relies on the coding sequences of different users being perfectly aligned in time (synchronous CDMA). Since such a level of synchronization is not easily achievable in certain situations (e.g. uplink connections from independent UEs), asynchronous CDMA makes use of Pseudorandom Noise (PN) sequences, which have better resistance to time shifts. For more information about spreading codes and other aspects of CDMA technology, the reader is referred to [11].

4.3.2 The Network

Figure 4.8 shows the UMTS network architecture as of Release 4 (published in April 2001) where the changes with respect to the GSM network of Figure 4.1 are already evident. The different parts of the UMTS network are explained in the following.

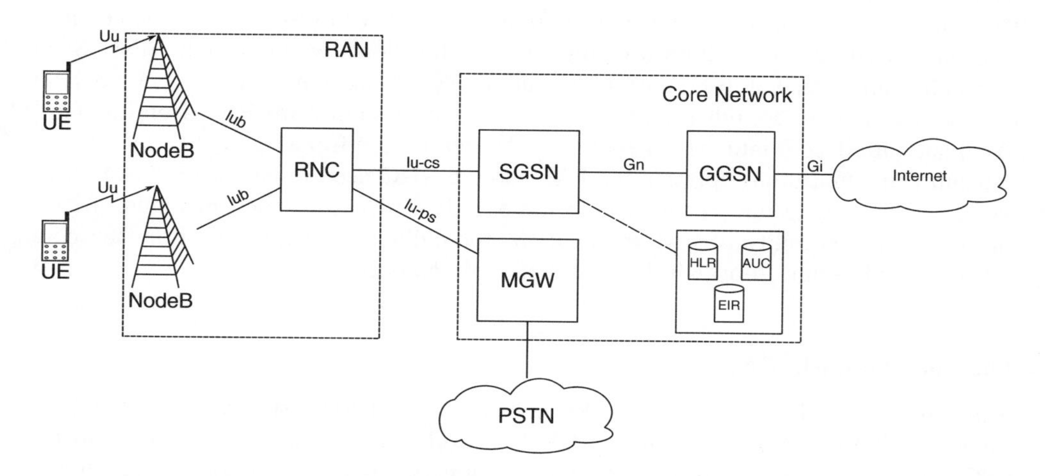

Figure 4.8 UMTS network architecture in Release 4

The Radio Access Network (RAN)

In UMTS, the RAN is the equivalent of the Base Station Subsystem (BSS) in GSM networks. The RAN in UMTS is called UMTS Terrestrial Radio Access Network (UTRAN), it connects the UEs with the core network and is responsible for handling the air interface. The only part of UTRAN visible to the mobile user is the NodeB, which handles a single cell throughout the WCDMA air-interface. The name *NodeB* (Node for Broadband Access) was given temporarily to this element during the standardization process at 3GPP. However, it soon became widely used and hence remained as the official term to denote UMTS base stations [12]. One of the main differences between UMTS and GSM is that UMTS introduced Asynchronous Transfer Mode (ATM) as the transport protocol to be used for carrying information and signalling in the backhaul. NodeBs are therefore connected to the RNC by ATM links through a logical interface called Iub [13], which is used for the negotiation of radio resources between the NodeB and the Radio Network Controller (RNC). In UMTS, NodeBs support cell sectorizing, i.e. each cell can be divided into several angular sectors served at different carrier frequencies.

The RNC is the equivalent of the BSC in GSM networks and is the decision-making element within UTRAN. The RNC controls several NodeBs, performs Radio Resource Management (RRM) and is capable of directly communicating with other RNCs through the Iur interface. Furthermore, the RNC communicates with the core network by means of a logical interface called *Iu*. The RNC performs Call Admission Control (CAC) and accepts or rejects calls depending on the level of interference present at the NodeB requesting the call. It also assigns CDMA codes to the UEs and determines the power control limits to avoid near – far problems (see Section 6.2.3).

In the case of femtocells, it is possible for the RAN to be fully integrated into the FAP device. However, there are different approaches where this could be done. If the RNC functionality is to be performed by the femtocell, then Iu messages to the core network need to be encapsulated into IP packets in order to be transmitted through the Internet. This configuration is called an Iu-tunnel or Iu-over-IP and it seems an appropriate architecture for small and medium businesses [14], where several users simultaneously access the femtocell. The elevated number of users in this case with respect to the home environment, introduces the need for bringing the RNC functionality closer to the HNB. Such an architecture is shown in Figures 4.2(a) and 4.9. Since the number of UEs in a SOHO environment is reduced, other approaches are needed to remove the RNC from the femtocell and introduce it into the core network. As can be seen from Figure 4.8, this implies that Iub communications between the FAP and the RNC need to be encapsulated on IP packets. This architecture is thus called Iub-over-IP and is illustrated in Figure 4.2(b). However, the feasibility of performing RRM for millions of femtocells from the core network is uncertain, Iu-over-IP being a preferred approach.

The Core Network (CN)

Although the RAN was remodelled, Release 99 of the UMTS standard kept the Core Network (CN) of GSM. MSCs in the GSM network used to transfer voice to/from the BSC via the A interface. However, UTRAN now uses ATM for transporting speech at a different rate and hence RNCs cannot talk directly to old MSCs. Due to this, a new network

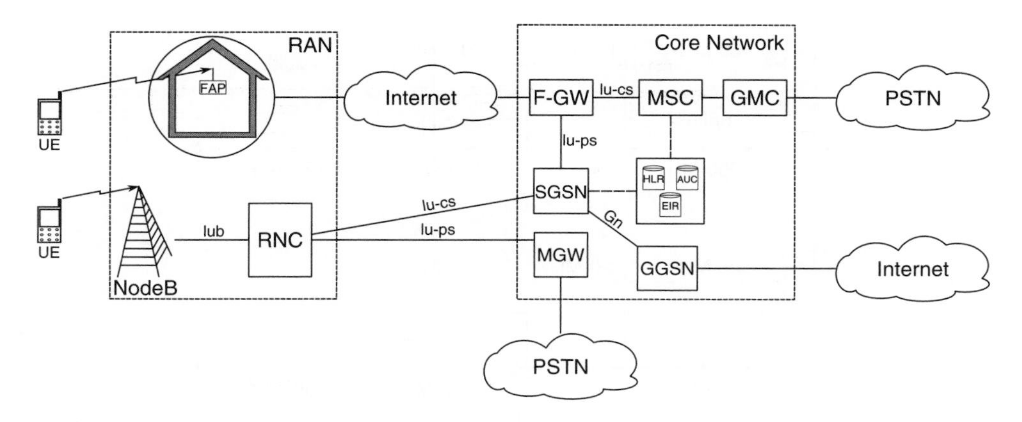

Figure 4.9 UMTS R4 femtocell network based on an Iu-over-IP architecture

element called the Media Gateway (MGW) was introduced to interface between UTRAN and GSM MSCs. As the specifications progressed, Release 4 changed the old GSM CN for an all-IP core and the MSC was finally removed from the network architecture (see Figure 4.8).

The SGSN is a network element inherited from the GPRS network that deals with data communications. It routes incoming packets to/from the appropriate RNC and it authenticates users into the network of the operator. The GGSN is nothing other than the entry point of the SGSN to the Internet.

4.3.3 The Air Interface

Figure 4.10(a) shows the frequency bands allocated to the FDD mode of UTRA. Similar to GSM, UMTS channels are *paired*, i.e. a single license is issued for a pair of uplink–downlink frequency channels. Each UMTS channel requires a bandwidth of 5 MHz, which means that there are only $60/5 = 12$ UMTS channels in the licensed band. However, in order for an operator to implement appropriate frequency planning and avoid interference between its NodeBs, more than one channel is necessary.

The UMTS protocol stack has two well defined parts:

- The Access Stratum (AS) comprises the layers that make up UTRAN plus lower layers that implement the ATM transport functionality.
- The Non-Access Stratum (NAS) includes the upper layers that communicate the UE with the CN.

Another key concept is that of the Radio Access Bearer (RAB), which is the means for transmitting information that the AS provides to the NAS. An RAB is basically the Service Access Point (SAP) that the Radio Link Control (RLC) layer provides to the upper layers in the UTRA protocol stack. This is shown in Figure 4.11, where the different elements of the RAN have been bundled together in the same network structural block to highlight the

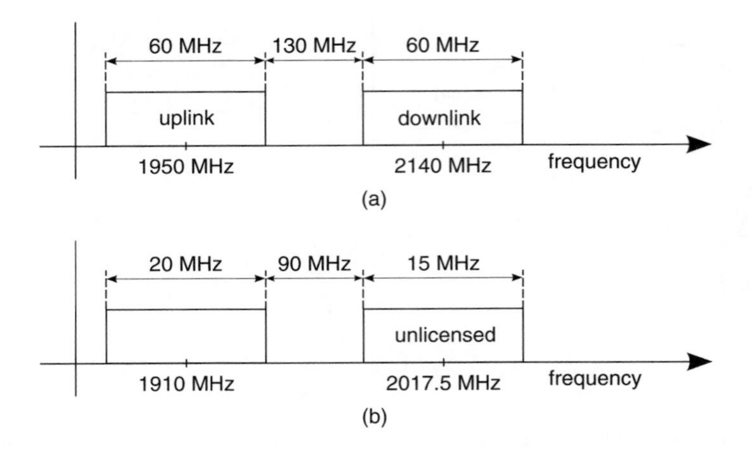

Figure 4.10 Operation bands of UTRA in Europe. (a) FDD. (b) TDD

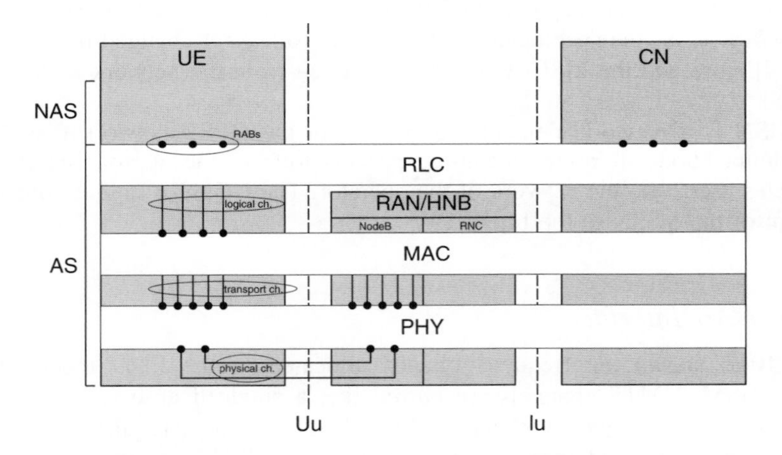

Figure 4.11 UTRAN channels and protocol stack

Iu-over-IP architecture of Figure 4.2(a). RABs are continuously established and released in order to provide transmission capabilities with different QoS to UMTS channels.

UTRA is thus one of the fundamental parts of the AS because it communicates the UE with the NodeB and it is the entry point of users into the UMTS network. The UMTS air-interface channels are SAPs provided by the lower layers and are classified into three categories:

- *Logical* channels are provided by the MAC layer to the RLC layer for the transmission of user information.
- *Transport* channels are provided by the Physical (PHY) layer.
- *Physical* channels are transmitted over the air and communicate the PHY layers of different UTRA elements.

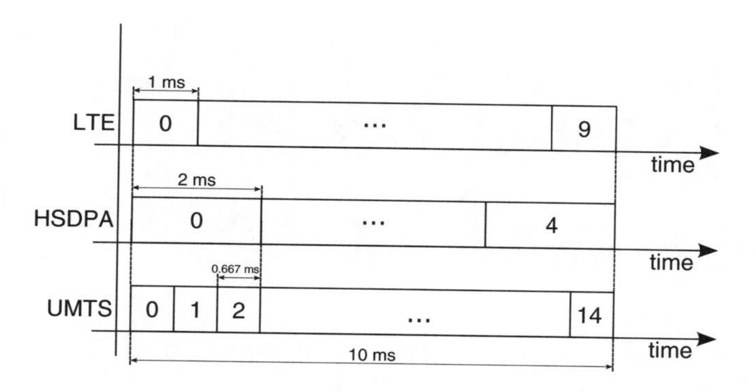

Figure 4.12 Radio frames

As explained in Section 4.3.1, the air-interface of UMTS is based on CDMA with a chip rate of $R_c = 3.84$ Mcps for a bandwidth of 5 MHz. The radio frame has a time duration of $t_f = 10$ ms and its structure is shown in Figure 4.12. Each frame is subdivided in $N_s = 15$ time slots or bursts of duration $t_s = t_f/N_s \approx 0.667$ ms, each containing $N_c = R_c \cdot t_s = 2560$ chips. The number N_b of bits carried by each time slot depends however on the CDMA Spreading Factor (*SF*) and the digital modulation being transmitted. In UMTS the modulation is Quadrature Phase Shift Keying (QPSK), which carries $N_{mod} = 2$ bits per symbol. Thus, N_b is easily computed by:

$$N_b = \frac{N_c}{SF} \cdot N_{mod} = \frac{2560}{SF} \cdot 2 \tag{4.3}$$

The *SF* is therefore a fundamental property of the physical channels and the bit rate R_b in one slot can be calculated using:

$$R_b = \frac{N_b}{t_s} \tag{4.4}$$

In UMTS, channels are often transmitted by mapping several upper-layer channels into one lower-layer subchannel. In a similar way, one upper-layer channel might be split across several lower-layer channels. Furthermore, some channels are unrelated to channels from upper layers and carry information that is significant only at the layer in which they are generated. Among other tasks, it is thus the responsability of each layer to *multiplex* and *segment* channels received through its SAPs into the appropriate subchannels of the layer beneath. Figure 4.13(a) illustrates those physical channels that are mapped to channels from upper layers and the different available channels are listed in the following.

Logical Channels

- As in GSM, the Broadcast Control Channel (BCCH) is a downlink channel containing the cell id and all necessary information for the UE to detect a UMTS cell.

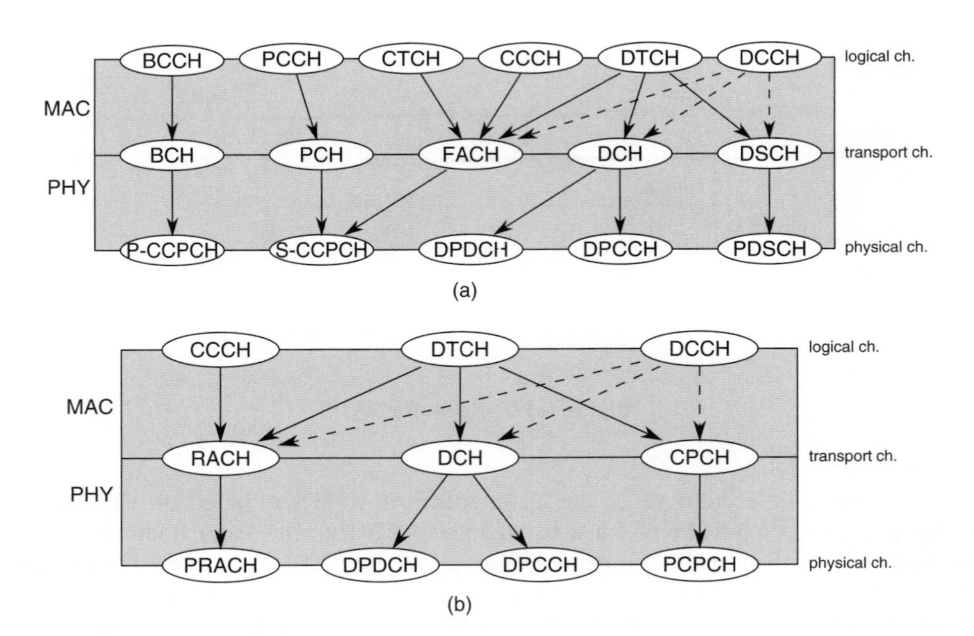

Figure 4.13 Allowed multiplexing of the FDD UTRA channels. (a) Downlink. (b) Uplink

- The Paging Control Channel (PCCH) is a downlink channel used to indicate the UE of an incoming connection.
- The Common Traffic Channel (CTCH) is a point-to-multipoint downlink channel to send the same data to a group of UEs.
- The CCCH is used to send control information to and from the UEs that have no assigned channels.
- The Dedicated Traffic Channel (DTCH) is the main traffic channel in UTRA and carries user data both in uplink and downlink.
- The DCCH is used to exchange control information between a given UE and the network.

Transport Channels

- As seen from Figure 4.13(a) the Broadcast Channel (BCH) carries the BCCH, which is to be decoded by every UE in the network.
- The PCH carries the PCCH in order to inform a UE without connection, of an incoming connection request.
- The Forward Access Channel (FACH) is a downlink channel able of carrying several types of traffic (common, dedicated, control information for connection set-up and small packets of user data).
- The Dedicated Channel (DCH) is the channel normally used to carry user data. However and as seen from Figure 4.13, user information can also be sent through other channels. This is decided by the RNC or the FAP depending on the data size.

- The Downlink Shared Channel (DSCH) can be used for the transport of data belonging to several users.
- The RACH is used by the UE to access the network for the first time in response to a network connection request. It can thus carry both user data as well as control information.
- The Common Packet Channel (CPCH) is used for transmitting power control commands as well as user data.

Physical Channels

- The Common Control Physical Channel (CCPCH) is a downlink channel used mainly for the transmission of control information (although small user data packets is allowed through the FACH). The CCPCH is however divided in two subchannels: the Primary Common Control Physical Channel (PCCPCH), which is always transmitted with a channelization code of $SF = 256$ and that carries the BCH, and the Secondary Common Control Physical Channel (SCCPCH).
- The Dedicated Physical Data Channel (DPDCH) carries user information from the DCH.
- The Dedicated Physical Control Channel (DPCCH) carries physical layer control information such as power control and data format.
- The Physical Downlink Shared Channel (PDSCH) is used to send the DSCH to different UEs.
- The Physical Random Access Channel (PRACH) carries control information that the UE sends in the RACH when accessing the network for the first time. It can also carry small user data.
- The Physical Common Packet Channel (PCPCH) is used for carrying the transport CPCH.
- Although not shown on Figure 4.13(a), the Common Pilot Channel (CPICH) is an important downlink physical channel transmitted continuously from the HNB. UEs read the power of this channel to detect available cells in its surroundings.

The PHY layer receives one block of data from the MAC layer each period known as the Transmission Time Interval (TTI). As mentioned earlier, the PHY layer manipulates the information carried by transport channels in order to transmit it successfully through the physical channels. The CRC and convolutional coding are performed at this level, as well as bit interleaving to distribute burst errors throughout different radio frames. The number of bits that can be interleaved depends on the TTI. On the one hand, longer TTIs allow more bits to be distributed throughout the frame and it hence lowers the impact of burst errors. On the other hand, shorter TTIs will reduce the effectivity of bit interleaving. The TTI value also impacts the frequency with which radio link adaptation can be applied. Short TTIs allow the Bit Error Rate (BER) to be estimated more often in reception and hence, quicker action can be taken in cases of link degradation. UMTS offers versatility in this field by allowing TTIs of 10, 20, 40 and 80 ms.

After interleaving, the bits of the resulting information blocks must be punctured or repeated in order for the block to match the bit rate of the channel. Such a procedure is known as *rate matching*. After that, the PHY layer can decide to multiplex all blocks with the same coding and interleaving into a single Coded Composite Transport Channel (CCTrCH)[‡]. The resulting CCTrCHs are then mapped to physical channels. Hereafter, physical channels are modulated with the WCDMA spreading function to reach the UMTS chip rate of $R_c = 3.84$ Mcps. Each HNB uses a spreading code of its own called a *scrambling* code. This allows UEs to differentiate the downlink signals arriving from different femtocells. Furthermore, each UE also has a scrambling code different from other users within the same femtocell. An intelligent management of the scrambling codes can be used as a means for reducing the impact of interference reaching the UE from surrounding femtocells. However, the UE needs to check all available scrambling codes when entering the femtocell for the first time. Therefore, and in order to reduce the registration time, the number of available codes must be kept low. Other type of codes named *channelization* codes are also used to distinguish the different physical channels on reception. Finally, the bit sequence is mapped to a QPSK modulation and modulated using a root-raised cosine filter with a roll-off factor of $\beta = 0.22$. A more in-depth study of UMTS can be found in [15].

4.3.4 HSPA Femtocells

As seen in Table 4.1, releases 5 and 6 of the 3GPP specifications introduced further enhancements to UMTS networks. These are detailed in the following.

HSDPA

The first version of the 3GPP Release 5 standard was published in March 2002 and it introduced several improvements to the downlink architecture. This release is commonly known as High Speed Downlink Packet Access (HSDPA) and, among other changes, it removes the transport channel DSCH from the specifications. In its place, its HSDPA equivalent is introduced: the High-Speed DSCH (HSDSCH), which is used in HSDPA to carry user data.

An option for using the higher order modulation 16-Quadrature Amplitude Modulation (QAM) instead of only QPSK is now supported (shown in Figure 4.14). By doing this, each modulation symbol carries $N_{mod} = 4$ bits and hence, the achievable throughput is much higher (see equation (4.3)). For instance, HSDPA is in theory capable of achieving 14.4 Mbps over a newly defined TTI of 2 ms. On the other hand, plain UMTS indoor systems only support speeds of up to 2 Mbps. However, this data rate is often limited by the Iub interface that connects the NodeB with the RNC. In femtocells designed according to the Iu-over-IP architecture, there is no such Iub interface because the RNC functionality is built into the FAP and hence, the Iub bottleneck is eliminated.

Hybrid Automatic Repeat reQuest (HARQ) is also included in HSDPA for an increased speed of packet retransmission in the physical layer. In HSDPA, data packets transmitted to the user in the downlink are kept in the buffer of the NodeB in order to accelerate packet retransmission when errors occur. This supposes a speed improvement with respect to plain UMTS macrocell systems, which only keep the data in the RNC. However, this

[‡] Note that CCTrCHs only exist within the physical layer.

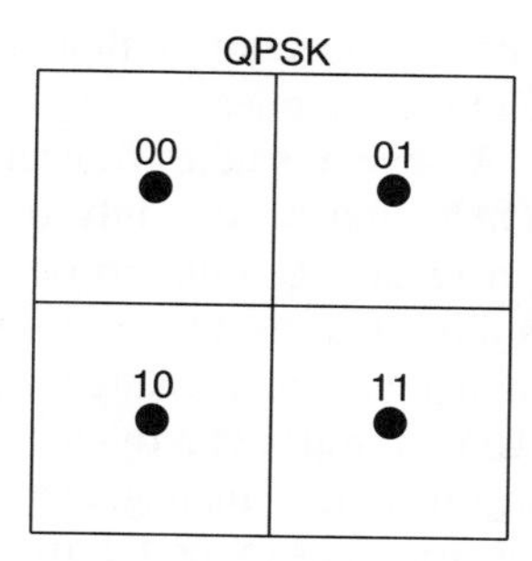

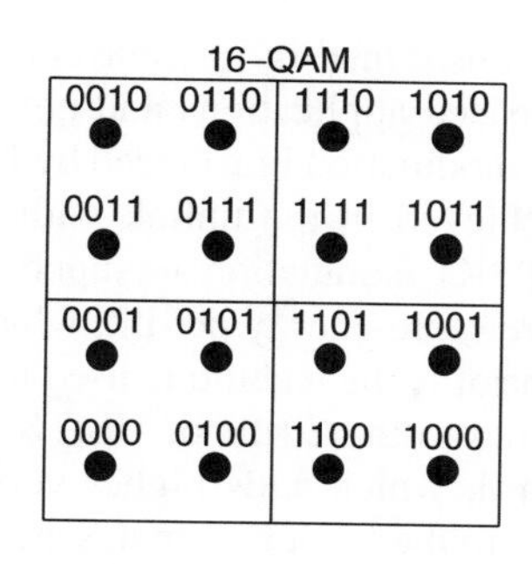

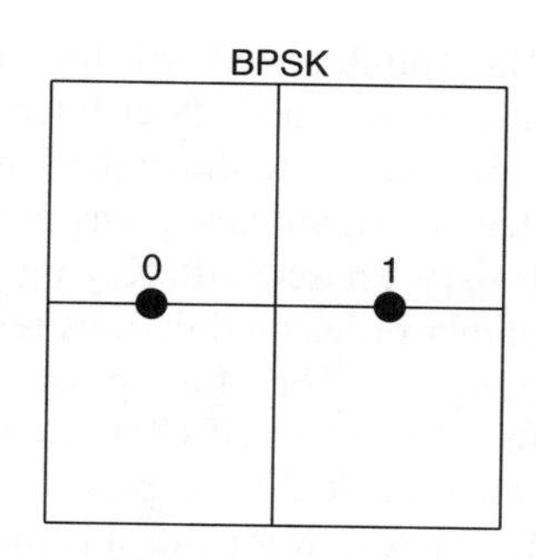

Figure 4.14 Gray coded digital modulations in UMTS, HSDPA and HSUPA

feature does not introduce significant gains in Iu-over-IP HSDPA femtocells with respect to UMTS because the RNC is built together with the HNB and they already share the same buffers. On the other hand, Iub-over-IP HNBs may highly profit from this approach because it avoids having to retransmit erroneous packets through the Iub tunnel and the Internet all the way from the RNC in the Core Network (see Figure 4.2). Besides, HSDPA reduces the TTI from 10 ms to 2 ms, which allows for a much faster scheduling and transmission of data over good radio links more often. Hence, less retransmissions are necessary. The scheduler in HSDPA is also removed from the RNC and located in the NodeB.

The DCH is also a key channel in HSDPA and it always carries signalling from the DCCH, which is always transmitted in parallel with other channels. Furthermore, the *SF* used to encode this channel is fixed in order to achieve the maximum available data rate. As mentioned above, the HSDSCH is now the transport channel carrying user data. However, instead of using power control, the RLC layer now uses Adaptive Modulation and Coding (AMC) to adapt to changes in the radio channel. The decision to change the Modulation and Coding (MC) is taken based on the Channel Quality Indicator (CQI) received in the uplink, which is an index pointing to the MC most suitable for the next transmission. Since there is no power control, the transmission power remains constant and the different digital modulations have to be selected in order to cope with channel variations. For more information on HSDPA, the reader is referred to [16].

HSUPA

Towards the end of 2004, release 6 of the 3GPP specifications, commonly known as High Speed Uplink Packet Access (HSUPA), was finally published and PicoChip was the first company to release a commercial HSUPA femtocell design [17]. This consisted of a software upgrade to their previous HSDPA product in order to make it compliant with 3GPP Release 6 and capable of delivering uplink speeds of up to 1.46 Mbps with HSUPA. Later, other companies such as Percello, Ubiquisys, Huawei or Ivy Network have also shown interest in such a design and started to develop their own HSUPA femtocell solutions. There are however concerns regarding the viability of supporting HSUPA technology in femtocell designs. Release 6 supports theoretical uplink speeds up to 5.76 Mbps, which is more than most of the ADSL connections in Europe can provide.

The uplink speed will thus be constrained by the current backhaul connection, although this would only affect Internet-based applications and not SOHO-centric ones.

In contrast to the higher order modulation introduced by HSDPA, several studies showed that no significant gains are achieved in the uplink with 16-QAM and hence, only the Binary Phase-Shift Keying (BPSK) modulation is supported in release 6. Furthermore, higher order modulations require more energy per bit, which is a limited resource in user terminals. Therefore, power control is the technique used to fight against quick changes of the channel in HSUPA. Compared to the 20 dB of UMTS, HSUPA introduces a dynamic range of 70 dB for power control, which adds higher versatility in fading management. Besides, power control is moved to the NodeB, which sends its instructions to the UE over several new dedicated channels (the Enhanced uplink Relative Grant Channel (ERGCH) and the Enhanced uplink Absolute Grant Channel (EAGCH)).

DCH is the transport channel where user data is typically carried in UMTS systems. However, this channel is replaced in HSUPA by the uplink Enhanced Dedicated Channel (EDCH), which is a channel dedicated to the transmission of information from the UE to the NodeB. HARQ is also supported in the PHY layer of HSUPA and implemented by means of the new EDCH HARQ Indicator Channel (EHICH), which is a downlink channel used for sending the acknowledgments corresponding to uplink transmissions.

HSUPA supports two TTI values: 2 ms and 10 ms. As with HSDPA, the lower value helps to increase the responsiveness against fast fading. However, the transmission of signalling information in the uplink, each 2 ms consumes a lot of power in those UEs located far from the NodeB. This motivated the introduction of support for TTIs of 10 ms which is today the most commonly supported TTI in user terminals. Femtocell users are usually much closer to their HNB than in the macrocell case. Furthermore, the femtocell edge is well defined and contained by the outer walls of the premises. This suggests that a TTI of 2 ms should be feasible in femtocell scenarios. However, this is an optional feature in most terminals [16] and not necessarily always supported.

4.4 OFDM-Based Femtocells

One of the main impairments of wireless channels is frequency selective fading. It is especially so in intense multipath environments where the behaviour of the channel differs between different frequencies (see Figure 4.15) and this is particularly true in indoor and urban environments. The distortion suffered by wideband signals (e.g. CDMA signals) when transmitted over such channels makes them difficult to recover and hence, narrowband signals are preferred for their higher resistance to these channels.

This led in the fifties to the development of multicarrier modulations, which consist on the transmission of information over several narrowband channels instead of one large wideband channel. However, this technology did not succeed until the 1990s, which is when electronics started to cope with its computational requirements. Furthermore, the subdivision of the data stream into several smaller ones allows for an efficient management of the radio resources and interference, which is the main problem of overlaid two-layer networks (see Chapter 6). Due to its highly efficient implementation by means of the Fast Fourier Transform (FFT), Orthogonal Frequency Division Multiplexing (OFDM) is the multicarrier technology selected for the PHY layer of IEEE Wireless Interoperability for Microwave Access (WiMAX) and 3GPP Long Term Evolution (LTE). These are some of

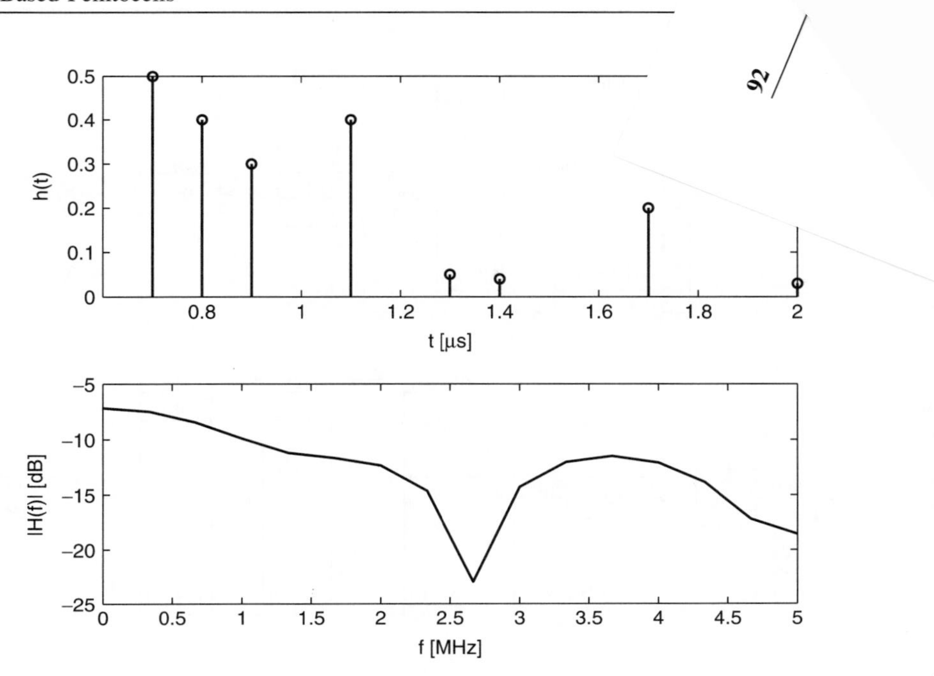

Figure 4.15 Channel impulse response and frequency response of a multipath channel

the candidate technologies to replace current UMTS networks. Since femtocells are also part of the RAN and need to interact with the outdoor macrocells (e.g. for performing handovers), FAPs based on these technologies are the natural evolution of current GSM or 3G NodeBs. Moreover, several manufacturers have already started the development of OFDM-based femtocells, which might become common the future.

In general terms, the femtocell concept applies equally to all RAN technologies. However, there are some fundamental differences between femtocells based on CDMA and OFDM. In scenarios where both femto and macrocells reuse the electromagnetic spectrum, interference averaging of CDMA helps to reduce the effects of interference. However, this does not happen in OFDM where one transmitter is enough to interfere completely with a given subcarrier. It is therefore assumed that OFDM femtocells will use some sort of *self-organizing* approach to cope with potential interference issues. The idea is that femtocells will scan the radio environment (either scanning it themselves or throughout measurement reports from the UEs) and use this information in a distributive manner (cooperatively or not) to choose the optimum subcarriers assignment.

In the following, an overview of WiMAX and LTE, which are candidate technologies in the evolution towards 4G, is presented.

4.4.1 OFDM Fundamentals

Figure 4.16 shows the fundamental structure of an OFDM transmission system. At the transmitter side, the data sequence is mapped into symbols X of a complex constellation

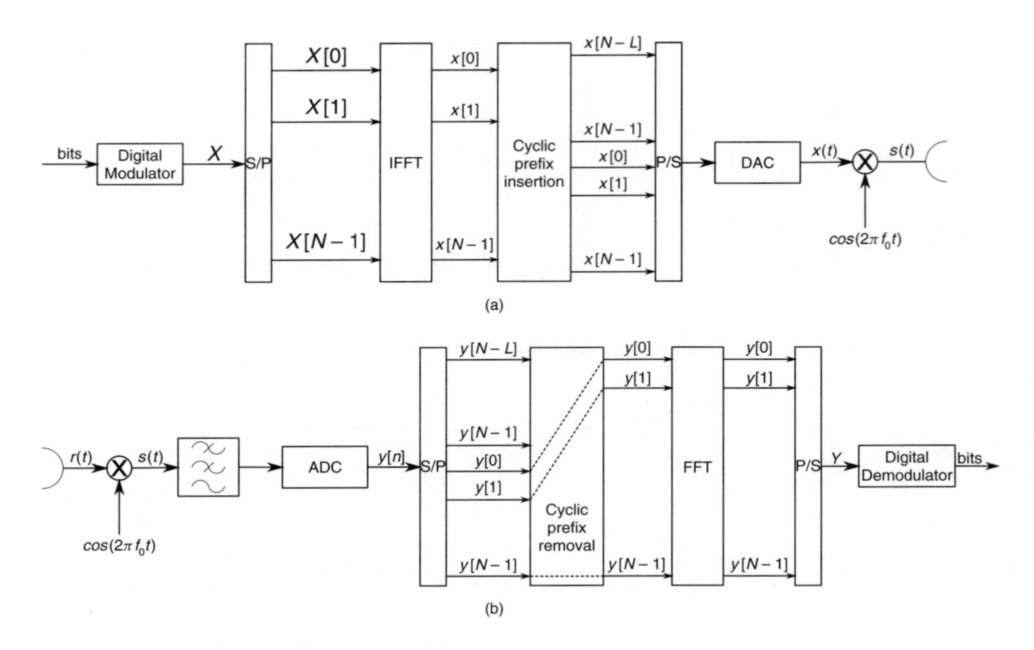

Figure 4.16 OFDM system with cyclic prefix. (a) Transmitter. (b) Receiver

(e.g. QPSK or 16-QAM) which are then input in groups of N to an Inverse Fast Fourier Transform (IFFT). N is typically a power of two and is a basic property of the system. The group of N samples output from the IFFT block is known as an *OFDM symbol* and it consists of the following sequence

$$x[n] = \frac{1}{\sqrt{N}} \sum_{k=0}^{N-1} X[k] e^{\frac{2\pi jnk}{N}}, \qquad n = 0, \ldots, N-1 \tag{4.5}$$

As seen, each complex symbol $X[k]$ is modulated by a complex exponential function. Each of these is called an *OFDM subcarrier* and they are arranged to be orthogonal relative to each other. Since the OFDM modulation is implemented via an FFT algorithm, the observed bandwidth of each subcarrier depends on the sampling frequency f_s and the FFT size N_{NFFT} throughout $\Delta f = f_s/N_{FFT}$. It should thus be noticed that the useful duration of the OFDM symbol (without cyclic prefix) is inversely related to the subcarrier spacing Δf by:

$$T_{OFDM} = \frac{1}{\Delta f} \tag{4.6}$$

There is thus an slight overlapping between adjacent subcarriers; however, orthogonality allows for an easy recovery of the original sequence through an FFT in the receiver side. The normalization factor $1/\sqrt{N}$ is used to guarantee an unitary transform of the signal going through both the transmitter and receiver parts of the system. Figure 4.17 illustrates an OFDM spectrum, where adjacent subcarriers have a spacing of Δf.

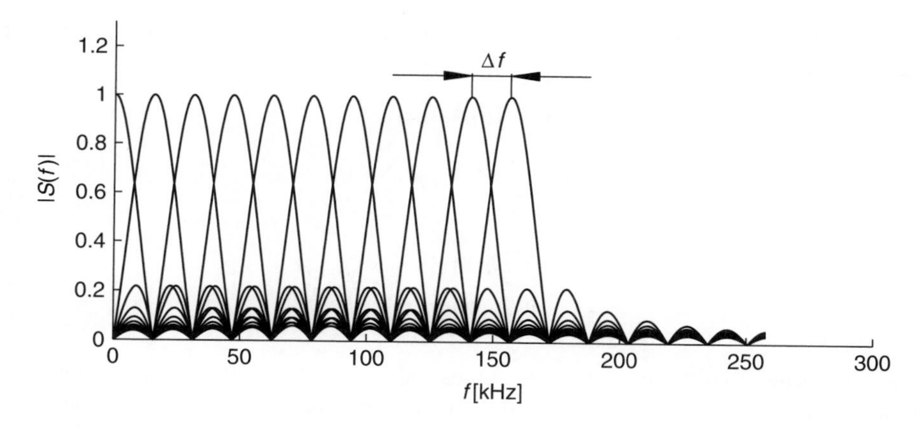

Figure 4.17 OFDM subcarriers modulating rectangular pulses and $\Delta f \approx 15.6\,\text{kHz}$

Due to the use of an IFFT, the OFDM signal can be thought of as being constructed in the frequency domain. This means that the energy of each symbol $X[k]$ translates directly into the energy carried by the kth subcarrier when transmitted over the air. This is useful, for instance, for avoiding transmitting data over certain bands suffering deep fading at a given instant. By exploiting the CQI information, the transmitter can decide, for example, not to transmit in subbands 10 and 47 just by doing $X[10] = 0$ and $X[47] = 0$. This frees the corresponding RF subcarriers and reduces the interference caused to other femtocells using those bands.

The basic principle for information recovery in OFDM is the orthogonality between subcarriers. This is typically guaranteed by highly accurate oscillators and timing between transmitter and receiver. However, it is the industry's objective to keep FAPs prices to a minimum, thus a risk exists of having low quality oscillators in OFDM femtocell access points that might degrade the performance of the transmission chain. This translates into subcarriers widening in frequency and causing interference to adjacent subcarriers in a phenomenon known as Intercarrier Interference (ICI), which must be properly calibrated by the FAP manufacturer. It has been shown [18] that some of the factors that increase ICI include large OFDM symbol durations and frequency offset errors, which cause a quadratic growth in ICI.

The Cyclic Prefix

As seen from Figure 4.15, a multipath channel causes the arrival of several signal echos at the receiver. The result of this is symbols widening in time and interfering with the next adjacent symbol in a phenomenon known as Intersymbol Interference (ISI). In order to reduce the impact of ISI, most OFDM systems make use of a technique called the *cyclic prefix*, which consists of repeating the last part of $x[n]$ at the beginning of the original sequence. This is illustrated in Figure 4.16(a), where the last L samples of $x[n]$ are introduced as a prefix before the DAC conversion, L being the length of the discrete multipath channel impulse response. The resulting sequence $x_{CP}[n]$ has thus a length of

$N + L$ and it follows that $x_{CP}[n] = x[n]_{\text{mod } N}$ at $-L \leq n \leq N - 1$. The received discrete signal is therefore:

$$\begin{aligned} y[n] &= x_{CP}[n] * h[n] \\ &= \sum_{k=0}^{L} h[k] x_{CP}[n-k] \\ &= \sum_{k=0}^{L} h[k] x[n-k]_{\text{mod } N} \\ &= x[n] \circledast h[n] \end{aligned} \tag{4.7}$$

where it is shown that the cyclic prefix forces the received signal $y[n]$ to be the circular convolution of the original signal with the channel impulse response $h[n]$. Following the properties of the circular convolution, (4.7) can be rewritten in the frequency domain as $Y[k] = X[k]H[k]$ for $0 \leq k \leq N - 1$. Therefore, as long as the channel response $h[n]$ is known, the original signal can be recovered free of *ISI* by:

$$x[n] = IFFT\left\{\frac{Y[k]}{H[k]}\right\} \tag{4.8}$$

OFDMA

Orthogonal Frequency Division Multiple Access (OFDMA) is a multiple users access scheme based on OFDM that exploits the spectrum arrangement in subcarriers to distribute users along the frequency domain. Instead of letting each user use all the spectrum all the time by transmitting through all the subcarriers, only a subgroup of those are assigned to each user. This maximizes the frequency reuse because it is unlikely that a user enjoys good channel conditions over all subcarriers at a given time instant. It is thus reasonable to let other users make use of those subcarriers. This approach introduces the need for dynamic frequency assignment algorithms that decide, for each OFDMA symbol, which users are assigned which subcarriers. Furthermore, in OFDMA systems, subcarriers are usually grouped into larger groups called subchannels for easier handling. For instance, Figure 4.18 shows a dynamic allocation of eight frequency subchannels that varies through time and exploits frequency diversity.

Single-Carrier FDMA

The peak-to-average-power-ratio *papr* of a signal $s(t)$ is defined as the ratio between its peak power and RMS value, i.e.

$$papr = \frac{\max[s(t)s^*(t)]}{E[s(t)s^*(t)]} \tag{4.9}$$

For instance, a constant DC signal has $papr = 1$, while a sinusoidal signal has $papr = 1.41$. However, in an OFDM signal with N subcarriers modulating the same constellation of complex symbols, it can be proved that $papr = N$ (this being the maximum *papr* of an

Figure 4.18 Allocation of OFDMA subchannels to different users

OFDM signal). Since OFDM signals pass through a power amplifier prior to transmission, a high *papr* value implies that the amplifier needs to have a large backoff in order to apply the same gain to all possible signal power values. Furthermore, signals with large *papr* also have large dynamic ranges, which implies that the D/A and A/D converters need high resolution to represent all amplitudes accurately.

These characteristics of OFDM signals might be hard to meet in certain mobile terminals due to low quality components chosen to minimize costs. Hence alternative transmission mechanisms had to be designed. The scheme shown in Figure 4.19 only differs from the OFDM system of Figure 4.16 in the Discrete Fourier Transform (DFT) and Inverse Discrete Fourier Transform (IDFT) blocks, which cancel each other out across the transmission chain. The objective of these blocks is to precode the symbols into groups of size $M < N$ before applying the N-size OFDM modulation. By doing this and completing with zeros to the IFFT size, the resulting coded signal $x_c[n]$ is an oversampled version of the original OFDM signal $x[n]$. This means that the power variations of $x_c(t)$ are much lower than those of $x(t)$ and hence, $x_c(t)$ is considered to behave as a single-carrier signal. The *papr* of such a signal is thus reduced, allowing for an increased efficiency of the power amplification.

All in all, the main advantages of multicarrier modulations such as OFDM can be summarized as:

- robustness against multipath;
- efficient implementation through FFT blocks;
- utilization of frequency diversity for multiple user access;
- high spectral efficiency;
- robustness against narrowband interference from nearby terminals.

4.4.2 WiMAX

Published in 2004 under the IEEE 802.16d standard, WiMAX is an interoperable wireless technology designed for the provision of *last mile* connectivity. However, the standard

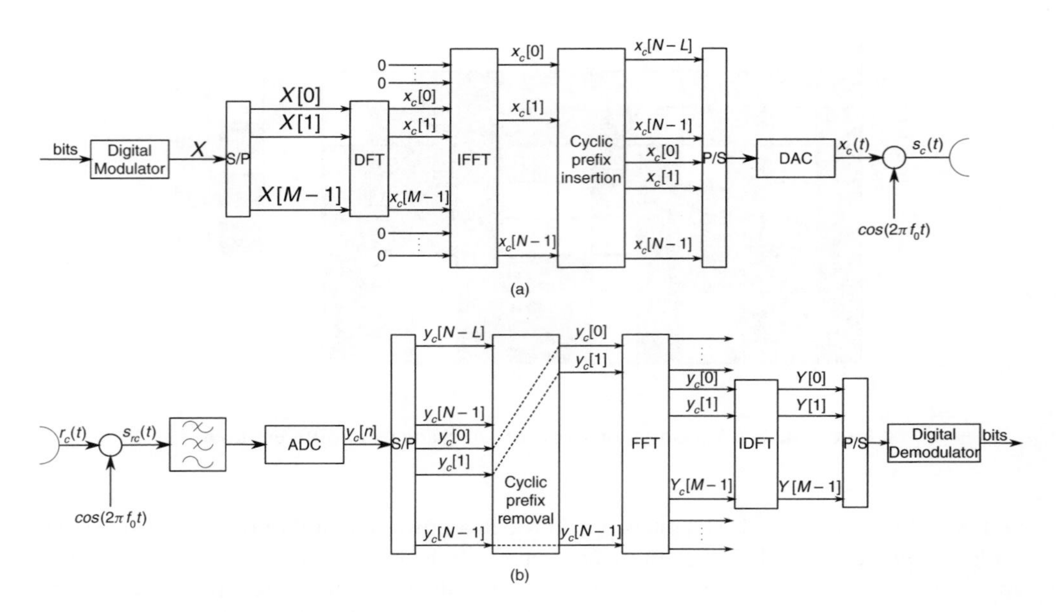

Figure 4.19 SC-FDMA system. (a) Transmitter. (b) Receiver

developed further as the 802.16e was released in 2005, which supports mobile connectivity and is thus known as mobile WiMAX. In contrast to 3GPP specifications, the WiMAX standard only defines the MAC and PHY layers for the radio access. Nevertheless, the Network Working Group (NWG) within the WiMAX Forum [19] developed an end-to-end network architecture supporting IP connectivity. This implies that a WiMAX Access Service Network (ASN) of any type (based on macrocells, femtocells, etc.) can also connect through an IP network (e.g. a DSL connection) to the CN of the WiMAX operator.

The fact that WiMAX is an all-IP technology makes it specially suitable for femtocell deployments. Furthermore, WiMAX supports theoretical symmetric data rates up to 70 Mbps, thus providing high QoS for mobile-centric femtocell applications. WiMAX emission licences have already been auctioned in most European countries and the USA. This adds to the publication of the standards in 2004 and 2005 and implies that other modern technologies such as LTE are 3 to 4 years behind in terms of marketability and equipment production. These facts have awakened the industry's interest in WiMAX femtocells as a feasible alternative to current UMTS and High Speed Packet Access (HSPA) HNBs. Some companies have even purchased frequency bands to be used exclusively by WiMAX femtocells, thus guaranteeing zero cross-layer co-channel interference from WiMAX macrocells. However, such an approach is costly as it does not maximize frequency reuse. Furthermore, self-organization features as well as distributed frequency allocation algorithms for OFDM cells are progressively emerging as promising solutions for coping with this type of interference. Network operators must hence carefully evaluate whether an orthogonal frequency assignment for WiMAX femto/macrocells affects their competitiveness. The industry is thus putting pressure on the WiMAX Forum to approve a WiMAX femtocell specification although, at the time of writting this book, nothing has so far been published.

Physical Layer

The PHY layer of 802.16d is based on OFDM, thus offering high resistance to multipath, which is an intrinsic characteristic of indoor and urban environments due to the abundance of reflections and scatterers. On the other hand, 802.16e is based on OFDMA, which exploits frequency diversity for the assignment of resources to different users. More specifically, the 802.16d version uses an FFT of size 256, while the PHY layer of 802.16e is scalable with FFT sizes between 128 and 2048.

Multiple Input Multiple Output (MIMO) techniques such as beamforming and transmit/receive diversity are also supported by WiMAX. Channel bandwidths of 1.25, 1.75, 3.5, 5, 7, 8.75, 10, 14 and 15 MHz are supported, being thus quite flexible in terms of bandwidth assignment for the different network layers. In the 802.16e standard, the available channel bandwidth relates to the OFDM modulation size that can be used, i.e. the FFT size. However, in the 802.16d version, the FFT is always constant and higher bandwidths translate into larger subcarriers spacing and thus shorter symbol period.

Mobile WiMAX uses the concept of subchannelization by grouping subcarriers (contiguously or not) in subchannels, which are the minimum frequency units that a WiMAX base station can allocate. Subchannels thus describe the granularity of the OFDMA frequency allocation. Permutation schemes based on distributed subcarriers, such as Full Usage of Subchannels (FUSC) or Partial Usage of Subchannels (PUSC), are suitable for fast motion users. This is because the spectrum changes rapidly over time and these schemes guarantee that not all subcarriers are affected in a negative way. Here, frequency diversity is exploited, averaging the overall network interference across multiple cells. On the other hand, permutation schemes based on contiguous subcarriers, are better suited for low motion users due to the fact that low fading spectral regions can be easily identified and remain constant during several subframes. In this case, multi-user diversity can be exploited by the system, providing each user with the subchannel that maximizes its SINR.

Further, slots define the granularity over the time domain and can have the duration of 1, 2 or 3 OFDM symbols. When a group of subchannels is assigned to a user over several contiguous time slots, the resulting resource is called the data region of the user.

Both WiMAX standards support TDD and FDD as a duplexing mechanism. However, the TDD mode is preferred by the industry due to the lower spectrum requirements. The WiMAX TDD frame (see Figure 4.20) has a variable duration between 2 ms and 20 ms and it distinguishes in time between downlink and uplink subframes. The first subframe is for downlink transmissions. Then, after a Transmit/Receive Transition Gap (TTG), the uplink subframe begins. The downlink subframe begins with a preamble spanned over all subcarriers, which is used for synchronization and channel estimation. After the preamble comes the Frame Control Header (FCH), which contains Media Access Protocol (MAP) messages indicating the burst profiles of each user. Only then data bursts to different users and allocated in independent data regions are transmitted. Similarly, the uplink subframe contains several data bursts as well as a ranging region used for the adjustment of power and other parameters.

Similarly to HSPA, WiMAX uses AMC to respond to the channel variations by supporting several modulation and coding schemes that vary from burst to burst. In the downlink, QPSK, 16QAM and 64QAM are supported, while the 64QAM modulation is optional

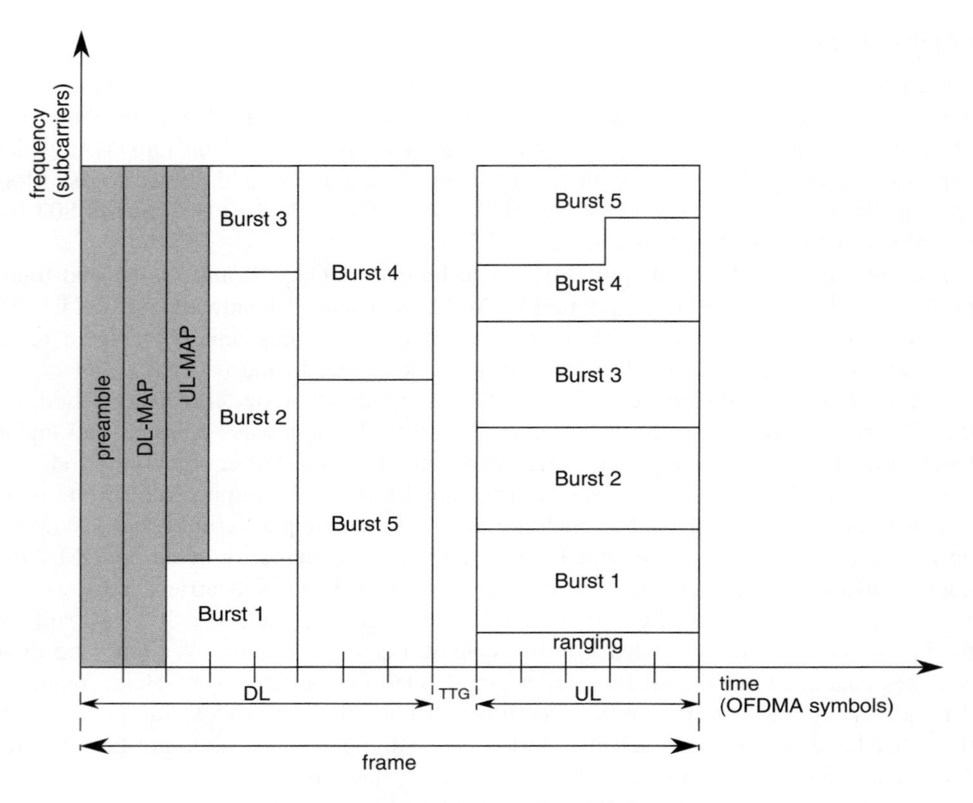

Figure 4.20 Example of a TDD WiMAX frame

in the uplink. Regarding error correction, Reed-Solomon codes are supported. Furthermore, the 802.16e standard included support for HARQ, thus providing high resistance to transmission errors while saving channel capacity.

MAC Layer

The MAC layer of WiMAX is connection oriented; it sits on top of the PHY layer and it converts MAC Service Data Units (SDUs) into MAC Packet Data Units (PDUs). Furthermore, there also exists a Convergence Sublayer on top of the MAC layer, which is used for interfacing with IP and Ethernet networks.

The way resources are allocated to users (called MS in the WiMAX terminology) is by means of a process called *polling*. This consists on periodically allocating slots to the MS. Then, if an MS needs resources, it can decide to request access to the assigned slot or not. The amount of resources allocated to each user depends on the QoS that the service of such an user requires. Hence, in WiMAX the following types of service are defined:

Unsolicited Grant Service (UGS) Used for applications that require a constant data rate such as VoIP without silence suppression.

Real-time Polling Service (rtPS) Used for real-time applications with packets of variable size, such as videoconference.

Non-real-time Polling Service (nrtPS) Used for applications in which delay is not an issue, such as file transfers.

Best Effort (BE) For applications without QoS requirements (e.g. web browsing).

Extended real time Polling Service (ertPS) For real-time applications with variable throughput requirements such as VoIP.

The MAC layer of WiMAX also supports a variety of power saving modes such as *idle* and *sleep* mode. These provide the MS with the possibility of turning itself off for determined periods, thus extending its battery life. WiMAX's MAC layer is large and complex. However, it is out of the scope of this book to enter in details and the reader is referred to [20] for more information.

4.4.3 LTE

In December 2008, release 8 of the 3GPP Technical Specifications reached its first stable version. This release is commonly known as the Long Term Evolution (LTE) and it introduces enhancements to previous specifications to achieve higher throughputs, spectral bandwidth, more flexible spectrum management, etc. Since GSM and UMTS are the more commonly deployed mobile networks worldwide, it is expected that LTE will be deployed by updating the existing networks and hence will become the most common mobile access technology worldwide. Furthermore, the LTE specifications introduce strong support for Home eNodeBs (HeNBs), which are the evolution of previous HNBs and are now considered as main points for radio access.

The industry has already commenced developing and testing LTE-based femtocells. For instance, in May 2008 picoChip announced the first microchip reference design for HeNBs based on the technical specifications available at the time. This led several FAP manufacturers to start developing their own products with the intention of being first on the market once the LTE licences are auctioned. However, at the time of writting this book, this has not yet taken place and therefore, the marketability of LTE femtocells is still uncertain. Furthermore, the data rates achieved by LTE are higher than those provided by most of today's DSL lines, which limits the advantages that femtocells based on this release can obtain. In the following, an overview of the main transmission schemes of the LTE radio interface is provided.

Evolved UTRAN (EUTRAN) Overview

In LTE, a Type 1 radio frame is defined having a duration of 10 ms (see Figure 4.12) and divided in ten subframes of duration $T_{sf} = 1$ ms. Furthermore and for legacy compatibility each subframe is divided into $N_{sl}^{sf} = 2$ slots of length $T_{sl} = T_{sf}/2 = 0.5$ ms. During each slot a total of $N_{sb}^{sl} = 6$ or $N_{sb}^{sl} = 7$ OFDM symbols are transmitted, being N_{sb}^{sl} dependent on the length of the selected cyclic prefix. LTE allows the use of an extended cyclic prefix (16.7 μs), which might be necessary to fight large delay spreads in some environments.

However, femtocell scenarios involve small areas with short delay spreads and hence, the normal length cyclic prefix ($T_{CP} = 5.2\,\mu s$) might be enough for HeNBs. In FDD mode, all subframes within the same band transmit either uplink or downlink information. However, in TDD mode all subframes, except 0 and 5 which carry system information, can be used for either downlink or uplink transmission. Furthermore, a *Type 2* radio frame is also defined for compatibility with previous TDD systems. Nevertheless in the following the use of *Type 1* frame is assumed. More information about the frame structures can be found in [21].

Downlink The radio access technology in the downlink of EUTRAN is OFDMA with a subcarrier spacing of $\Delta f = 7.5\,\text{kHz}$ for multicast transmissions, and $\Delta f = 15\,\text{kHz}$ for all other cases. OFDMA subcarriers are bundled together in groups of $N_{sc} = 12$ adjacent subcarriers forming what is known as a resource block. Since a single time slot carries $N_{sb}^{sl} = 7$ OFDM symbols with a normal cyclic prefix, the total number of subcarriers contained in one resource block during one time slot is $N_{sc}^{rb} = N_{sc} \cdot N_{sb}^{sl} = 12 \cdot 7 = 84$. The size N_{FFT} of the FFT defines the total number of subcarriers of the system, where the centre one carries no information. This is due to the fact that this is the DC subcarrier and the local oscillator at the UE might cause it high interference due to leakage. Such a subcarrier is thus left unused.

Another advantage of LTE is its flexibility with respect to spectrum management. This translates into allowing the downlink to be composed of an arbitrary number of resource blocks between 6 and 110, thus allowing network operators to use bandwidths between 1 MHz and almost 20 MHz. However, not all subcarriers within a resource block carry useful information. In LTE, a set of complex reference symbols modulates certain subcarriers throughout the OFDM grid. These are transmitted between the first and fifth OFDM symbols of a time slot and have a frequency separation of six subcarriers. Hence, each resource block carries $N_{rs}^{rb} = 4$ reference symbols per time slot[§]. The objective of reference symbols is to serve as a means for cell identification throughout predefined reference symbol sequences, as well as for channel sounding.

LTE supports QPSK, 16QAM and 64QAM as modulation schemes. Therefore, the minimum usable data rate of a resource block with normal cyclic prefix occurs for the case of QPSK ($N_{bit}^{sb} = 2$ bits per symbol). Furthermore, each subframe can use up to $N_{sig} = 3$ OFDM symbols for carrying L1/L2 signaling channels, implying thus a throughput of:

$$\begin{aligned} R_{min}^{rb} &= \frac{N_{bit}^{sb} \cdot (N_{sc} \cdot (N_{sl}^{sf} \cdot N_{sb}^{sl} - N_{sig}) - N_{rs}^{rb} - N_{rs}^{rb}/2)}{T_{sf}} \\ &= \frac{2 \cdot (12 \cdot (2 \cdot 7 - 3) - 4 - 2)}{1\,\text{ms}} \\ &= 252\,\text{kbps} \end{aligned} \tag{4.10}$$

where $N_{rs}^{rb}/2$ subcarriers have to be subtracted from the first slot of the subframe because the other $N_{rs}^{rb}/2$ are already included in the first N_{sig} signalling symbols (see Figure 4.21). This value represents the minimum achievable downlink data rate of LTE in one resource block when one single antenna is used. In the case that the network operator has a

[§] The number of reference symbols per slot might be different in Multicast-Broadcast Single-Frequency Network (MBSFN) subframes, which have a different structure.

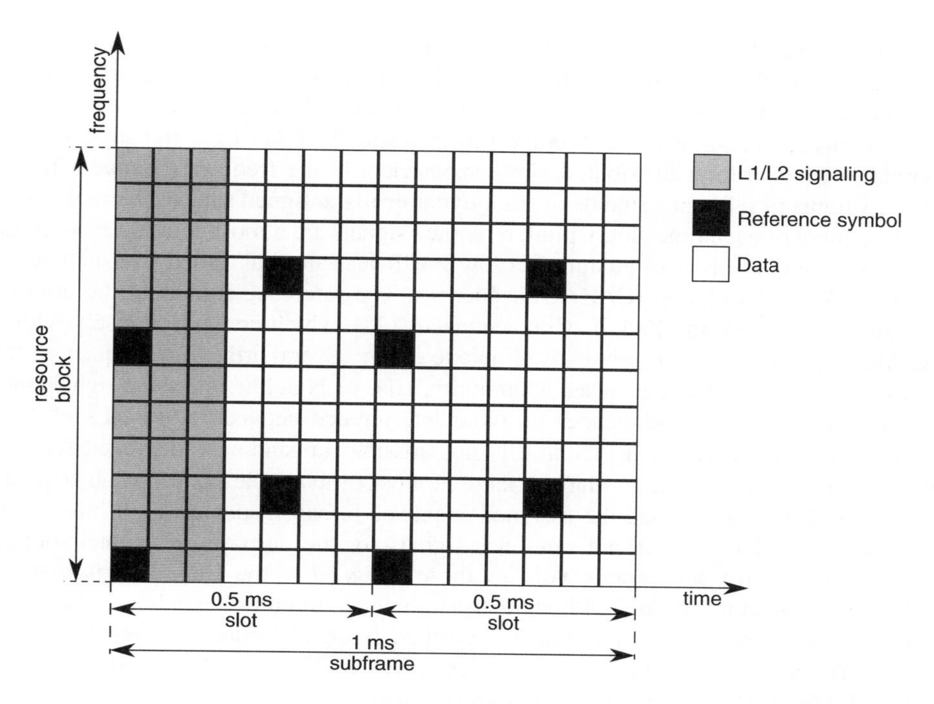

Figure 4.21 Downlink resource block and subframe structure in downlink LTE

bandwidth of $BW = 20\,\text{MHz}$ available, a total of $N_{rb}^{BW} = 110$ resource blocks can be utilized and the minimum downlink throughput is thus $N_{rb}^{BW} \cdot R_{min}^{rb} = 27.72\,\text{Mbps}$. The maximum downlink throughput for a single transmitting antenna is estimated with (4.10), $N_{bit}^{sb} = 6$ (64QAM) and $N_{sig} = 1$, giving a result of $R_{max}^{rb} = 900\,\text{kbps}$. However, it must be noted that these throughput values include redundant bits due to coding. The usable data rate is thus lower.

The packets of data received in the PHY layer through the transport channels are called transport blocks. These undergo several processes before modulation which include Cyclic Redundance Check (CRC) calculation, channel coding (Turbo codes), bits interleaving, scrambling and HARQ. This is how codewords are formed. Furthermore, LTE introduces several MIMO schemes with support for up to four transmitting antennas, which include open-loop MIMO, beam-forming and spatial multiplexing. However, a detailed description of all these systems is beyond the scope of this book and the reader is referred to [22].

Uplink The radio transmission technology in LTE is Single Carrier FDMA (SCFDMA) and it has been designed to resemble as much as possible the downlink OFDMA scheme. However, some fundamental differences exist. For instance, the centre DC subcarrier is not left unused in the uplink frame. Furthermore and in contrast to LTE uplink, when a user is assigned several resource blocks, these must be adjacent in the frequency domain during a given time slot, although the frequency allocation between adjacent time slots is allowed to change.

Reference signals are also necessary in the uplink to provide the HeNB with channel knowledge and allow coherent demodulation. However, these signals are different from those in the downlink case. Since the objective of the uplink radio interface is to minimize the instantaneous Peak-to-Average Power Ratio (PAPR) of the transmitted signal, reference signals are not distributed along subcarriers in the frequency domain. Instead, LTE UEs transmit reference signals over all dynamically assigned subcarriers in the fourth OFDM symbol of each time slot. Uplink reference signals are almost constant in amplitude along subcarriers in order to guarantee a low PAPR and are thus called Constant Amplitude Zero Auto-Correlation (CAZAC) sequences. The reference signals in the uplink of LTE are derived from the Zadoff–Chu sequences [23], which comply precisely with the CAZAC property. Since it is relatively simple to derive several orthogonal sequences from the original Zadoff–Chu sequences, neighboring LTE cells define reference signals based on different Zadoff–Chu sequences to avoid interference between close UEs of neighbouring cells. However, such planning is not possible in randomly deployed femtocell networks. It is thus important to have a large number of available Zadoff–Chu sequences so that the HeNB can change the assigned sequence in case it detects high interference. LTE also allows UEs to transmit reference signals for the purpose of channel sounding over all the system's subcarriers (not just those assigned to the UE). This is true even for UEs not transmitting data, although these channel sounding signals are transmitted with less frequency than the previously described frequency allocated signals. Channel sounding is one of the significant technologies in LTE and it allows the HeNB to schedule the uplink based on an accurate channel knowledge.

In LTE, different frequency resource blocks are allocated to different users for each OFDM symbol. However, the OFDM modulation scheme implies that all frequency resources are to be demodulated simultaneously throughout the FFT block. It is thus imperative that resource blocks transmitted from different UEs do not overlap each other in time so that they can be properly demodulated. Since the propagation delays for different users might vary, transmissions from different users need to be coordinated by the network. This is done in the uplink using *timing advance*, which consists of sending uplink transmissions before the start of the corresponding uplink subframe. After the reception of a random access burst from a UE, the HeNB decides if the UE requires a timing advance correction depending on the burst arrival time. Then, an appropriate offset is estimated and sent to the UE.

The rest of the PHY layer functionality in uplink is basically the same as in downlink. Once again, this includes CRC insertion, channel coding, interleaving, scrambling, HARQ and data modulation. As with the downlink case, the supported modulations in uplink are QPSK, 16QAM and 64QAM. Similarly, L1/L2 control signalling is transmitted in the uplink to carry CQI information and HARQ acknowledgments among others.

References

[1] 'UMA Today,' http://www.umatoday.com/.

[2] Ericsson. (2008, Sep.) Press Release. [Online]. Available: http://www.ericsson.com/ericsson/press/releases/20080911-1250616.shtml

[3] (2009, Jan.) HSL 2.75G Femtocell. [Online]. Available: http://www.haysystems.com/mobile-networks/hsl-femtocell/webcast

[4] V. Chandrasekhar and J. G. Andrews, 'Uplink Capacity and Interference Avoidance for Two-Tier Femtocell Networks,' *IEEE Transactions on Wireless Communications*, February 2008.
[5] J. Bannister, P. Mather, and S. Coope, *Convergence Technologies for 3G Networks IP, UMTS, EGPRS and ATM*. J. Wiley & Sons, Inc., 2004, ch. 3, p. 46.
[6] 'Universal Mobile Telecommunications System,' 3GPP, 2008. [Online]. Available: http://www.3gpp.org/article/umts
[7] 'TSG Radio Access Network,' 3GPP, 2008. [Online]. Available: http://www.3gpp.org/RAN
[8] 'User Equipment (UE) radio transmission and reception (FDD),' Feb. 1999. [Online]. Available: http://www.3gpp.org/ftp/Specs/html-info/25101.htm
[9] 'User Equipment (UE) radio transmission and reception (TDD),' Feb. 1999. [Online]. Available: http://www.3gpp.org/ftp/Specs/html-info/25102.htm
[10] 'Service requirements for Home Node B (HNB) and Home eNode B (HeNB),' Dec. 2008. [Online]. Available: http://www.3gpp.org/ftp/Specs/html-info/22220.htm
[11] V. P. Ipatov, *Spread Spectrum and CDMA: Principles and Applications*. John Wiley & Sons, Chichester, Mar. 2005.
[12] B. Walke, P. Seidenberg, and M. P. Althoff, *UMTS The Fundamentals*. John Wiley & Sons, Ltd, Chichester, 2003.
[13] 'TS 25.430 UTRAN Iub Interface: general aspects and principles (Release 4),' Sep. 2002.
[14] S. Rao and R. R. Bhat, 'Assessing Femtocell Network Architecture and Signaling Protocol alternatives,' http://www.embedded.com, Feb. 2008.
[15] H. Holma and A. Toskala, Eds., *WCDMA for UMTS*, 3rd ed. John Wiley & Sons, Chichester, 2004.
[16] ———, *HDSPA/HSUPA for UMTS*. John Wiley & Sons, Chichester, 2006.
[17] picoChip, 'picoChip announces industry's first HSUPA-femtocell reference design,' http://www.picochip.com/pr/first-hsupa-femtocell-reference-design, 2007.
[18] A. Goldsmith, *Wireless Communications*. New York, NY, USA: Cambridge University Press, 2005.
[19] 'WiMAX Forum,' http://www.wimaxforum.org/.
[20] J. G. Andrews, A. Ghosh, and R. Muhamed, *Fundamentals of WiMAX Understanding Broadband Wireless Networking*. Prentice Hall, 2007.
[21] P. Lescuyer and T. Lucidarme, *Evolved Packet System (EPS) The LTE and SAE evolution of 3G UMTS*. John Wiley & Sons, Chichester, Mar. 2008.
[22] E. Dahlman, S. Parkvall, J. Sköld, and P. Beming, *3G Evolution. HSPA and LTE for Mobile Broadband*, 1st ed. Elsevier, Jul. 2007.
[23] D. C. Chu, 'Polyphase Codes with Good Periodic Correlation Properties,' *IEEE Transactions on Information Theory*, vol. 18, no. 4, pp. 531–532, Jul. 1972.

5

System-Level Simulation for Femtocell Scenarios

David López Pérez, Guillaume de la Roche and Hui Song

The current wireless communication systems have a high degree of freedom, thus allowing a large number of parameters to be configured. For example, scheduling and resource allocation mechanisms are left to the free implementation of vendors and operators, as well as antenna, power and frequency configurations. Therefore, based on the experience of the Radio Frequency (RF) engineers, it is improbable that the optimal configuration of a network can be found.

In order to analyse the performance of these wireless systems and find their optimal configuration, *network planning and optimization tools* based on analytical models or simulations have been widely used by both industry and research communities [1, 2].

In the case of *femtocells*, since the number and location of these devices are unknown by the operator, and because they can be switched on/off or moved at any time, self-organization techniques will play a very important role to the detriment of classic network design. However, these planning and optimization tools will still aid vendors and operators:

- To understand the *impact* that the deployment of a new femtocell layer will cause in the existing macrocell network in terms of coverage, capacity and interference.
- To develop new *algorithms* to face the technical challenges imposed by femtocells, such as for example, synchronization, access method, spectrum allocation, interference avoidance, etc. [3]

In this chapter, a simulation tool able to cope with both aspects taking macrocell and femtocell hybrid scenarios into account is presented. In addition, experimental evaluations that will help the reader to comprehend the advantages and drawbacks that femtocells will bring are given. The rest of the chapter presents the following:

Femtocells: Technologies and Deployment Jie Zhang and Guillaume de la Roche

- In Section 5.1, the need for simulation tools to analyse the performance of the network.
- In Section 5.2, the differences between a Link-Level Simulation (LLS) and System-Level Simulation (SLS).
- In Section 5.3, the models employed to emulate the radio channel.
- In Section 5.4, the differences between a *static* and *dynamic* SLS.
- In Section 5.5, the methodology used to generate the system-level performance results of a femtocell network using static SLSs.
- In Section 5.6, an analysis of the coverage and capacity enhancement achieved due to a femtocell deployment.
- Finally, in Section 5.7, an overview of a dynamic SLS module.

5.1 Network Simulation

Network planning and optimization tools able to evaluate the overall performance of a network are needed to aid vendors and operators in the deployment of new systems and the refinement of existing algorithms.

These tools are currently used by the operators to fine-tune their networks and to analyse the possible advantages and disadvantages of any modification of their systems [4, 5]. Nevertheless, there is a large variety of these tools, which are normally specialized in a certain area and based on a given approach.

In the following, the main ways to evaluate the overall performance of a network, for example, when deploying a new cell or testing a new algorithm are summarized (Table 5.1).

The first approach would consist in the *real implementation* of the new cell or algorithm in the existing network. This approach is the most accurate and reliable, but also the most complex and expensive. Therefore, this approach would not be considered by the network managers, since a malfunctioning of the system would result in the lost of millions of dollars.

The second approach would consist in the use of *analytical models*. In this way, different constraints and formulas could be used to model the relationship between several parameters of the network. Analytical models are often used to derive bounds on the capacity of simple systems with well-defined statistical properties, e.g. Point to Point (P2P) communications. However, due to the size, complexity and nonlinearity of the current wireless systems, it is extremely difficult to determine the exact capacity and performance of a network using these models. Just to make the problem solvable a high number of assumptions and simplifications that will in turn reduce the accuracy and reliability of the solution must be made.

Table 5.1 Performance and capacity evaluation approaches

Approaches	Accuracy	Reliability	Complexity	Cost
Implementation	High	High	High	High
Analytical Model	Low	Low	Medium	Low
Simulation	Medium	Medium	Medium	Low

The third approach is based on *computer simulation*. In this way, the elements and operations of the wireless network are modelled by software. This option is the most suitable for the analysis of complex wireless systems, since it is simpler and cheaper than the real implementation and more accurate and reliable than the analytical model. The number of assumptions and simplifications made depends on the complexity of the computer simulation, but it is usually less than in the case of employing analytical models. In addition, computer simulations not only provide more accurate results, but can also model more complex systems. However, in order to get a statistically representative result for the performance of the network, this option needs a high number of computer iterations. Due to this fact, in order to avoid prohibitive computational costs, a trade off between accuracy and complexity is needed.

In the rest of this book, the evaluation of the performance of macrocell and femtocell hybrid wireless networks will be analysed using computer simulation. In the rest of this chapter, the models employed to carry out these simulations are presented.

5.2 Link and System Level Simulations

Since the simulation of the transmission of all the bits between all the Base Stations (BSs) and User Equipments (UEs) of a given wireless communication system is prohibitive due to the large computational cost, the simulation of a wireless network is usually separated into two levels: *link-level simulation* and *system-level simulation*.

At the *link level*, the behaviour of the radio link between a single transmitter and a single receiver is studied bit by bit, taking the radio propagation phenomena on a small temporal scales (symbol duration) into account. At this level, all relevant aspects of the Physical (PHY) and Medium Access Control (MAC) layer must be considered (Figure 5.1):

- At the PHY layer, functions such as coding, interleaving and modulation must be carefully modelled according to the given standard in order to achieve accurate and reliable results.
- At the MAC layer, the impact of different techniques, e.g. Hybrid Automatic Repeat reQuest (HARQ), should also be considered, since they affect the performance of the system.

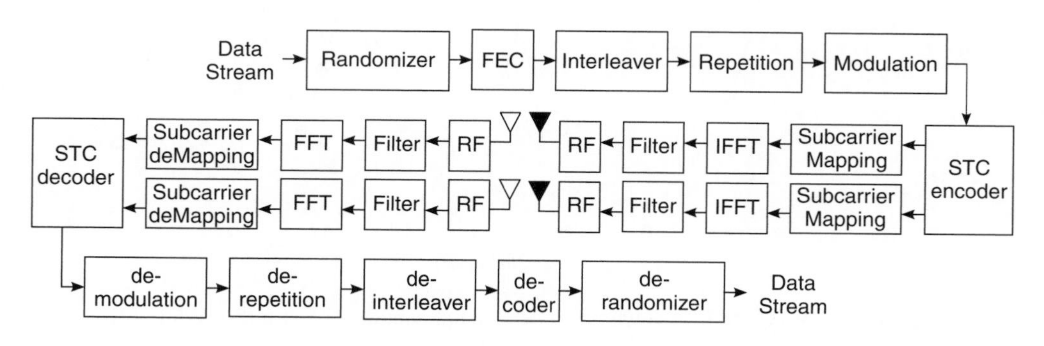

Figure 5.1 WiMAX link-level simulator block diagram

In Wideband Code Division Multiple Access (WCDMA) systems, LLSs have been broadly used to analyse the advantages and drawbacks of Power Control (PC) algorithms [6], while in Orthogonal Frequency Division Multiple Access (OFDMA) systems, they have been widely used to study the performance of Adaptive Modulation and Coding (AMC) and subcarrier mapping techniques [7].

At the *system level*, the behaviour of a network as a set of BSs and UEs is analyzed, considering MAC related issues such as mobility, scheduling and radio resource management. Now, the target is not to study the performance of a single link over small temporal scales, but the performance of the overall system (Quality of Service (QoS), capacity, delay) over larger periods of time.

As mentioned above, link-level and system-level simulations work on very different time scales, and they are executed independently due to computational cost issues. The interaction between both levels is obtained by means of different interfaces called *Look Up Tables* (LUTs). The LUTs are a set of tables that represent the results of the LLS in a simplified way (Figure 5.2), relating the signal quality experienced in the radio link, for example, in terms of bit energy to noise density (E_b/N_0) or Signal to Interference plus Noise Ratio (SINR), to a given parameter of link performance, e.g. Bit Error Rate (BER) or BLock Error Rate (BLER).

This set of tables are consulted during the SLS to obtain the necessary link-level information generated in the LLS. This procedure represents a simple, but efficient way

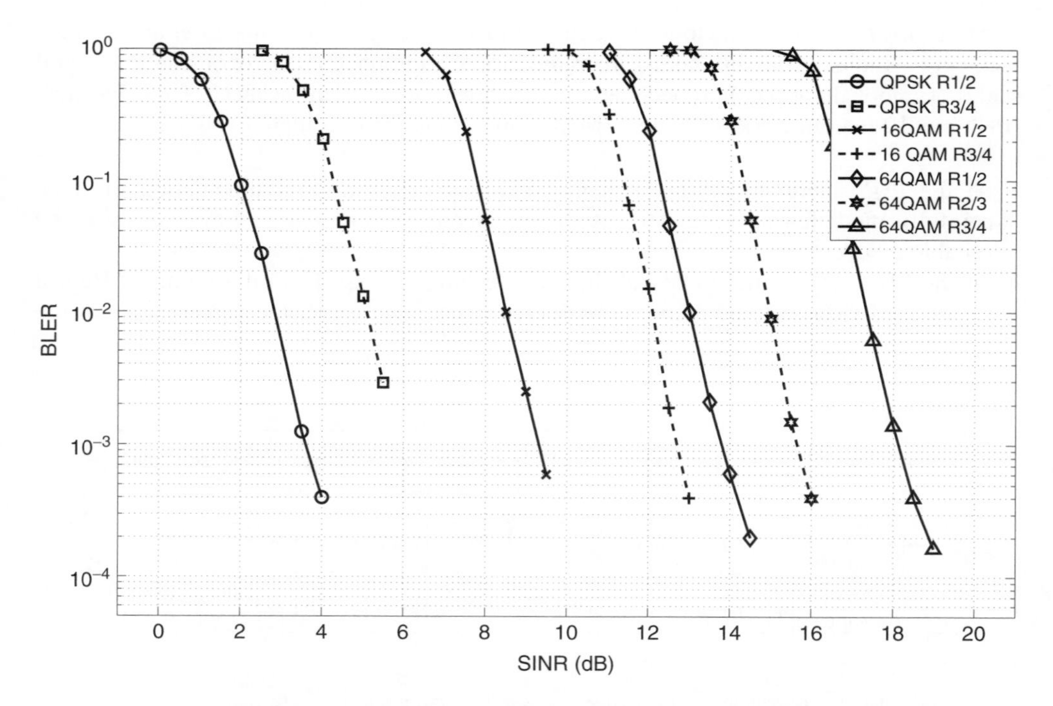

Figure 5.2 WiMAX LUT under AWGN channel conditions and turbo coding [7]

of including the fluctuations of the transmission channel, which take place on small time scales, into the SLS, whose effects are related to larger periods of time (hours, days).

The information derived from these LUTs can be used by the network designer in different ways. First of all, they can be used to benchmark the performance of a given wireless system, by analysing the spectral efficiency in bps/Hz over a single connection. Given the signal quality (SINR) of a certain radio link, the spectral efficiency can be computed as follows:

$$C = C_{max} \cdot (1 - BER) \quad (5.1)$$

where C_{max} corresponds to the symbol efficiency of the modulation and coding scheme used in the transmission (Table 5.2), and the BER can be derived from the SINR using LUTs.

Moreover, the LLS results can also be used by the network designers to derive the appropriate AMC thresholds in link-adaptation systems, as for example, High Speed Downlink Packet Access (HSDPA), Wireless Interoperability for Microwave Access (WiMAX) or Long Term Evolution (LTE). This is important because if these AMC thresholds are underestimated, the system will suffer from packet loss, whereas if they are overestimated, the system will suffer from low efficiency. Table 5.2 illustrates an example of these thresholds for a WiMAX system under AWGN channel conditions.

In order to predict correctly the performance of a wireless system by using LLSs or SLSs, it is not only necessary to accurately simulate the behaviour of the PHY and MAC layers, but also to model precisely the radio channel. Since the behaviour of the radio link strongly depends on the scenario (rural, urban, indoor), as well as the traffic (service, applications) and mobility (fixed, nomadic, mobile) conditions of the users, the *channel model* used in the simulation should be selected according to these features.

In order to model precisely the effects of the attenuation, fading and interference in femtocell scenarios, not only indoor propagation and fading models taking into account the effects of a large number of diffractions and reflections will be needed, but also accurate *indoor-to-outdoor* and *outdoor-to-indoor* models able to capture the particularities of different household structures will be required (Section 5.3).

Table 5.2 WiMAX radio access bearers

RAB	Modulation	Coding	Efficiency (bits/symbol)	SINR threshold (dB)
RAB1	QPSK	1/2	1	2.88
RAB2	QPSK	3/4	1.5	5.74
RAB3	16QAM	1/2	2	8.79
RAB4	16QAM	3/4	3	12.22
RAB5	64QAM	1/2	3	13.23
RAB6	64QAM	2/3	4	15.88
RAB7	64QAM	3/4	4.5	17.50

5.3 Wireless Radio Channel Modelling

Network planning and optimization tools will help femtocell operators to understand the impact of the femtocells in the existing macrocell networks and to develop new inbuilt femtocell algorithms. These tools rely on accurate descriptions of the underlying physical channel in order to perform trustworthy link- and system-level simulations with which to study the femtocells' behaviour. To increase the reliability of these tools, accurate radio wave propagation models are thus necessary. Such models aim at taking into account efficiently the physical effects that have an influence on the propagation of the signal.

5.3.1 Physical Effects

When a femtocell is installed inside a house, its radiated power will produce waves emitted in all the directions according to the antenna pattern that define the part of the energy released in those different directions. Then, a part of the signal will remain indoors, but another part will go outside the building and affect the quality of signal of outdoor users because of the interference. Moreover, the signal of a macrocell installed outside will not only cover the exterior environment, but a part of it will enter the buildings and interfere with the femtocells. That is why identifying the physical effects that occur on the waves can help in understanding how a system will behave.

Reflection When a wave arrives at an obstacle, depending on the physical coefficients of the material, one part of the signal will pass through this obstacle (the refracted part) whereas another part will be reflected from the material.

Diffraction If a wave encounters an obstacle, of which the size is in the order of its wavelength, some diffraction phenomena might appear. This is described by the Huygens–Fresnel principle, and will produce many outgoing waves in various directions depending on the shape of this obstacle.

Diffusion When the surfaces of the obstacles are perfectly planar, only reflections and diffractions will occur. In reality most materials, like concrete, are rough. Depending on the roughness factor of the material, and the wavelength of the material, a wave reaching an obstacle will not produce a single reflected ray, but might produce many smaller rays corresponding in fact to numerous reflections due to the irregularities of the surface.

For a specific user in an environment, the received signal is equal to the sum of the reflected, diffracted and diffused rays. In indoor environments, reflection, diffraction and diffusion will typically occur from the walls and the furniture. Outdoors these phenomena will mainly occur from the buildings. In general, when the outdoor antennas are located at high levels, the diffraction effects that occur on the roofs of the buildings, called roof-top diffraction, are important. When a user is in direct visibility of the emitter, the main component is the direct path called Line Of Sight (LOS). When there are obstacles between the emitter and the receiver the resulting signal is the sum of indirect paths which is called Non Line Of Sight (NLOS).

Different models that deal with the propagation problem can be found. Some are called empirical because they are based on statistics and measurements but do not aim at simulating all the physical phenomena. Others are called deterministic because they are based on deterministic laws that try to estimate all these phenomena.

5.3.2 Propagation Models

Empirical Models

Empirical models are based on statistics and measurements, they do not require the knowledge of the geometry of the environment because the reflexions and diffractions are not computed. In this approach only the distance between the emitter and the receiver have to be taken into account. The loss L in dB at a distance d of the emitter can be estimated with:

$$L = 10 \cdot n \cdot log(d) + C \tag{5.2}$$

Both the constant value C which represents the system losses, and the parameter n which characterize the path loss exponent, depend entirely on the kind of environment, and some typical values can be found for typical kinds of environments. Very often these parameters are very dependent from the scenario, that is why doing measurements will help to find the best parameters.

Semi-Empirical Models

Because of the low accuracy of empirical models, some other models have been proposed. They are also empirical because they are based on only one path between the emitter and the receiver, and because they do not simulate the physical phenomenas. But some deterministic parts can be added to the model, so that obstacles are taken into account. With such models, an attenuation value (depending on the material) is added to the path loss each time the ray crosses an obstacle. At the Universal Mobile Telecommunication System (UMTS) frequencies, typical values for these parameters are as in the example represented in Table 5.3.

Table 5.3 Typical values of attenuations in dB

Material	Attenuation
Main wall (concrete)	15 dB
Inner wall (plaster)	5 dB
Window (glass)	1.5 dB
Door (wood)	0.5 dB

Ray-Tracing-Like Models

Ray-tracing-like models are based on the laws of Descartes. According to this, radio wave propagation can be approximated using geometric waves. With this kind of model, geometrical rays are computed in all directions from the emitter, and each time a ray encounters an obstacle, the reflected ray is computed so that the angle of the reflected ray and the incident ray are symmetrical to the normal vector. Two main methods have been deployed:

- Ray launching: this method launches numerous rays from the emitter in all directions and computes all the reflections and diffractions.
- Ray tracing: this method considers all the receiving points and computes, thanks to the method of images, which are the possible rays that contribute to the received power at this point.

On the one hand, ray launching models are generally less accurate because it is difficult to reach all the points of the environment due to the angular dispersion. Indeed, due to complexity reasons, a limited number of rays has to be launched (typically every 1 or 2 degrees), which is why when the distance increases some receiving points can be missed. On the other hand, ray tracing is more accurate because it computes exactly the rays reaching the considered receiving points, but this method is more memory and time consuming.

In Figure 5.3, the radio coverage of a femtocell inside an office is plotted. The received signal has been computed with a 3D ray launching model [8]. To estimate the diffractions on the corners with such models, the general theory of diffraction and uniform theory of diffraction are used. These consist in isolating the small objects that will be considered as new sources, the coefficients to apply to each ray depending on the direction. Because of the complexity, the number of reflections and diffractions to be computed for each ray has to be limited (e.g. three reflections and three diffractions are typical values in complex environments). To take into account the diffusion, some coefficients of roughness can also be applied depending on the materials.

FDTD-Like Models

Finite-Difference Time-Domain (FDTD) methods propose solving the Maxwell's equations on a discrete spacial grid. The Maxwell equations describe the properties of the

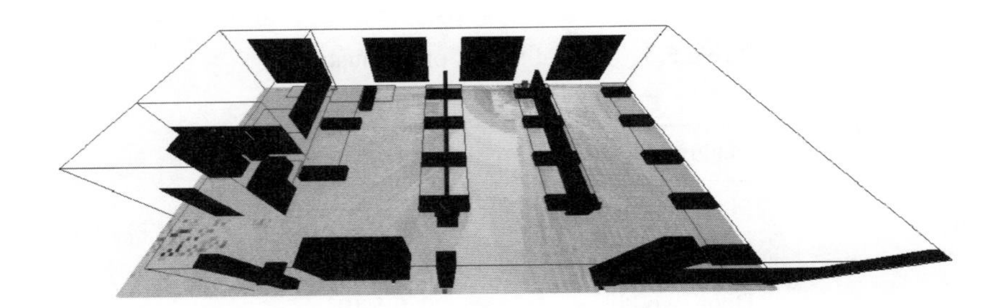

Figure 5.3 Received signal strength prediction in a 3D scenario

electric and magnetic fields and relate them to their sources, charge density and current density. That is why they are very accurate because they implicitly take into account all the reflections and diffractions. However, because the continuous problem of propagation is made discrete, the spatial step and the time step have to be as small as possible, thus making these models very memory and time consuming. Hence these methods have only recently started to be used, as more powerful computers has been developed.

Normally, in order to have a good approximation, FDTD-like models should use a spatial step Δ very small compared with the wavelength λ. A typical value is to use $\Delta = \lambda/10$ making the size of the matrices to consider very large.

In Figure 5.4 a 2D radio coverage of a femtocell has been computed with an FDTD method in an urban scenario. Due to complexity, some simplifications have been proposed and the method has been implemented on a Graphics Processing Unit (GPU) parallel processor [9]. Indeed, FDTD-like models are well adapted to parallel computing, because at each iteration each pixel has to update its energy depending on the neighbouring pixels, which can be computed independently from one pixel to another. Some FDTD-like models, like the one presented in [10], have also been proposed in the frequency domain, making possible to solve the problem in a multi-resolution approach, with a complex preprocessing but a very quick propagation phase.

5.3.3 Choice of a Model

The literature in the area of propagation is very large and choosing the best model is not always an easy task. Here some few recommendations to help the researchers are given.

Accuracy vs Complexity

Empirical models are very easy to implement and have the advantage of being very fast because only one path is computed at each receiving point. However they suffer from a

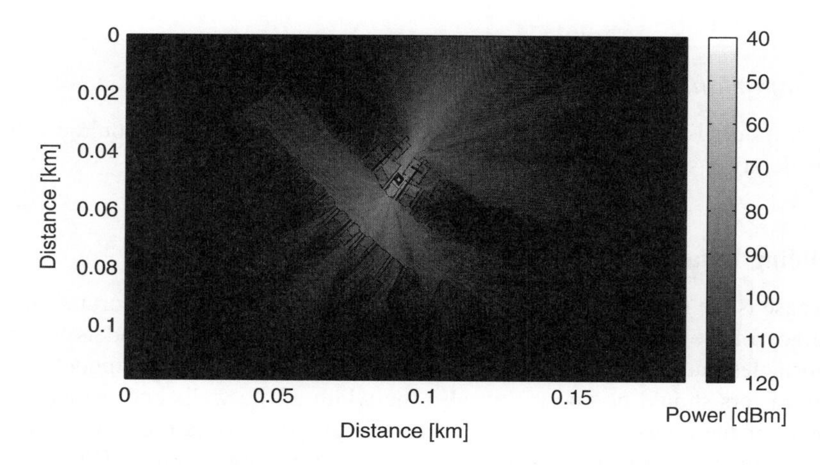

Figure 5.4 Received signal strength prediction in a 2D scenario

lower accuracy because do not take the obstacles into account efficiently. They are a good choice when a simple model is required and when the accuracy is not the most important factor. Moreover, when the database representing the obstacles and the materials is not available or accurate enough, they are the only option.

When a high accuracy is required, deterministic models are the best choice but this cannot be done without a complexity cost. In general, the most complex models are also the most accurate models. Of course this can be discussed because the implementation also has an impact and how to optimize the memory and reduce the simulation time, being these the main challenges. When a choice has to be made between ray tracing or FDTD models, it is generally observed that ray tracing is more complex to develop but is less memory and time consuming compared with FDTD approaches, which are easier to implement but very high time and memory demanding. The accuracy of these models will always depend on the approximations done during the implementation. For example in ray tracing a high number of rays increases both the complexity and the accuracy, and so it is for FDTD-like models with the use of small spatial or time steps.

2D vs 3D Model

All propagation models have been implemented in 2D or 3D, the 2D being easier to implement and less demanding of resources, but also less accurate. Even if 3D-like models perform more accurately, its high complexity sometimes leads to try to use 2D-like approaches. For example in a flat scenario with small buildings, and considering that most antennas have a pattern that radiates in the horizontal plane, the choice of a 2D model can be a judicious approach. Moreover, in a multifloored building it is also sometimes possible to estimate the radio coverage between the different levels by using quasi 3D models, also called 2.5D models. However some complex environments where the antennas are located at different heights, like for example a city with high buildings, and base stations on the roofs of the buildings, and femtocells at different levels of buildings, will absolutely require the use of full 3D propagation models.

5.3.4 Important Factors

Whichever model is chosen, developers of radio propagation models should carefully take into consideration some important factors.

The Building Database

The database is the input of the model and its accuracy will have an important influence on the accuracy of the simulation results. When using semi-empirical models, the positions of the obstacles and the material should be known. If a deterministic model is used, the same parameters should be known but also the width of the walls and the exact physical coefficients of the constituting materials. For this purpose, some radio propagation tools available on the market use the Autocad Drawing Interchange Format (DXF) format that often contains this information.

The Antennas

The shape of the antenna pattern has an important impact on the resulting signal in a given environment. Usually antenna manufacturers provide a diagram in both the horizontal and vertical planes. This diagram illustrates which quantity of energy radiates in which direction. This information has to be integrated into the propagation tool so that the simulations are accurate.

The Calibration

Because no model can perfectly simulate the real life situation, there are always some errors between simulations and measurements. Moreover, some errors in the database or in the real knowledge of the antenna pattern will have a negative impact on the quality of the results. Furthermore, it is not possible to take into account all the elements in the simulations, like for example, the furniture inside a home, the trees in the streets, the passing users and cars who modify the signal, or the weather, which also have an influence on the coefficients of the materials. To all these factors, fading effects due to shadowing and multipath must also be added, which create constructive or negative sums of waves and then modify the resulting signal.

That is why it is important to notice that there will always be some difference between the simulation and the measurements. However, it is possible to make them fit as much as possible by doing a calibration of the model. The calibration consists in making real measurements in the considered scenario, to adapt the parameters of the model, so that the errors between simulations and measurements are reduced. A higher number of measurements will help to reduce the errors but will have a cost because measurement campaigns are often complex. Usually, with correctly calibrated models, it is common to obtain a Root Mean Square Error (RMSE) between prediction and measurements between 5 dB and 10 dB. An accuracy of less than 5 dB would not really make sense because the fading effects are usually in the same order of values.

5.3.5 Simulation of the Fading Effects

The behaviour of the radio signals cannot only be simulated by taking into account an empirical or deterministic model representing the path loss. There are always small statistical variations of the channel that also need to be taken into account.

Shadowing

Shadowing corresponds to the large-scale variations of the channel produced by large obstructions. Indeed, depending on the chosen propagation models, some obstacles will be more or less correctly taken into account in the simulation. That is why the shadowing can be simulated by adding a log-normal variation of the received power around the predicted median value. The shadow fading corresponds to the value of received power that will be added or removed to the received power, depending on a certain probability. This value and the associated probability depend mainly on both the radio propagation model and the environment. Therefore, in femtocell scenarios, measurements must be performed in order to model shadow fading.

Fast Fading

Fast fading corresponds to the small-scale local variations of the channel produced by multi-path. These varations are mainly due to the different rays that reach the receiver (reflections and diffractions in the environment). The sum of all these rays, because of their phases, can have a destructive or constructive effect on the resulting signal, therefore producing small variations in the received power. Different fast fading models have been used like Rayleigh or Rician fading. The power delay profile, representing the power and time of arrival of each ray, is often used to characterize fast fading at a specific location. That is why ITU has defined different profiles (pedestrian, vehicular) [11] that should be carefully included in the simulation in order to take into account the numerous reflections that usually occur in femtocell scenarios.

5.4 Static and Dynamic System-Level Simulations

Although the general target of system-level simulations is to characterize the overall performance of a wireless network, they can be applied with different objectives. In the case of femtocells, SLSs may be used to analyse the advantages and disadvantages of the deployment of a large femtocell layer over the existing macrocell network, or to accurately evaluate the impact of new inbuilt femtocell algorithms in the overall system performance, e.g. self-organization techniques.

Since SLSs may be applied for different purposes, different SLS models tailored to such targets should be used. In this book, we are differentiating between static and dynamic system-level simulation.

In a static SLS (Section 5.5), the aim of the simulation is to study the average performance of the system over large areas and for long periods of time. In this case, the simulation is normally based on multiple and independent Monte Carlo snapshots [12], which is a widely used approach in network planning and optimization [13, 14]. In a Monte Carlo snapshot, the time domain is neglected and several users with different features and QoS requirements are randomly spread over the area of study. The performance of the overall network is then analysed in terms of coverage and capacity. The Key Performance Indicators (KPIs) that are normally used are the cell/user throughput, as well as the number of success and outage users. In order to get a statistical representative result of the average performance of the system, a static Monte Carlo simulation needs a large number of snapshots (Figure 5.5). Therefore, fast simulation algorithms able to support thousands of SLSs within a short period of time are necessary to predict accurately the overall behaviour of the studied system. As a result, time plays a very important role to the detriment of accuracy.

In a dynamic SLS (Section 5.7), the target of the simulation is to accurately model the functioning of the system with a high degree of detail. In this case, the evolution of the network over time is taken into account, and the simulation allows the network to live as a function of the time or events. To capture the end-to-end behaviour of a network,

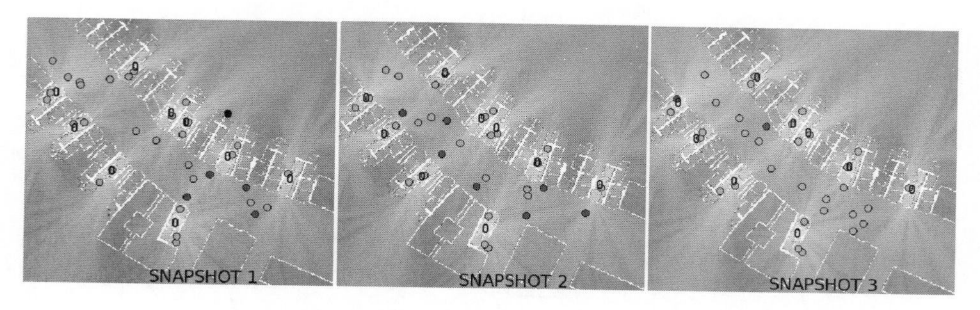

Figure 5.5 Monte Carlo snapshots

the dynamic features of the users (mobility models) and the traffic (traffic models) must be simulated, as well as the fluctuations of the channel over time and frequency (shadow and multipath fading) [15]. In this way, the behaviour of different techniques, e.g. power allocation strategies and radio resource management techniques, can be analysed in detail over time. As a result of the higher level of detail, the running time significantly increases, and therefore, only small areas can be analysed for shorts periods of time. The network performance is assessed by various measures such as cell and user throughput, call blocks and drops, end-to-end delay and jitter, packets loss and/or retransmission ratio.

5.5 Static System-Level Methodology for WiMAX Femtocells

The design and evaluation of Monte Carlo SLSs able to model the behaviour of wireless cellular networks is not a new topic. In fact, this issue has been widely investigated and commercialized [13, 14] in the past. However, as new technologies emerge, these simulation tools must be reviewed and adapted. This is the case with femtocells, whose particular features require the use of tailored simulation engines that model the innovative characteristics of these devices.

In this section, the Monte Carlo SLS methodology used in this book to carry out coverage and capacity performance analyses of macrocell and femtocell hybrid networks is introduced. Furthermore, the achievable performance of co-channel femtocell deployments, and their interference impact on an existing macrocell layer are discussed, based on simulation results.

5.5.1 Network Characterization

In the following, the main elements and parameters of the Centre for Wireless Network Design (CWiND)'s Monte Carlo SLS (Figure 5.6) for WiMAX networks are introduced for illustration purposes. However, this SLS structure could be applied to other technologies such as Global System for Mobile communication (GSM), UMTS, High Speed Packet Access (HSPA) or LTE taking the different radio access network interfaces into account.

Cell Site or Base Station A cell site, also called a base station, refers to a geographical point where one or several transmitters/receivers equipped with one or several antennas are located.

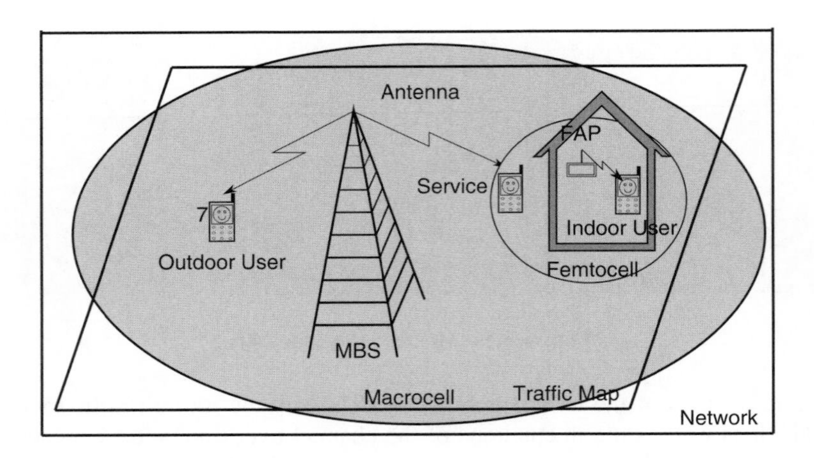

Figure 5.6 System-level simulation elements

In this simulation platform, two different types of cell site are distinguished:

- *Macrocell Base Stations* (*MBSs*) that are designed to provide radio coverage outdoors,
- *Femtocell Access Points* (*FAPs*) that are designed to provide radio coverage indoors.

Cell or Sector A cell C_i, also called sector, refers to a geographical area covered by one or several transmitters/receivers.

In this simulation platform, two different types of cell are distinguished:

- *macrocells* C_i^{macro}, which are covered by a MBS,
- *femtocells* C_i^{femto}, which are covered by a FAP.

User A user UE_x refers to a network subscriber, and is characterized by different parameters such as its geographical position, as well as its equipment, service and mobility type. Although in a Monte Carlo SLS the movement of the users is not considered, their mobility features can be used to model the behaviour of their channel. In this way, the fading of the channel can be estimated according to the speed of the users, taking the coherence bandwidth and Doppler effects into account.

In this simulation platform, two different types of user are distinguished:

- *non-subscribers* UE_x^{ns}, which are users not registered in any nearby femtocell,
- *subscribers* UE_x^{s}, which are users registered in a nearby femtocell.

Subscribers are thus defined as the rightful users of the femtocell, and they are usually mobile terminals of the femtocell owner and close family and friends.

When using Closed Subscriber Group (CSG) access, non-subscribers UE_x^{ns} are only allowed to connect to the macrocells; meanwhile, subscribers UE_x^{s} are allowed to connect to their own femtocell or to the macrocells. Conversely, when using open access, any user can connect to any macrocell or femtocell regardless of their user type.

Traffic Map A traffic map provides information about the number of users and the types of service available in a specific area.

In this simulation platform, two different types of traffic map are distinguished:

- *outdoor traffic maps* that are placed outdoors, and generate non-subscribers,
- *indoor traffic maps* that are placed indoors, and generate subscribers.

A given indoor traffic map is always associated with a femtocell. In this way, the subscribers generated by this traffic map are the rightful users of this femtocell.

Subcarrier, Subchannel, Orthogonal Frequency Division Multiplexing (OFDM) Symbol and Slot OFDMA/Time Division Duplexing (TDD) (Figure 5.7) is a multicarrier technology where:

- the available radio spectrum (wireless channel) is formed by R orthogonal subcarriers, which in turn are combined into K groups called subchannels,
- the time domain is segmented into consecutive frames of a given duration T_{frame}, which in turn are divided into T time slots called OFDM symbols.

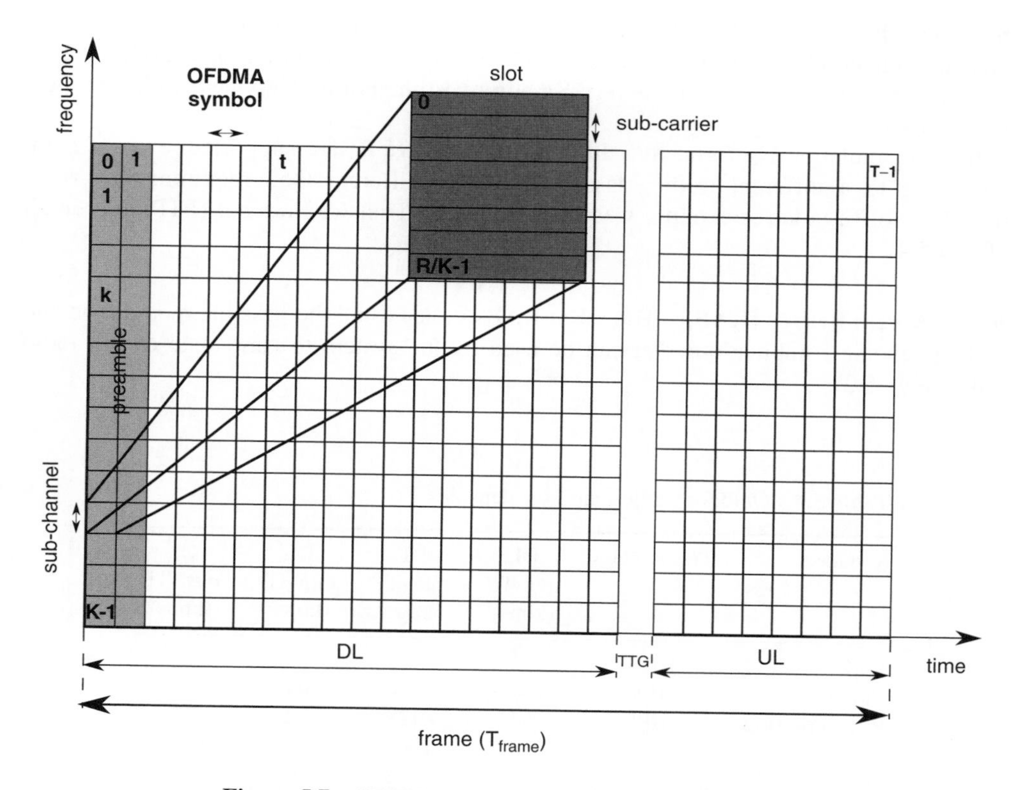

Figure 5.7 OFDMA/TDD frame structure of WiMAX

The number and exact distribution of the subcarriers that constitute a subchannel depend on the subcarrier permutation mode. Subchannels may be built of either contiguous or pseudo-random distributed subcarriers across the spectrum (Section 4.4.2).

In WiMAX [16], the slot is the minimum frequency-time resource that can be allocated by the cells. A slot is composed of one subchannel, and one, two or three OFDM symbols, depending on the network configuration. Several contiguous slots can be allocated to different users in the form of bursts as a multiple-access mechanism. The bandwidth allocated to a user is a function of its bandwidth demand, QoS requirements and channel conditions.

Radio Link A radio link models the propagation phenomena between transmitters and receivers $C_i \leftrightarrow UE_x$. It characterizes the fluctuations of the received signal strength taking the effects of the path loss attenuation, shadow fading and multipath fading into account. This data is extracted from the radio coverage simulations presented in Section 5.3.

Antenna Cells C_i and users UE_x are equipped with at least one antenna, which is a device that is made to transmit and receive electromagnetic waves efficiently, and is characterized by different parameters such as radiation patterns, polarization, azimuth, tilt and gain.

Services In WiMAX, five different classes of traffic with different priorities of service are defined [17]: Unsolicited Grant Service (UGS), real-time Polling Service (rtPS), Extended real time Polling Service (ertPS), non-real-time Polling Service (nrtPS) and Best Effort (BE).

In this simulation platform, five different services SER_s are supported, each of them belonging to a different traffic class, and having different QoS requirements: Voice over IP (VoIP), video streaming, web browsing, File Transfer Protocol (FTP) and email (Table 5.4).

Radio Access Bearer (RAB) The set of RABs indicates all the possible modulation and coding scheme combinations that can be used by the system to carry data over a radio link (Table 5.2).

Table 5.4 Outdoor traffic map user densities

Service	Traffic class	DL min TP (kbps)	DL max TP (kbps)	UL min TP (kbps)	UL max TP (kbps)
VoIP	UGS	12.2	12.2	12.2	12.2
Video	rtPS	64	64	64	64
Web browsing	nrtPS	64	128	32	64
FTP	nrtPS	0	1000	0	100
Email	BE	0	128	0	64

5.5.2 Static SLS Methodology

The static SLS methodology presented here takes multiple independent Monte Carlo snapshots of the network to determine the average behaviour of the system over long time scales. Within each Monte Carlo snapshot, the simulator takes several steps, which can be divided into three layers: Network, MAC and PHY layer, to compute the final performance of the network. This process is illustrated in Figure 5.8 and Figure 5.9.

First of all, let us define a WiMAX network as:

- a set of N cells $\{C_0, C_i, C_j, \ldots, C_{N-1}\}$ and M users $\{UE_0, UE_x, UE_y, \ldots, UE_{M-1}\}$,
- with K sub-channels $\{0, k, \ldots, K-1\}$ and T OFDM symbols $\{0, t, \ldots, T-1\}$,
- where R RABs $\{0, RAB_r, \ldots, RAB_{R-1}\}$ are available for transmission,
- and S services $\{0, SER_s, \ldots, SER_{S-1}\}$ are defined.
- $slot_{i,k,t}$ is formed by the kth subchannel and Tth OFDM symbol of the ith cell.

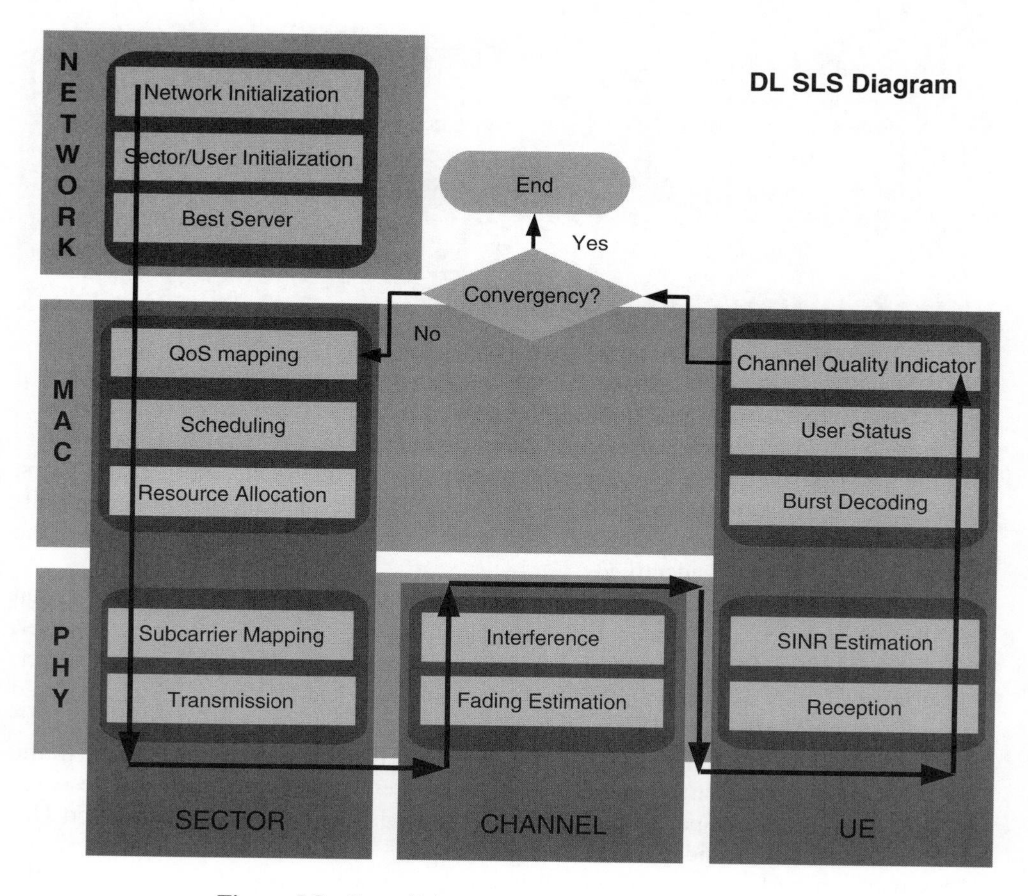

Figure 5.8 Downlink system-level simulation diagram

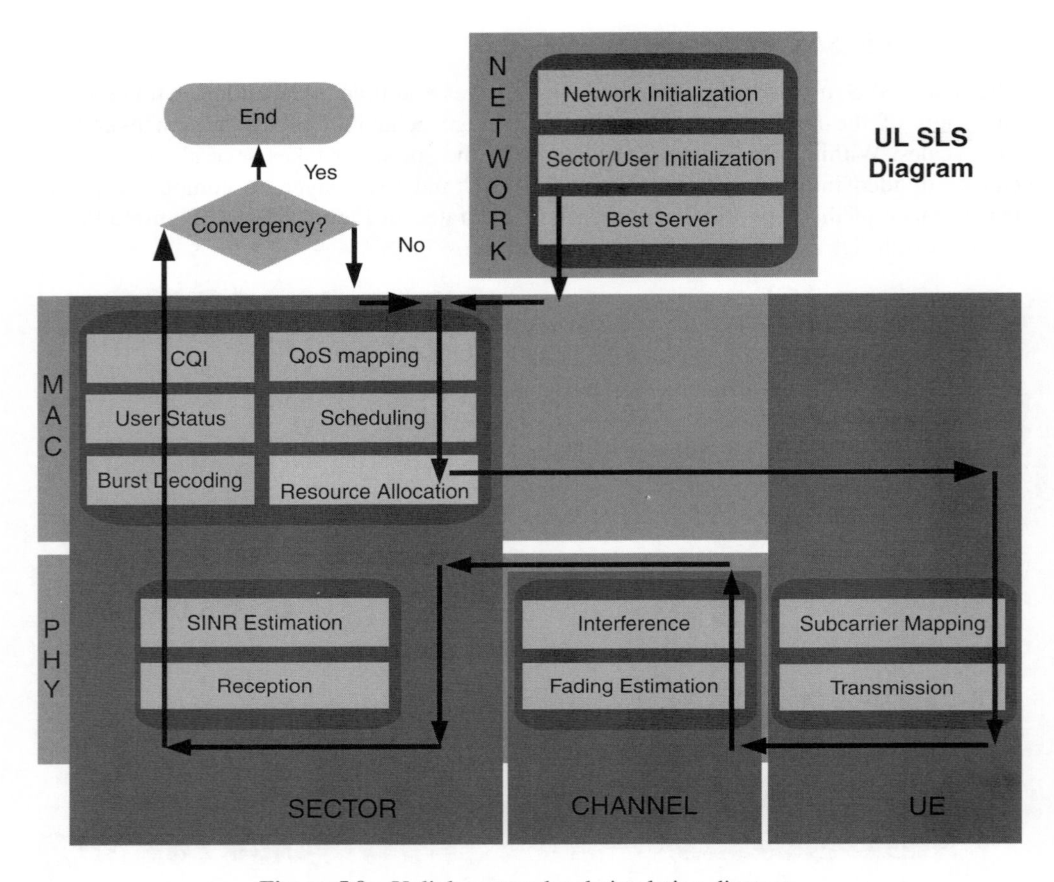

Figure 5.9 Uplink system-level simulation diagram

At the network layer, first, the network configuration is fed in (cells, traffic maps, users, services and bearers). Then, the path loss between the users and the neighbouring cells is computed, using one of the propagation models presented in Section 5.3.2. Finally, the best server of each user is identified.

The UE selects as a best server the cell that provides the largest received pilot signal strength, where this parameter must be larger than the sensitivity of the antenna of the user. Note that non-subscribers of a CSG femtocell cannot select this femtocell as best server.

Afterwards, in this simulation platform, DownLink (DL) and UpLink (UL) are analysed separately at the MAC and PHY layer.

At the MAC layer, initially, the scheduler classifies the users according to their traffic class (UGS, rtPS, ertPS, nrtPS and BE), and subsequently, the scheduling and resource allocation procedures are performed. Link adaptation is supported in both DL and UL, and power control is considered when needed.

This platform performs power control in the UL. It works in such a way that if the signal quality of the user is larger than is necessary to achieve the best RAB defined in the system, the user reduces its transmitted power just to get the required signal quality to keep this RAB.

At the PHY layer, the transmission of the information between the transmitter and receiver is simulated. At this stage, the interference between the slots of the frames of the different cells is computed, and the average signal quality of each radio link is estimated.

Multiple iterations are performed over the same snapshot to reach a stable solution. The stability of the snapshot is achieved when the RABs allocated to the users remain constant for a number of iterations. The RAB allocated to each user is selected according to the channel state information that is fed back from one iteration to the next. This channel state information indicates the average SINR suffered by the user over all its allocated slots in the previous iteration. Once the solution is stable, the KPIs of the users and the results are extracted.

Since the position of the users in a Monte Carlo snapshot is randomly generated, to get a statistically representative result of the performance of the system, a large number of snapshots must be considered. In this way, the correlation between the position of the users and the results is reduced.

Now that the structure of the simulation tool has been presented, the DL and UL iterations are analysed in more detail, paying attention to six different aspects: user generation, scheduling, resource allocation, interference estimation, throughput calculation and user state.

User Generation Within each snapshot, the simulator randomly distributes several users over the area of study according to the features of the traffic maps. Since each traffic map could be associated with several service types, the number of users generated in each snapshot depends on two parameters: the number of service types per traffic map, and the density of users per service.

On the one hand, outdoor traffic maps indicate the density of users (non-subscribers) per service according to the ratio $user/km^2$, which depends on the type of service and area. Table 5.5 and Table 5.6 indicate values for these densities.

On the other hand, indoor traffic maps indicate the density of users (subscribers) per service according to the ratio *user/house*, which is randomly selected between $[0 - UE_{femto}^{max}]$, where UE_{femto}^{max} is the maximum number of simultaneous users supported per femtocell.

Note that macrocell and femtocell users are uniformly distributed inside the traffic maps.

As mentioned above, a traffic map can support different services SER_s, e.g. VoIP, video, web browsing, FTP, email, which are defined by certain requirements of QoS and throughput. For example, a service such as VoIP is described by its traffic class UGS,

Table 5.5 Outdoor traffic map user densities

Area type	Dense urban	Urban	Rural
User density ($user/km^2$)	80	45	10

Table 5.6 User traffic mix [18]

Service type	VoIP	Video	Web browsing	FTP	Email
Percentage of users	35	20	20	10	5

and its DL and UL minimum 12.2 kbps and maximum 12.2 kbps demanded throughput among other parameters. Such service requirements will be used to decide when and what resources should be allocated to each user, as well as to determine their final user state (successful, outage, etc.). In Table 5.4, the traffic class and throughput requirements of different services are illustrated.

Scheduling Within an snapshot, the cell layout is fixed and the channel undergoes multi-path fading according to the position and motion of the users (Section 5.3.5). Channel state information is then fed back from the users to the cells in terms of SINR (Section 5.5.2). Using this information, power control algorithms and link adaptation techniques can be used to overcome the problems imposed by the fluctuations of the radio channel. Moreover, this information can also be used to exploit multi-user diversity (Figure 5.10), considering sophisticated scheduling techniques, where the users that have favourable radio channel conditions are scheduled in first place.

In this simulation platform, the users are scheduled independently in each cell. First, the MAC scheduler classifies the different user connections into different traffic classes, each one with its own service priority and QoS requirements (UGS, rtPS, ertPS, nrtPS and BE). Then, the MAC scheduler classifies the user connections within one traffic class according to a given strategy. We are differentiating between three policies: First In–First Out (FIFO), best SINR and Proportional Fair (PF) [19]. When using the first policy, the users are scheduled according to their time of arrival, independently of the Channel Quality Indicator (CQI) fed back from the user to the cell in previous iterations. Conversely, best SINR and PF make use of this information to classify the connections within one service class. The best SINR policy tends to increase the network throughput, since the users with larger SINR (close to cell) are scheduled first. The PF policy tends to increase the fairness between the users, since the user with a larger ratio current to average SINR (better channel conditions) has higher priority.

Radio Resource Management Radio resources are allocated by the cells to the users in order to satisfy their bandwidth demands and QoS requirements, while taking the traffic and channel conditions into account.

The radio resources will be assigned to the users in the order indicated by the scheduler. Services belonging to traffic classes with a higher priority will be scheduled first. The connections within a traffic class will be served according to the scheduling policy.

In WiMAX, the resources are allocated in the form of burst, which is a two-dimensional (frequency/time) data region of contiguous slots using the same modulation and coding.

In this simulation tool, it is considered that in the buffer of each user the data queue is always full of packets so that we can focus on the analysis of the system performance.

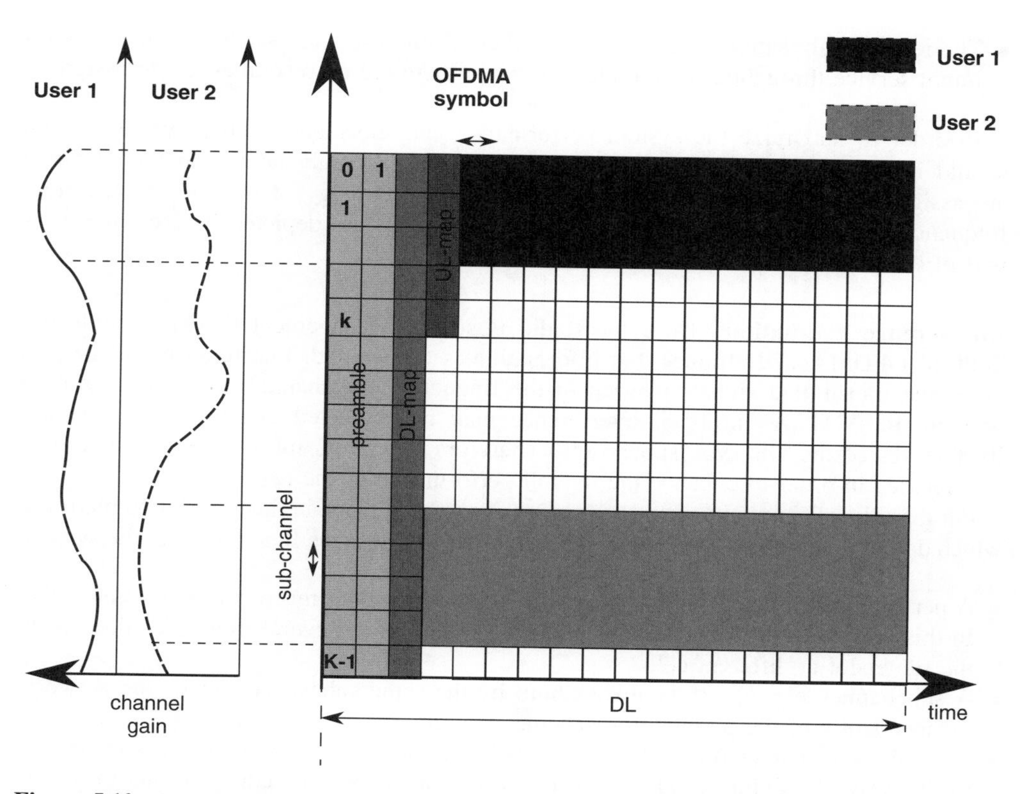

Figure 5.10 In OFDMA, different subchannels are allocated to different users, preferably in the range where they have a high channel gain (multi-user diversity)

The RAB (user profile) is selected according to the CQI fed back in previous iterations. The higher the RAB, the higher the bearer efficiency (*bits/symbol*) and bit rate of the slot. Seven different user profiles are defined in this WiMAX platform (Table 5.2).

Therefore, since the cell knows the RAB to be assigned to the user, as well as its requested throughput, the cell can compute the number of slots that are needed to serve this user, and then build its burst. The number of slots can be calculated as follows:

$$n_{slots} = \frac{TP_{requested} \cdot T_{frame}}{\frac{C}{K} \cdot RAB_{eff}} \tag{5.3}$$

where $TP_{requested}$ denotes the service throughput requirement, T_{frame} indicates the frame duration, C/K represents the number of data subcarriers per subchannel, and finally, RAB_{eff} is the RAB efficiency ($bits/symbol$).

Then, the resource allocation procedure is as follows:

- The cell assigns as many slots as needed to the user appointed by the scheduler to satisfy the minimum service throughput requirement.
- After serving all the users of all traffic classes with this requirement, if there are radio resources left, the cell assigns one more slot to each user to enhance its throughput.

- The resource allocation procedure stops when all the users are satisfied with the maximum service throughput requirement or when there are no resources left to assign.

In order to maximize the system performance, the subchannel and power allocation should be based on the channel conditions. The specific subchannels and power that are assigned to each user depend upon the allocation strategy of the system. Different frequency and power allocation strategies for femtocells are depicted in Section 8.6, as part of the self-organization procedure of the femtocell.

Interference Estimation Once the Radio Resource Management (RRM) module has built all OFDMA/TDD frames, the information is transmitted. Interference will happen when the transmitted signals overlap in the frequency (subchannel) and time (symbol) domain. In WiMAX, intra-cell interference may be neglected due to the orthogonality features of the OFDMA subcarriers. Therefore, operators must cope with inter-cell interference in order to enhance the overall performance of the network.

For the sake of simplicity, four assumptions have been made in this simulation platform, which do not involve any loss of generality in order to assess the functioning of the system:

- A perfect synchronized WiMAX network is assumed (both macrocells and femtocells). In this way, inter-cell interference will occur only when several users are allocated to the same slot at different cells.
- A subchannel, e.g. $k = 0$, is always built by the same subcarriers across the network, independently of the permutation scheme selected.
- The coherence bandwidth of the channel is larger than the bandwidth of the subchannel. In this way, the fading of all subcarriers within a slot will be constant, and they may change only from subchannel to subchannel.
- The coherence time of the channel is larger than the duration of the OFDMA/TDD frame. In this way, the fading of all OFDM symbols within a subchannel will be constant, and they may change only from one frame to another.

Concerning the second assumption note that, in AMC schemes, a given subchannel $k = 0$ is always built by the same subcarriers across the network, while in Full Usage of Subchannels (FUSC) and Partial Usage of Subchannels (PUSC) schemes, this maybe or maybe not true. However, it has been proved that in FUSC and PUSC systems in which logical subchannels are identically built in all cells of the network perform better [20].

As a result of the third and fourth assumptions, the effective SINR of a slot is equal to the SINR of one of its subcarriers (flat subchannel) since all of them suffer the same fading conditions. However, if the fading of all subcarriers within a slot would not be constant, different metrics can be used to calculate the effective SINR of a slot from the SINR of its subcarriers, e.g. Mean Instantaneous Capacity (MIC), Exponential Effective SINR Mapping (EESM) or Effective Code Rate Map (ECRM) [7].

The model used to compute the interference in a given slot $slot_{i,k,t}$ is depicted in Figure 5.11.

Downlink Case In downlink, where the interference is suffered by the users, it can be said that a certain user UE_x, whose best server is C_i, suffers from the interference of cell

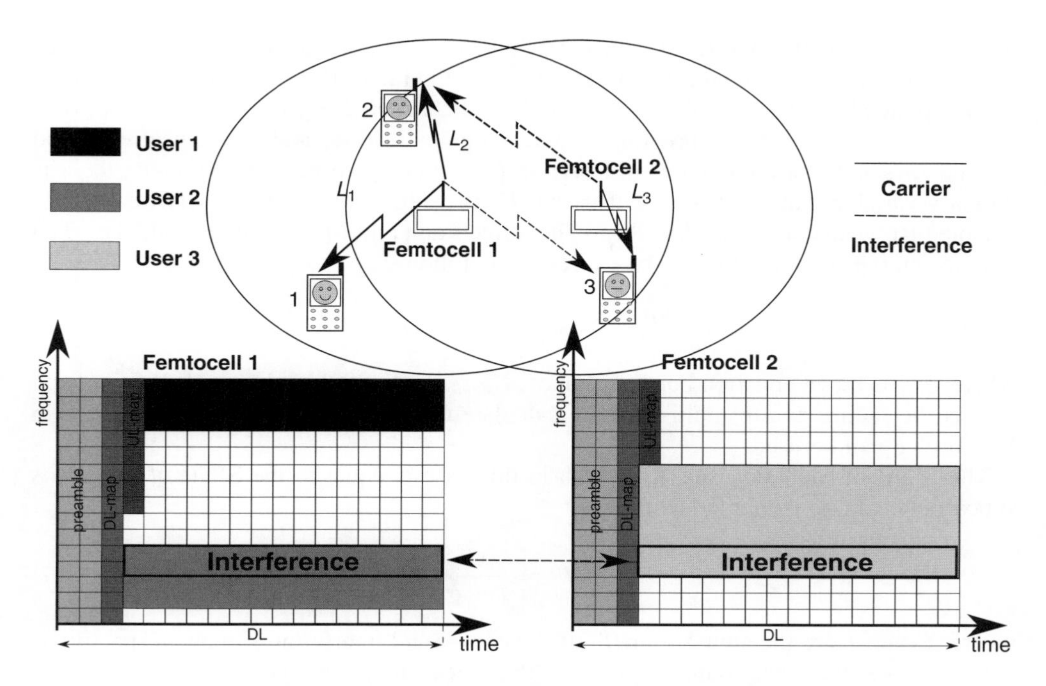

Figure 5.11 Interference in OFDMA/TDD femtocell scenarios: downlink case

C_j, if and only if, C_i and C_j are using the same subchannel for DL transmission at the same OFDM symbol. The final interference suffered in DL by UE_x at slot $slot_{i,k,t}$ will be the sum of the interferences coming from all neighboring cells C_j.

$$I_{x,k,t}^{DL} = \sum_{j=0, j \neq i}^{N-1} \sum_{t=0}^{T-1} (P_{j,k} \cdot G_j \cdot L_j \cdot Lp_{j,x} \cdot G_x \cdot L_x) \cdot \phi_{j,k,t} \tag{5.4}$$

where, x indicates the interfered user, UE_x; k indicates the kth subchannel and t indicates the tth symbol; i is the serving cell, C_i; j is an interfering cell, C_j; $P_{j,k}$ is the power applied by C_j in a subcarrier of the kth subchannel; $Lp_{j,x}$ is the path loss between C_j and UE_x; G_j and G_x stands for the antenna gains in C_j and UE_x, respectively, and L_j and L_x stands for the equipment losses in C_j and UE_x, respectively. Finally, $\phi_{j,k,t}$ is a binary variable that is equal to 1 if cell C_j is using slot $slot_{j,k,t}$, or 0 otherwise.

Uplink Case In uplink, where the interference is suffered by the cells, it can be said that a certain cell C_i, whose connected user is UE_x, suffers from the interference of user UE_y, if and only if, UE_x and UE_y are using the same subchannel for the UL transmission at the same OFDM symbol. The final interference suffered in UL by C_i at slot $slot_{i,k,t}$ will be the sum of the interferences coming from all neighbouring users UE_y.

$$I_{i,k,t}^{UL} = \sum_{y=1, y \neq x}^{M-1} \sum_{t=0}^{T-1} (P_{y,k} \cdot G_y \cdot L_y \cdot Lp_{y,i} \cdot G_i \cdot L_i) \cdot \theta_{x,k,t} \tag{5.5}$$

where, i indicates the interfered cell, C_i; k indicates the kth subchannel and t indicates the tth symbol; x is the connected user, UE_x; y is the interfering users, UE_y; $P_{y,k}$ is the power applied by UE_x in a subcarrier of the kth subchannel; $Lp_{y,i}$ is the path loss between UE_y and C_i; G_y and G_i stands for the antenna gains in UE_y and C_i, respectively, and L_y and L_i stands for the equipment losses in UE_y and C_i, respectively. Finally, $\theta_{x,k,t}$ is a binary variable that is equal to 1 if user UE_y is using $slot_{x,k,t}$, or 0 otherwise.

Note that when computing Lp, shadowing effects and multi-path fading should be taken into account. Therefore, Lp can be expressed as follows:

$$Lp = L_{att} \cdot L_{shadow} \cdot L_{ff} \tag{5.6}$$

where L_{att} is the attenuation due to distance, L_{shadow} is the fading due to shadowing, and L_{ff} is the fading due to multi-path. To calculate these values, the models presented in Section 5.3 can be used.

The SINR of each slot, $slot_{i,k,t}$, which in this case is equal to the SINR of one of its subcarriers, can be computed as follows:

$$SINR = \frac{C}{I + \sigma} \tag{5.7}$$

where, C and I are the signal strength of the carrier and interfering signals, respectively, and σ denotes the background noise. Note that linear units must be used.

The interfering signal power I can be derived from the interference model presented above, while the carrier signal power C can be similarly estimated using:

$$C_{x,k}^{DL} = P_{i,k} \cdot G_i \cdot L_i \cdot Lp_{i,x} \cdot G_x \cdot L_x \tag{5.8}$$

$$C_{i,k}^{UL} = P_{x,k} \cdot G_x \cdot L_x \cdot Lp_{x,i} \cdot G_i \cdot L_i \tag{5.9}$$

The background noise, on the other hand, can be approximated by:

$$\sigma_n = n_0 + nf_{eq} \tag{5.10}$$

$$n_0 = -174 \frac{dBm}{Hz} \cdot 10 \log \left(F_{sampling} \cdot \frac{SC_{used}}{SC_{total}} \right) \tag{5.11}$$

where, n_0 stands for the thermal noise, and nf_{eq} for the noise figure of the terminal (DL case) or the cell (UL case). In addition, $F_{sampling}$ denotes the sampling frequency of the system, while SC_{used} and SC_{total} represents the number of considered and total subcarriers, respectively.

Once the SINR of all slots allocated to a user are known, the effective SINR of the user, which will be fed back to the cell, is computed by using a MIC metric [7]. This information will be used in the next iteration of the snapshot to select the adequate RAB for transmission.

Throughput Calculation Taking into account the assumption that the fading of all subcarriers within a slot is constant, the bit rate of a given slot, $slot_{i,k,t}$, can be calculated as the bit rate of one of its subcarriers multiplied by the number of subcarriers within the slot as shown in the following equation:

$$BR_{slot}^{i,k,t} = \frac{RAB_{eff}^{i,k,t}}{T_{frame}} \cdot \frac{C}{K} \tag{5.12}$$

where $RAB_{eff}^{i,k,t}$ denotes the efficiency of the subcarriers (*bit/symbol*) within the slot $slot_{i,k,t}$. This value can be assessed by using the SINR of the subcarrier and Table 5.2. Moreover, T_{frame} indicates the duration on the OFDMA/TDD frame, and C/K represents the number of subcarriers per subchannel.

Once the bit rate of the slot $slot_{i,k,t}$ is known, the throughput of this slot can be derived as follows:

$$TP_{slot}^{i,k,t} = BR_{slot}^{i,k,t} \cdot [1 - BLER_{i,k,t}(SINR, RAB)] \tag{5.13}$$

where $BR_{slot}^{i,k,t}$ denotes the bit rate of the slot $slot_{i,k,t}$; $BLER_{SINR}^{i,k,t}$ indicates the BLER of the slot, which is function of the suffered SINR and the used RAB.

Finally, the total user UE_x throughput can be calculated as the sum of the throughput of all slots assigned to the user as shown in the following equation.

$$TP_x = \sum_{k=0}^{K-1} \sum_{t=0}^{T-1} TP_{i,k,t} \cdot \phi_{i,k,t} \tag{5.14}$$

where $\phi_{i,k,t}$ is a binary variable that is equal to 1 if cell C_i is using slot $slot_{i,k,t}$, or 0 otherwise.

Moreover, the throughput of the cell can be computed as the sum of the throughput of all its slots.

User State Finally, the state of the user is estimated. The possible states of the users are defined as follows:

- No coverage: A user is considered to be without coverage, when the power received coming from the user's best server is smaller than the user's equipment sensitivity.
- No resource: If the user has a user profile (RAB), but the cell has not sufficient resources to satisfy the minimum requested service throughput.
- Outage (No RAB): If the SINR reported by the CQI is smaller than the SINR required to get the minimum user profile (RAB).
- Outage (Tx failure): When the user is transmitting, but the user has achieved the minimum requested service throughput.
- Successful: When the user is transmitting, and the user has achieved the minimum requested service throughput.

5.6 Coverage and Capacity Analysis for WiMAX Femtocells

It is expected that femtocells will benefit both users and operators:

- Users may enjoy better signal qualities due to the proximity of transmitters and receivers, the result of this being communication with greater reliabilities and throughputs, as well as power and battery savings.
- From the operator's point of view, femtocells may extend indoor coverage and enhance network capacity. Femtocells will help to manage the exponential growth of traffic within the macrocells, due to the handover of the indoor traffic to the backhaul connection. They will also reduce the capital and operational expenditure of the network, since they are paid for and maintained by the user.

However, these benefits are not easy to realize, since the network managers must face different challenges before femtocells can become widely deployed. To achieve these benefits, for example, the management of electromagnetic interference between the macrocell and femtocell layer, and between femtocells plays an important role. This interference could counteract the above mentioned advantages provided by the femtocells, and severely downgrade the performance of the entire network.

In the following, WiMAX SLSs based on Monte Carlo snapshots will be used to extract realistic statistics about the impact that the deployment of a wide femtocell layer will have over the existing macrocell networks. In this way, the benefits and drawbacks that femtocells might bring will be assessed, and some of the challenges that the operators must face will be addressed. The analysis presented here will be divided into three different subsections: coverage, signal quality and capacity.

5.6.1 Scenario Description

First of all, let us describe the scenario used to perform these experimental evaluations.

Figure 5.12(a) and Figure 5.12(b) presents an aerial coverage view of a residential area within Luton, a town located to the north of London (UK), where the study has been carried out.

In this scenario, femtocells operate in the same frequency band as the existing macrocells (co-channel deployment). Femtocells operating in a dedicated and separate frequency band (orthogonal deployment) is a possible and optimal solution for the avoidance of interference between macrocells and femtocells. However, this approach will drive the operators to a reduced spectral efficiency (*bps*/Hz) usage, which is extremely expensive and undesired. Therefore, femtocells operating in a co-channel frequency band with existing macrocells seems to be a more appropriate solution, although technically far more challenging due to interference.

In this experiment, two macrocells and a large number of CSG femtocells were used. Their position can be identified from the coverage plot of the above mentioned figures. In this scenario, femtocells have been deployed according to a femtocell penetration of 10%. This means that 1 of every 10 households is equipped with a FAP. This value is equivalent to half of all subscribers of a market-leading UK mobile operator deploying femtocells [22]. This translates to 100 femtocells deployed in this simulation, whose position was randomly selected.

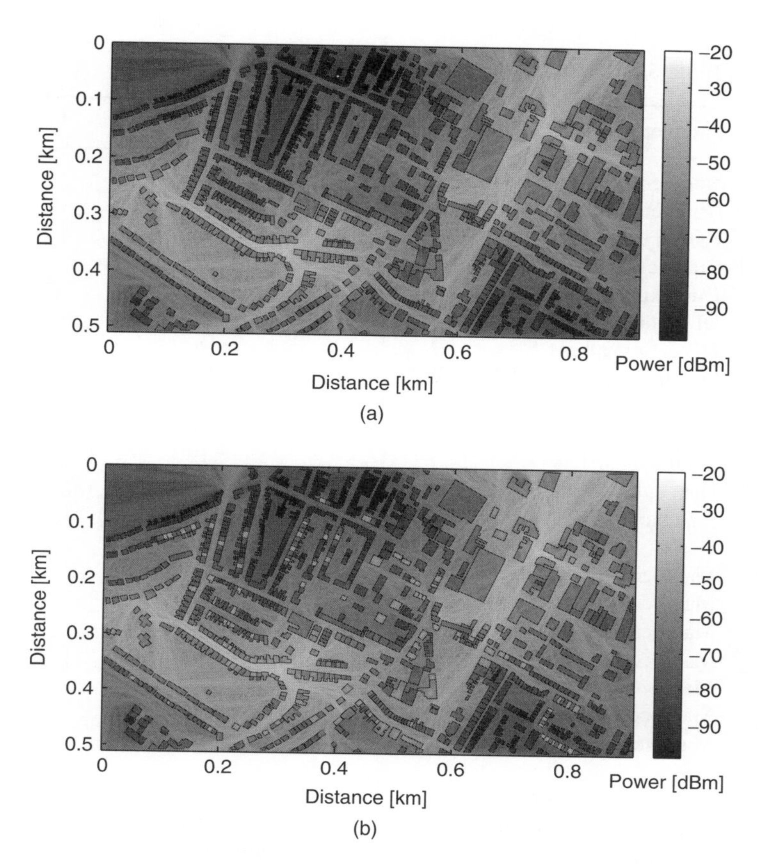

Figure 5.12 Coverage plot for Luton, UK. (a) Deployment without femtocells. (b) Deployment with femtocells

With regards to the user distribution, different traffic maps have been used to simulate the generation of subscribers and non-subscribers. On the one hand, there is an outdoor traffic map, containing 45 $user/km^2$. On the other hand, there is an indoor traffic map per household equipped with a FAP, hosting up to four users. Subscribers and non-subscribers are placed randomly within the traffic maps.

The parameters of this SLS are shown in Table 5.7.

5.6.2 *Coverage*

A signal strength analysis within the scenario presented above is carried out in this section. The objective of this study is to give an example of the coverage extension provided by femtocells.

To carry out this analysis, FDTD coverage predictions were performed by means of computer simulation. These coverage predictions have been calibrated and compared with measurements [23]. The RMSE of these predictions has been estimated to be around 6 dB.

Table 5.7 System-level simulation parameters

Parameter	Value
Number of macrocells	2
Number of femtocells	100
Carrier frequency	3.5 GHz
Channel bandwidth	5 MHz
Duplexing scheme	TDD 1:1
DL data symbols	19
UL data symbols	18
Overhead symbols (DL + UL)[a]	11
Permutation scheme	AMC
Frame duration	5 ms
Total subcarriers	512
Data subcarriers	318
Subchannels	8
Macrocell TX power	33 dBm
Macrocell antenna gain	18 dBi
Macrocell antenna pattern	Omnidirectional
Macrocell antenna height	30 m
Macrocell antenna tilt	3
Macrocell noise figure	4 dB
Macrocell cable loss	3 dB
Femtocell TX power	10 dBm
Femtocell antenna gain	0 dBi
Femtocell antenna pattern	Omnidirectional
Femtocell antenna tilt	0
Femtocell antenna height	1 m
Femto noise figure	4 dB
Femto cable loss	0 dB
UE Tx power	23 dBm
UE antenna pattern	Omnidirectional
UE antenna height	1.5 m
UE noise figure	8 dB
UE cable loss	0 dB
Density of outdoor users	45 *user*/km^2
Density of outdoor users	Up to four *users/house*
Scheduling algorithm	best SINR
Outdoor users service	VoIP
Minimum VoIP throughput	12.2 Kbps
Maximum VoIP throughput	12.2 Kbps

Table 5.7 *(continued)*

Parameter	Value
Indoor users service	Data
Minimum data throughput	128 Kbps
Maximum data throughput	320 Kbps

[a] In a 5 ms OFDMA/TDD WiMAX frame there are 48 OFDM symbols available. According to [21], when using a DL to UL ratio of 1:1, it can be supposed that on average three and seven OFDMA symbols will used by overhead in DL and UL, respectively. Moreover, one OFDMA symbol is employed as a guard time to switch from DL to UL (TTG). This overhead information contains the preamble, cyclic prefix, etc. This way, we can concentrate on the analysis of data transmission.

In Figure 5.12(a), the coverage provided by the two existing macrocells in terms of received signal strength (dBm) is illustrated. In this plot, the effect of the interference is not taken into account, and only the best received signal strength (best server) at each pixel is shown. These values were computed using the formulation introduced in Section 5.5.2. This prediction simulates the behaviour of the preamble information transmitted by every cell, which is used by the UEs as a pilot signal to conduct network entry and synchronization. If our UE is to have an antenna sensitivity of −100 dBm, it can be seen that the coverage provided by the two macrocells is acceptable outdoors, but sometimes insufficient indoors. The shadowing effect of the buildings, and the attenuation caused by thick walls make the coverage unavailable within some areas and households in this scenario.

Femtocells were designed in the first instance as a solution to extend indoor coverage. In Figure 5.12(b), it can be seen how, by deploying femtocells, the network coverage is enhanced in those areas where the FAPs are located. Households, in which the coverage was insufficient before, are covered now by a radio signal with a high strength, by around −45 dBm. In more detail, Figure 5.13 shows the histogram of received signal strength considering every pixel within the scenario. From this histogram, it can be seen how, using femtocells, the samples shift to the right-hand side of the histogram towards higher received signal strength.

In conclusion, it can be stated that femtocells increase indoor network coverage. However, this is at the expense of increasing interference outdoors due to the uncontrolled deployment of new cells by the users. The power leakage from indoors to outdoors by the femtocells will be seen as interference for non-subscribers passing close to the femtocells.

5.6.3 *Signal Quality*

A signal quality analysis within the scenario presented above is carried out in this section. The objective of this study is to give an example of the signal quality (SINR) enhancement provided by the femtocells.

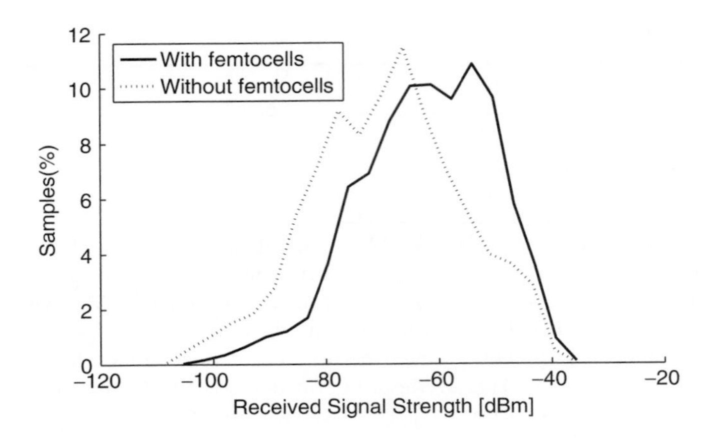

Figure 5.13 Received signal strength histogram for the two scenarios

Figure 5.14(a) shows the signal quality *SINR* provided by the two existing macrocells. In this case, the SINR level of the preamble, also known as pilot signal, is used as KPI. To estimate this value at every pixel of the scenario, the received signal strength of the best server has been considered as the carrier C, and the sum of the received signal strength coming from all neighbouring cells has been considered as interference I. These values and the noise level N are computed using the formulation presented in Section 5.5.2. The resulting SINR value corresponds to a worse case scenario in the case of data transmission, since it is considered that all users operate in the same frequency band and interfere with each other.

In the deployment not using femtocells, it can be seen that the SINR levels are large in those areas close to the macrocells, but they drop rapidly on the boundary between the macrocells, where the carrier and interference signals are received at similar strengths. This effect will not be a problem for those users located on the cell edge using different subchannels, but it may jam their data communication in the case of subchannel collision (interference).

Figure 5.12(b) demonstrates how deploying femtocells, the SINR is enhanced in those areas where the femtocells are located. Households, in which the SINR was low before, are covered now by a radio signal with a large SINR, around 40 dB.

In conclusion, it can be stated that femtocells will increase the indoor signal quality. This opens up the possibility for operators to offer new services with larger throughput requirements (high speed Internet, gaming, etc.). However, this is at the expense of raising the interference outdoors due to the uncontrolled deployment of new cells by the customers. Non-subscribers do not benefit from the larger SINR provided by CSG femtocells, and they will suffer from greater interference instead.

5.6.4 Performance

In this section, a performance analysis within the scenario presented above is carried out. The objective of this study is to determine the performance enhancement provided by the femtocells using SLSs.

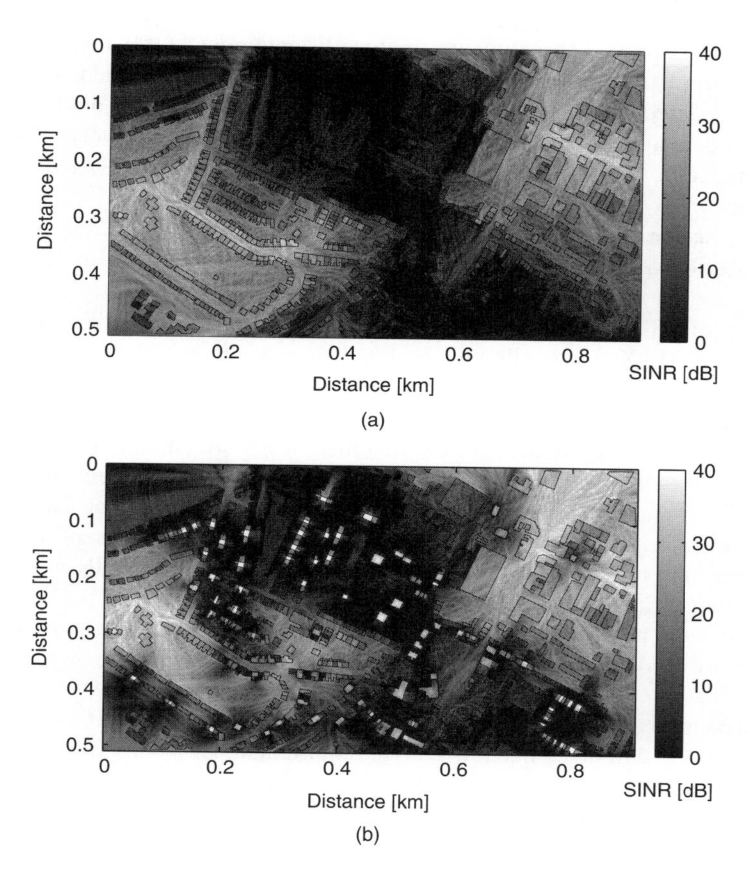

Figure 5.14 Received signal quality plot for Luton, UK. (a) Deployment without femtocells. (b) Deployment with femtocells

The simulation parameters are the same as suggested in Table 5.7, and the performance of the network is assessed by different measurements such as the cell throughput, and the percentage of successful and outage users.

Different network deployments within the given scenario have been proposed in order to explore the limits of operation of the macrocells and observe the benefits provided by femtocells. These network deployments are summarized in the following:

- The first scenario consists of two macrocells, and no femtocells. Both outdoor and indoor users attempt to carry a VoIP service, whose minimum throughput requirement is 12.2 kbps. Note that when using no femtocells, the 2 macrocells have 20 MHz bandwidth, (32 subchannels).
- In the second scenario, the deployment layout is unchanged, and no femtocells are used. In this case, outdoor users keep demanding VoIP services, while indoor users change their demand to web browsing, requesting a minimum throughput of 64 kbps.
- In the third scenario, the indoor users upgrade their service demand to 256 kbps.

- In the fourth scenario, an extensive femtocell deployment (100 FAPs) is carried out on top of the existing macrocell layout. A co-channel deployment of CSG femtocell is considered across the entire network. In this case, both outdoor and indoor users keep demanding VoIP and data (256 kbps) services, respectively. Note that when using femtocells, the two macrocells and the femtocell have 5 MHz bandwidth, (eight subchannels).
- In the fifth scenario, the deployment layout is unchanged, and 100 femtocells are used. In this case, however, the indoor users upgrade is service demand to 320 kbps.

In the rest of this section, only DL is considered. The reason behind this is that the prime use of WiMAX femtocells will be for data services, which are asymmetrical, i.e. more demands on the DL than UL. The results of the simulation (averaged over 100 snapshots) are presented in Table 5.8.

In the first scenario Figure 5.15(a), the two macrocells are enough to serve the majority of both outdoor and indoor users. Table 5.8 shows that the percentage of successful users is 91.75% and that the percentage of users that suffers from a lack of resources is 0%. However, as can be seen in Figure 5.15(a), several indoor users (8.25%) located in areas where the radio coverage is limited are in outage. Femtocells would help these users, providing a larger indoor coverage and conducting their data packets over the IP backhaul connection. In addition, it can be seen from the results that the indoor traffic is greater than outdoor (7.468 versus 1.838 Mbps). There are more users indoors than outdoor.

In the second scenario, the demands of the indoor users grow (64 kbps), and more resources are needed to meet this new service requirement. As a consequence of the growth of indoor traffic, the throughput of the indoor layer increases from 7.468 to 22.483 Mbps. However, although the two macrocells still have enough resources to manage the request,

Table 5.8 System-level simulation results

Scenario outdoor/indoor service	Success users (%)	Outage users (%)	No resources users (%)	Outdoor throughput (Mbps)	Indoor throughput (Mbps)
Two macrocells VoIP/VoIP	91.75	8.25	0	1.838	7.468
Two macrocells VoIP/Data(64 kbps)	87.94	12.06	0	1.780	22.483
Two macrocells VoIP/Data(256 kbps)	44.44	1.27	57.29	1.089	41.707
Two macrocells/ 100 femtocells VoIP/Data(256 kbps)	98.41	1.59	0	2.107	60.681
Two macrocells/ 100 femtocells VoIP/data(320 kbps)	98.41	1.59	0	2.059	84.302

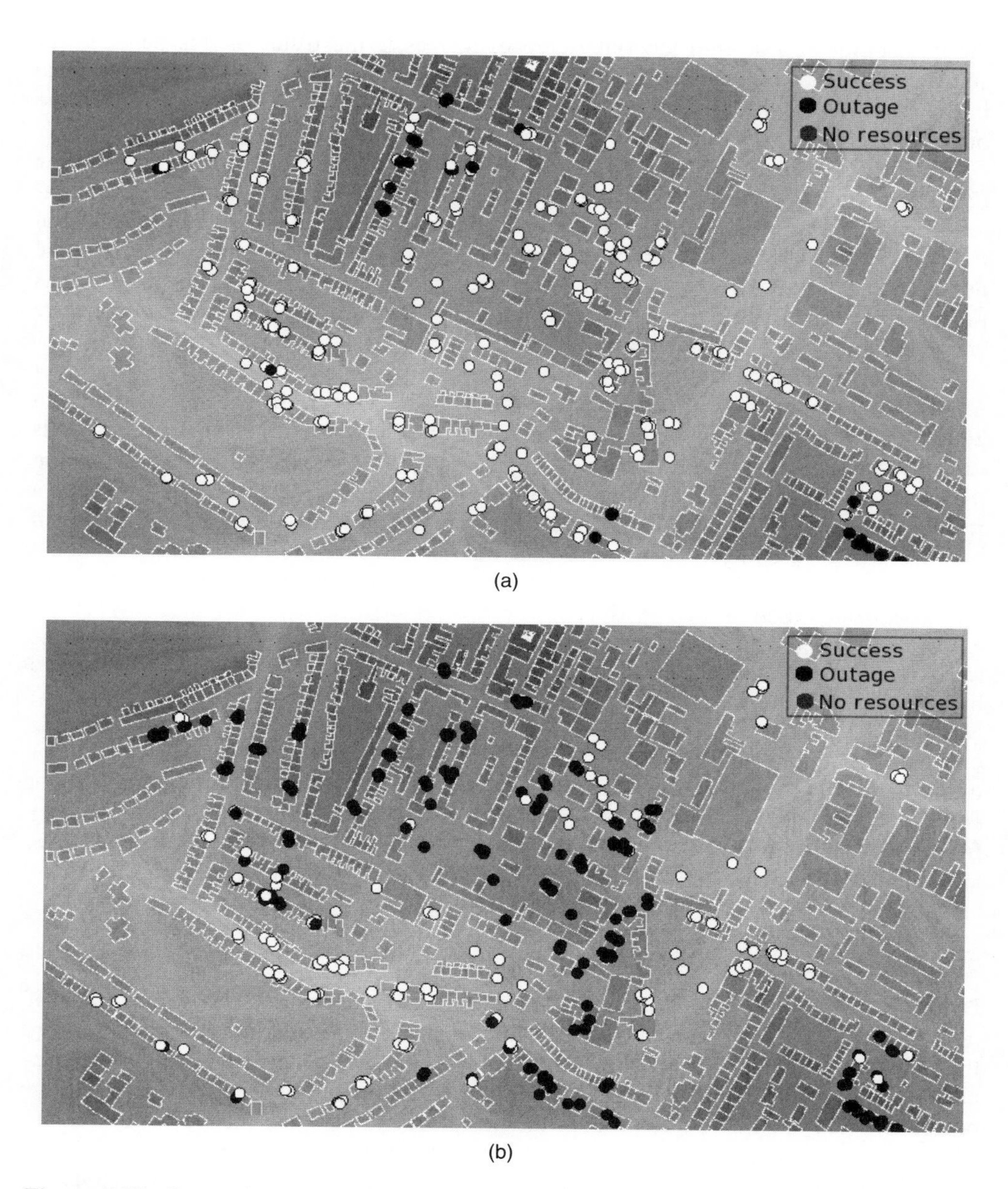

Figure 5.15 System-level simulation plot in Luton, UK. (a) Deployment with two macrocells. Outdoor and indoor users carry VoIP services. (b) Deployment with two macrocells. Outdoor and indoor users carry VoIP services, while indoor users carry data (256 kps) services

the probability of slot collision (interference) increases, the result of this being lower signal qualities for the outdoor users or even outages in some cases. Due to the increase in the interference, the percentage of outages increases from 8.25% to 12.06%, and the outdoor layer throughput decreases from 1.838 to 1.780 Mbps.

In the third scenario (Figure 5.15(b)), the demands of the indoor users again grow (256 kps), and more slots are needed to meet this new service requirement. As a consequence of the growth of the indoor traffic, the throughput of the indoor layer again increases from 22.483 to 41.707 Mbps. However, the resources of the two macrocells are not enough to serve the user demands now, and the number of users with no resources rapidly grows to 57.29%. In this case, the macrocells are overloaded, and users that before were able to transmit, are now blocked due to the lack of resources.

In the fourth scenario (Figure 5.16(a)), an extensive use of femtocells is considered. Let us remember that in this scenario the bandwidth of the macrocells have been reduced from 20 to 5 MHz. Despite everything, due to the handover of indoor traffic from the macrocells to the backhaul connection, the problems of lack of resources disappear (0%), reducing the need for large bandwidths in the macrocells and enhancing their reliability. As a result, the operator can reduce the capital and operational expenditure of the macrocell network. Moreover, Table 5.8 shows that the number of successful users (subscribers and non-subscribers) is high (98.41%) and that the overall throughput carried by the network increases from 42.796 to 62.788 Mbps.

In the fifth scenario (Figure 5.16(b)), the demands of the indoor users again grow (320 kps). However, it is not a problem for the femtocell layer, since this traffic is absorbed by the back-haul link, increasing the indoor layer throughput from 60.681 to 84.302 Mbps. Note that the number of both successful and outage users remained unchanged, but the outdoor layer throughput slightly decreases. This is due to interference. Since the indoor traffic increases, the number of occupied slots increases and thus the probability of slot collision. Due to this fact, non-subscribers in areas close to femtocells are likely to suffer from greater interference. This interference will downgrade the SINR of non-subscribers and decrease their RAB and throughput.

5.7 Overview of Dynamic System-Level Simulation

Dynamic system-level simulations will help to analyse accurately the performance of new in-built femtocell algorithms in different scenarios. In this kind of simulation, the time domain is not neglected, and the evolution over time of the network can be precisely examined. This feature is particularly useful for analysing the behaviour of those procedures that must take into account the movement of the user and fluctuation of the radio channel over time, e.g. handover algorithms and feedback techniques. Moreover, it will help to understand the behaviour of different strategies, e.g. power and frequency allocation, in the presence of several traffic types with very different features such as real time services and best effort.

In the case of femtocells, novel self-organization techniques that will allow the femtocell to sense its environment and tune its parameters must be investigated, before being implemented. This type of simulation can cover a large variety of topics. It can be used to analyse the effect of an inaccurate synchronization, and derive bounds in the performance of the oscillators. Conversely, it can also be used to understand the advantages

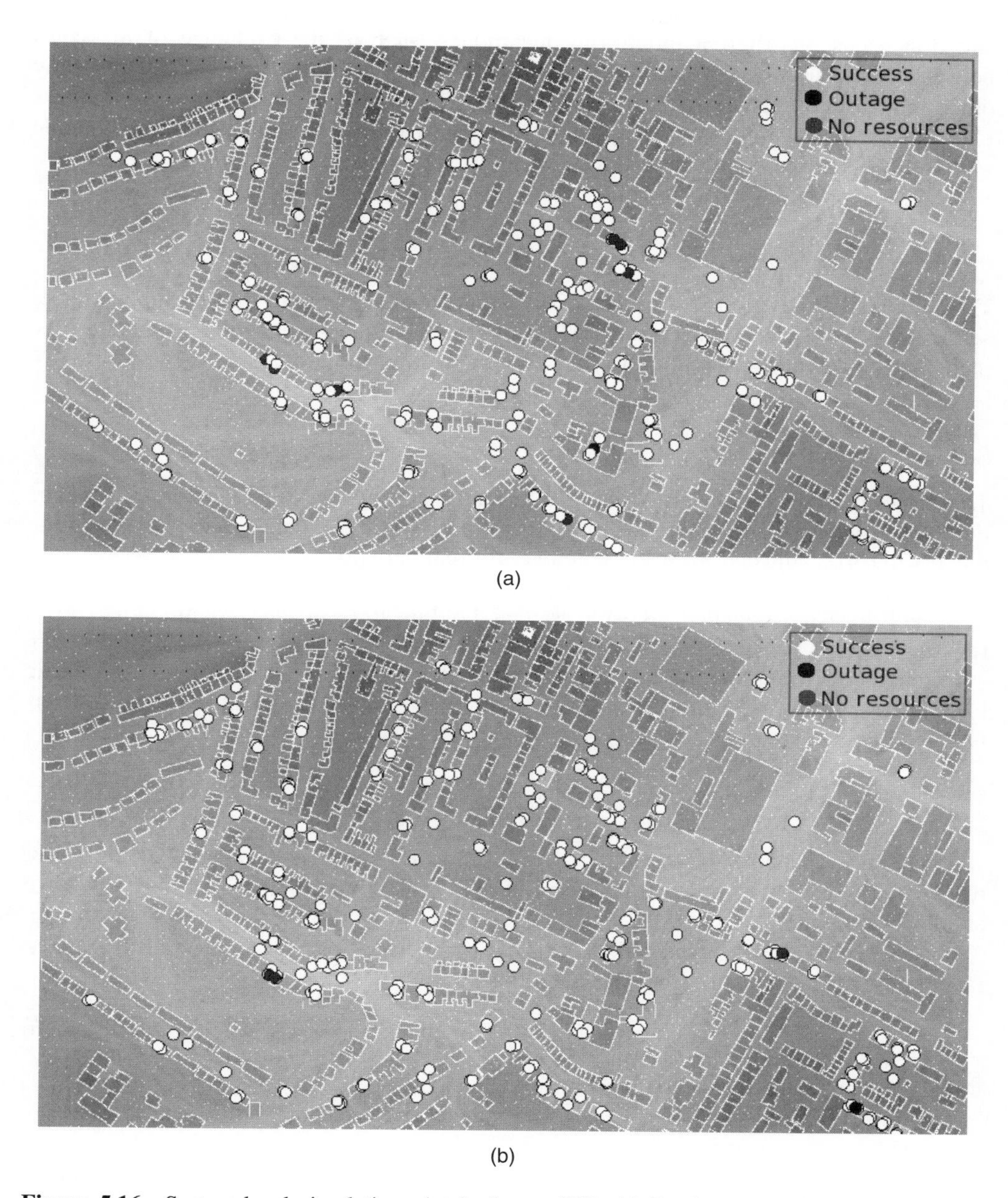

Figure 5.16 System-level simulation plot in Luton, UK. (a) Deployment with two macrocells and 100 femtocells, outdoor and indoor users carry VoIP services, while indoor users carry data (256 kps) services. (b) Deployment with two macrocells and 100 femtocells, outdoor and indoor users carry VoIP services, while indoor users carry data (320 kps) services

and drawbacks of different monitoring techniques, such as the exchange of messages between femtocells or the use of measurement reports coming from the users. Many other applications exist.

In this section, the different key features that must be taken into account in a dynamic system-level simulation are introduced.

5.7.1 Traffic Modelling

Traffic models are used in SLSs in order to simulate the behaviour of a given application or service, e.g VoIP, video conference, FTP. This behaviour is considered at different levels, including the session generation process and the data generation within each session.

Session Generation

The target of traffic modelling is to create analytical models that statistically describe the most representative parameters of a given traffic type. The parameters of these traffic models must be adjusted to fit the features of a real network.

First of all, in order to capture the dynamics of the end customer, the arrival rate of the users must be modelled, as well as their holding time in the system. Moreover, a femtocell user is likely to carry multiple sessions simultaneously (call, email, FTP).

In macrocell scenarios, the Poisson process has been widely used to model the traffic load of the network over time. In this case, the probability of the arrival of n new sessions is given by:

$$p_n(T) = \frac{(\lambda T)^n}{n!} \mathrm{e}^{(-\lambda T)} \tag{5.15}$$

where T and λ denote the sampling period, and the average arrival rate of the users, respectively. Note that these sessions are normally distributed within a defined coverage area or traffic map.

For each new session, the holding time t_H is described by an exponential distributed random variable generated from the next PDF:

$$f_{t_H}(t_H) = \mu \mathrm{e}^{(-\mu t_H)} \tag{5.16}$$

where μ denotes the mean holding time.

To simulate different service densities within a given scenario, different values of λ and μ can be defined per service and/or area.

In the case of femtocells, a common approach is to assume that the femtocell is occupied at all times from one up to the maximum number of allowed users in the femtocells [24]. Although, this assumption is appropriate for analysing capacity and interference in femtocells, it might be not realistic. Therefore, new models that capture the behaviour of the femtocell users taking the femtocell applications into account must be investigated.

Data Generation

The dynamics of the traffic generated by a session should be accurately modeled over time in order to derive the performance of the system. In the following, two different traffic models (data generation) are described for illustration purposes, VoIP and FTP [25].

VoIP A typical VoIP call can normally be divided into two states: active and inactive. In the active state, the user is transmitting packets, while in the inactive, the user is receiving information. Therefore, a two-state Markov chain can be used to model this application.

The probability of moving from the active to the inactive state is equal to 0.6, while the probability of changing from the inactive to the active state is 0.4. Moreover, the period of time that the call is active or inactive is defined by an exponential distribution of mean 1.0 s and 1.5 s, respectively.

When the user is in the active state, packets of fixed size are generated at regular intervals. The values of these parameters depend on the voice codecs and compression scheme used. If a simplified AMR (Adaptive Multi Rate) audio data compression technique is used, the payload of the AMR blocks is 33 bytes, and they are generated at a rate of 20 ms.

When the user is in the inactive state, a comfort noise is generated in order to avoid confusing the customer between being disconnected or awaiting for new data packets. The payload of these packets is 7 bytes, and they are generated at a rate of 160 ms.

FTP In FTP applications, a session consists of a sequence of file transfers, separated by reading times. The main parameters of this model are the size of the file to be transferred, and the length of the reading time. The reading time is modelled by a exponential distribution of mean 180 s. The size of the file follows a log-normal distribution of mean 2 MByte and standard deviation 0.72 MByte. The size of the file must not exceed 5 MByte.

The two traffic models presented here are just an example of a large variety of models. This kind of model allows a more exhaustive evaluation of the performance of the system. For example, the analysis of scheduling techniques that takes the size of the queues where data is store into account, or the allocation of resources between real and non-real time services, etc.

In the case of femtocells, vendors and operators are working towards the creation of novel applications that will generate the need for femtocell in the market. As a result, new models that capture the particularities of these new services will be needed. Using these models, new optimization techniques can be investigated to enhance the performance of the femtocells.

5.7.2 Mobility Modelling

Mobility models are used in SLS in order to simulate the movement of the users across the scenario, e.g. position, speed, directions.

A large variety of mobility models exists. Some of them model the movement of the users in a simplistic way, and some others with a large level of detail.

A straightforward way of modelling the user motion is by using Gaussian distributions [26]. Let us denote the future position of a user as:

$$(x_t, y_t) = \begin{cases} x_{t-1} + v_t\,\Delta t\,cos(\alpha_t) + n_{t-1} \\ y_{t-1} + v_t\,\Delta t\,sin(\alpha_t) + n_{t-1} \end{cases} \tag{5.17}$$

where (x_{t-1}, y_{t-1}) denotes the previous user position, v_t indicates the user velocity at time t. α_t represents the user direction at time t, and n_{t-1} is a noise with Gaussian distribution. Moreover, Δt denotes the period of time between two consecutive updates of the model.

The velocity and direction can be derived as follows:

$$v_t = N\left(v_{t-1}, 1\frac{m}{s^2}\Delta t\right) \tag{5.18}$$

$$\alpha_t = N\left(\alpha_{t-1}, 2\pi - a\tan\left(\frac{\sqrt{v_t}}{2}\right)\Delta t\right) \tag{5.19}$$

where v_{t-1} and α_{t-1} denote the previous user velocity and direction, respectively, and $N(a, b)$ indicates a Gaussian distribution of mean a and standard deviation b.

The user movement must also take into account the environment description, i.e. wall, room and corridor. Users can not move across a wall or another solid object. For this, a mobility graph that constraines the user movements according to the obstacles in the environment can be used. Vertices represent possible destinations, and edges correspond to physically valid paths over which the users can move [27].

On the other hand, more complex models can be used to simulate the mobility of the customers. In [28], a model based on measurements is presented. In this work, for example, tracks of always on WiFi devices are collected, and the distribution of the speed and pause times are derived. This kind of model provides a more accurate simulation of the user behaviour.

In the case of femtocells, indoor mobility models taking the scenario features into account might be used to examine accurately the performance of subscribers and non-subscribers. These models can be important for studying the handover procedures in the femtocell case. However, the behaviour of the femtocell users must be analysed, since it can be different from the models used in ad-hoc or sensor network simulations. Therefore, research is needed in this area.

References

[1] E. Amaldi, A. Capone, and F. Malucelli, 'Planning UMTS base station location: Optimization models with power control and algorithms,' *IEEE Transactions on Wireless Communications*, vol. 2, no. 5, pp. 939–952, September 2003.

[2] F. Gordejuela-Sanchez, A. Jüttner, and J. Zhang, 'A multiobjective optimization framework for IEEE 802.16e network design and performance analysis,' *IEEE Journal on Selected Areas in Communications. Special issue on Broadband Access Networks: Architectures and Protocols*, April 2009.

[3] V. Chandrasekhar and J. G. Andrews, 'Femtocell networks: A survey,' *IEEE Communication Magazine*, vol. 46, no. 9, pp. 59–67, September 2008.

[4] A. Eisenblätter, 'Frequency assignment in GSM networks: Models, heuristics, and lower bounds,' Ph.D. dissertation, Technische Universität Berlin, Fachbereich Mathematik, Berlin Germany, 2001.

[5] A. Eisenblätter, H.-F. Geerdes, T. Koch, A. Martin, and R. Wessäly, 'UMTS radio network evaluation and optimization beyond snapshots,' *Mathematical Methods of Operations Research*, vol. 63, no. 1, pp. 1–29, February 2006.

[6] J. Laiho, A. Wacker, and T. Novosad, *Radio Network Planning and Optimisation for UMTS*, 2nd ed. John Wiley & Sons, Ltd, Chichester, December 2006.

[7] J. G. Andrews, A. Ghosh, and R. Muhamed, *Fundamentals of WiMAX Understanding Broadband Wireless Networking*. Massachusetts, USA: Prentice Hall, February 2007.

[8] Z. Lai, N. Bessis, G. de la Roche, H. Song, J. Zhang, and G. Clapworthy, 'An intelligent ray launching for urban propagation prediction,' in *3rd European Conference on Antennas and Propagation (EuCAP)*, Berlin, Germany, March 2009.

[9] A. Valcarce, G. de la Roche, and J. Zhang, 'A GPU approach to FDTD for radio coverage prediction,' in *11th IEEE International Conference on Communication Systems*, Guangzhou, China, November 2008.

[10] J.-M. Gorce, K. Jaffres-Runser, and G. D. L. Roche, 'Deterministic approach for fast simulations of indoor radio wave propagation,' *IEEE Transactions on Antennas and Propagation*, vol. 55, pp. 938–942, March 2007.

[11] 'Guidelines for Evaluation of Radio Transmission Technologies for IMT-2000,' ITU-R M.1225, 1997.

[12] U. Türke, Ed., *Efficient Methods for WCDMA Radio Network Planning and Optimization*. DUV, December 2007, Monte Carlo Snapshot Analysis, pp. 45–66.

[13] Forsk. Forsk atoll–global rf planning solution.

[14] AWE Communications. Awe communications–wave propagation and radio network planning.

[15] T. Kwon, H. Lee, S. Choi, J. Kim, and D.-H. Cho, 'Design and implementation of a simulator based on a cross-layer protocol between mac and phy layers in a wibro compatible IEEE 802.16e OFDMA system,' in *IEEE Communications Magazine*, vol. 43, no. 12, December 2005, pp. 136–146.

[16] IEEE Std 802.16e-2005, 'IEEE standard for local and metropolitan area networks: Air interface for fixed broadband wireless access systems–physical and medium access control layers for combined fixed and mobile operation in licensed bands,' February 2006.

[17] C. Huang, H. Juan, M. Lin, and C. Chang, 'Radio resource management of heterogeneous services in mobile wimax systems,' in *IEEE Wireless Communications*, vol. 14, 2007, pp. 20–26.

[18] NGMN, 'NGMN radio access performance evaluation methodology,' A White Paper by the NGMN Alliance, January 2008.

[19] V. Hassel, 'Design issues and performance analysis for opportunistic scheduling algorithms in wireless networks,' PhD dissertation, Norwegian University of Science and Technology, Faculty of Information Technology, Mathematics and Electrical Engineering, Department of Electronics and Telecommunications, January 2007.

[20] H. Jia, Z. Zhang, G. Yu, P. Cheng, and S. Li, 'On the performance of IEEE 802.16 OFDMA system under different frequency reuse and subcarrier permutation patterns,' in *IEEE International Conference on Communications*, vol. 24, june 2007, pp. 5720–5725.

[21] ATDI, 'Mobile WiMAX from OFDM 256 to s-OFDMA,' ATDI White paper, 2007.

[22] United Kingdom Office of Communications (Ofcom), 'The communications market 2006, annual communications market report,' http://www.ofcom.org.uk/research/cm/cm06, August 2006.

[23] A. Valcarce, G. D. L. Roche, A. Jüttner, D. López-Pérez, and J. Zhang, 'Applying FDTD to the coverage prediction of WiMAX femtocells,' *Eurasip Journal of Wireless Communications and Networking. Special issue on Advances in Propagation Modelling for Wireless Systems*, February 2009.

[24] H. Claussen, L. T. W. Ho, and L. G. Samuel, 'An overview of the femtocell concept,' *Bell Labs Technical Journal*, vol. 3, no. 1, pp. 221–245, May 2008.

[25] W. Lee, J. Choi, K. Ryu, and R. Kim, 'Draft IEEE 802.16m evaluation methodology and key criteria for 802.16m – advanced air interface,' IEEE 802.16 Broadband Wireless Access Working Group, March 2007.

[26] Widyawan, M. Klepal, and D. Pesch, 'Influence of predicted and measured fingerprint on the accuracy of RSSI-based indoor location systems,' in *4th Workshop on Positioning, Navigation and Communication, 2007. WPNC '07.*, March 2007, pp. 145–151.

[27] A. Cavilla, G. Baron, T. Hart, L. Litty, and E. de Lara, 'Simplified simulation models for indoor manet evaluation are not robust,' in *IEEE Sensor and Ad Hoc Communications and Networks*, vol. 4, no. 7, October 2004, pp. 610–620.

[28] M. Kim, D. Kotz, and S. Kim, 'Extracting a mobility model from real user traces,' in *25th IEEE International Conference on Computer Communications. INFOCOM.*, April 2006, pp. 1–13.

6

Interference in the Presence of Femtocells

Alvaro Valcarce and David López-Pérez

6.1 Introduction

The deployment of femtocells introduces some changes in the topology of conventional macrocellular networks. The new network architecture is composed of two clearly separated layers, the macrocell layer and the femtocell layer. Such a network architecture is therefore called a two-layer or two-tier network. The first layer comprehends the plain old traditional cellular network, while the second one incorporates several shorter range cells that can be planned (e.g. microcells) or distributed in a random manner (e.g. femtocells). In the case of home base stations, they are randomly located inside of the same geographic area covered by the larger cellular network (also called the umbrella macrocell) and they can exploit the same spectral frequencies as the umbrella. Therefore, one of the advantages of deploying smaller cells in such a network structure is increased coverage in not fully covered areas within the umbrella macrocell. Furthermore, and since femtocells are used by a reduced number of users, higher data rates are also achieved.

The topology of two-layer networks brings, however, new problems and creates new design challenges. When several transmitters emit their signals in the same frequency band and within the same geographic location, a receiving system sensing that frequency band will not be able to distinguish which transmitter it is listening to. This is a very elementary description of the problem of interference in telecommunications systems and it is one of the main challenges that the deployment of femtocells will face. Interference-limited systems such as Code Division Multiple Access (CDMA) will be greatly affected by the presence of femtocells and will recquire the introduction of interference avoidance techniques such as time-hopping or power control. Furthermore, capacity-limited systems like Orthogonal Frequency Division Multiple Access (OFDMA) will need to adopt intelligent

Femtocells: Technologies and Deployment Jie Zhang and Guillaume de la Roche

frequency planning strategies to cope with the presence of interference along subcarriers due to the presence of a femtocell layer.

Femtocells will provide higher spectrum efficiency, spatial frequency reuse and better coverage in areas not fully covered by macrocells, e.g. indoor rooms or locations near the cell edge. However, if interference cancellation or avoidance techniques are not applied, dead zones can appear within the macrocell, disrupting its service in the proximity of a femtocell. Furthermore, the opportunistic location of femtocells suggests that randomness must be taken into account when analysing femtocell related issues. This chapter covers the problems caused by interfering femtocells and gives specific examples of how randomnly located interferers affect CDMA and OFDMA systems.

6.2 Key Concepts

6.2.1 Co-Layer Interference

Co-layer interference is described as the unwanted signal received at a femtocell and sent from other femtocells, decreasing thus the quality of its communication. The name *co-layer* makes reference to the fact that all femtocells belong to the same network layer, unlike other elements like base stations, NodeBs and so on, which belong to the macrocell layer. Co-layer interference occurs mainly between inmediate neighbours due to low isolation between houses and appartments. This problem is thus independent of the disruption caused to the macrocell layer. A diagram summarizing the main problems originating from co-layer interference is presented in Figure 6.1, which are described in more detail in the following paragraphs.

Since the deployment of femtocells is opportunistic, it is likely that several femtocells would be installed in locations close to each other, for example, horizontally in adjacent terraced houses or vertically in blocks of appartments, thus interfering one another. To illustrate this, let us assume that a Global System for Mobile communication (GSM) femtocell f_a illuminates an arbitrary location L_a. It can then be said that L_a belongs to the coverage area of f_a. If there are additional signals from surrounding femtocells using the same frequency, location L_a is said to suffer from co-channel interference and the system's performance suffers. Furthermore, when signals from several femtocells are present at

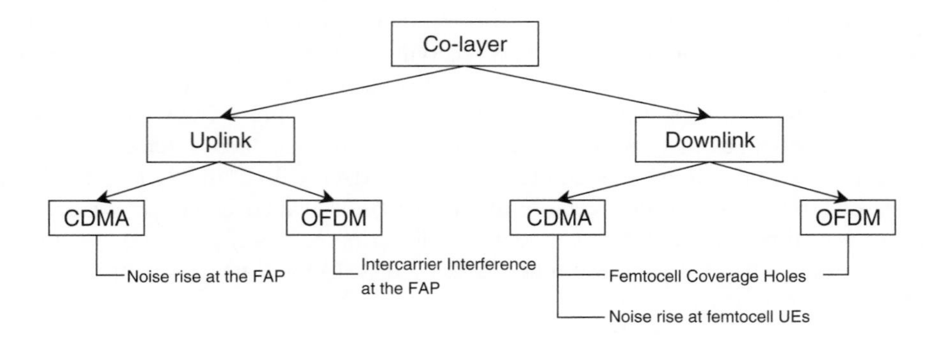

Figure 6.1 Main problems caused by co-layer interference

a given location, the overall interference can be higher than any of the independent femtocell power levels. If the Carrier to Interference and Noise Ratio (*CINR*) value is too low, it could be impossible to establish a communication through any femtocell and such a location would be considered a dead zone. This should not be confused with the concept of coverage holes, which, according to [1], are regions with low pilot *CINR* due to path loss, hence being network entry not possible for the User Equipment (UE). The degradation of the communication varies between RF technologies due to their different performances in the presence of interference. As will be shown in the following, the geometry of dead zones changes between different air-interface technologies.

Since femtocells can be deployed in Closed Subscriber Group (CSG), open-access or in a hybrid access mode, the impact of the co-layer interference will be different depending on the access method. To illustrate this, in Figure 6.2 a realistic femtocells scenario is presented under different Quality of Service (QoS) requirements. For instance, Figure 6.2(a) shows the coverage areas of nine randomly deployed femtocells when the coverage area is defined as those locations where the main femtocell signal is strictly higher than the overall interference plus noise if only one carrier is assumed. If these were open-access femtocells, users could move freely from their femtocells to others as long as they do not cross one of the scenario's dead zones (pictured as black areas in Figure 6.2). As mentioned earlier, the geometry of dead zones depends on the *CINR* requirements of the air-interface technology in use. In telecommunications systems, users typically require a minimum quality of the main signal in order to use certain services. For example in GSM and for the performance of handoffs, it is not enough to have a carrier signal slightly larger than the interference plus noise level ($CINR > 0\,\text{dB}$). For instance, Figure 6.2(b) presents the same scenario with the added restriction of having the dominant signal at least 10 dB above the interference-plus-noise level ($CINR > 10\,\text{dB}$). It is then clear how dead zones grow larger as the QoS requisites increase and it could be that, even with open-access femtocells, a user would not be able to move freely from one femtocell to another due to the presence of these gaps.

In the case of CDMA systems, the conditions for service impairment can be further relaxed. For instance, it can be shown that the requisite of having $CINR > -20\,\text{dB}$ in downlink is achieved by all femtocells in the whole of the previous scenario, which is enough for the performance of standard speech connections [2]. It must be, however, mentioned that such a signal quality is heavily dependent on the density of femtocells per unit area. Nevertheless, the modulation process in CDMA spreads the signal power across the spectrum, hence reducing the power spectral density and also the received *CINR*.

In OFDMA, dead zones are seen differently by each user. This is dependent on the subchannel allocation, which might differ between various femtocells. Therefore, an OFDMA femtocell user might perceive a dead zone for certain subchannels, whilst other subchannels remain interference free. To illustrate this, let us suppose that at a given moment in time, the allocation of subchannels decided by the OFDMA femtocells of the previous scenario is that shown in Figure 6.3. The dark areas in Figure 6.3(a) represent the regions where $CINR < 10\,\text{dB}$ for subchannels 1 to 4, i.e. dead zones for those subchannels. On the other hand, Figure 6.3(b) shows the regions where $CINR < 10\,\text{dB}$ for subchannels 5 to 8, which might differ from the previous ones.

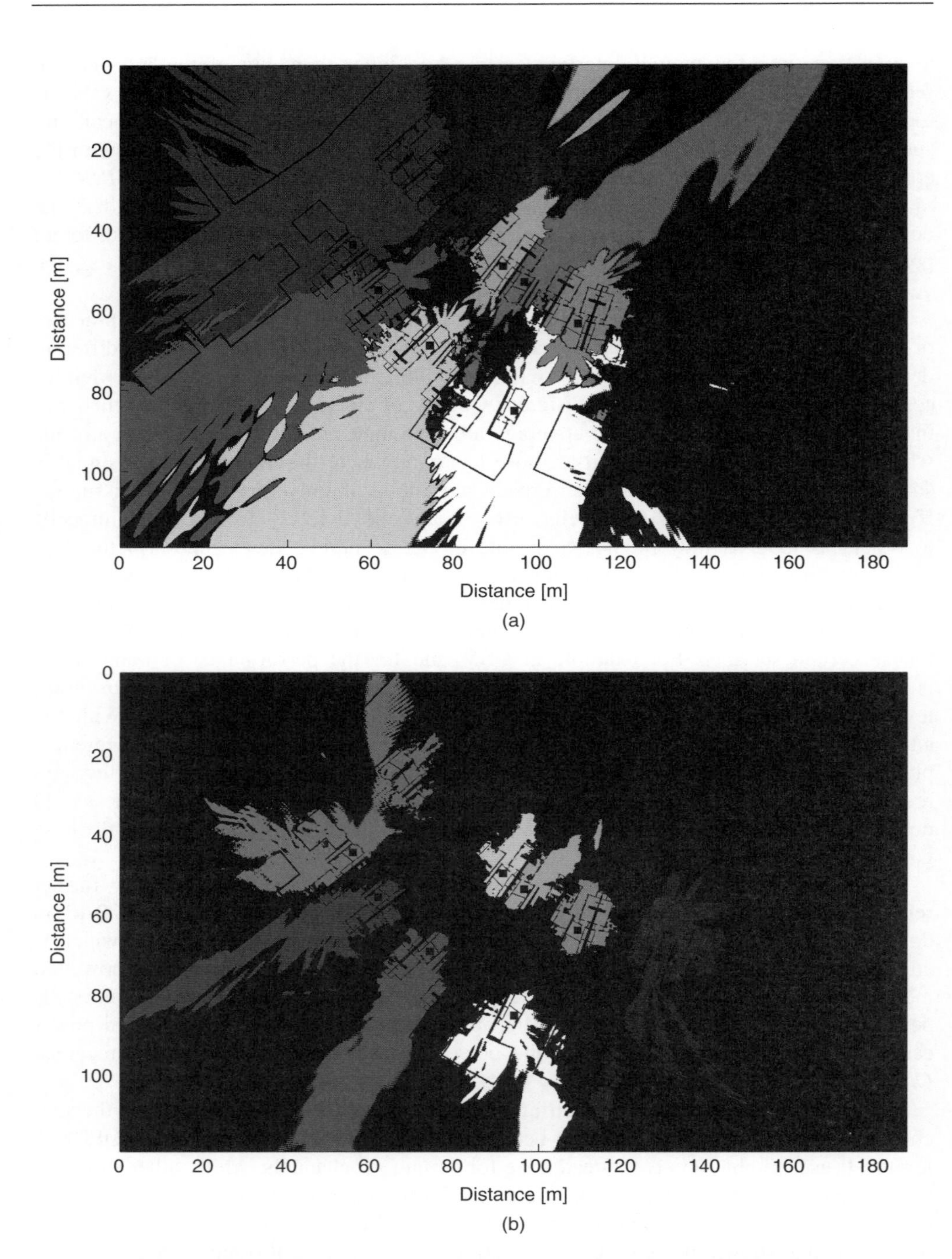

Figure 6.2 Downlink coverage areas of femtocells in a residential environment with absence of macrocell coverage. The dark squares indicate the location of the FAPs and the black areas are regions where the *CINR* requisite is not met by any femtocell. (a) Coverage areas if $CINR > 0\,\text{dB}$. (b) Coverage areas if $CINR > 10\,\text{dB}$

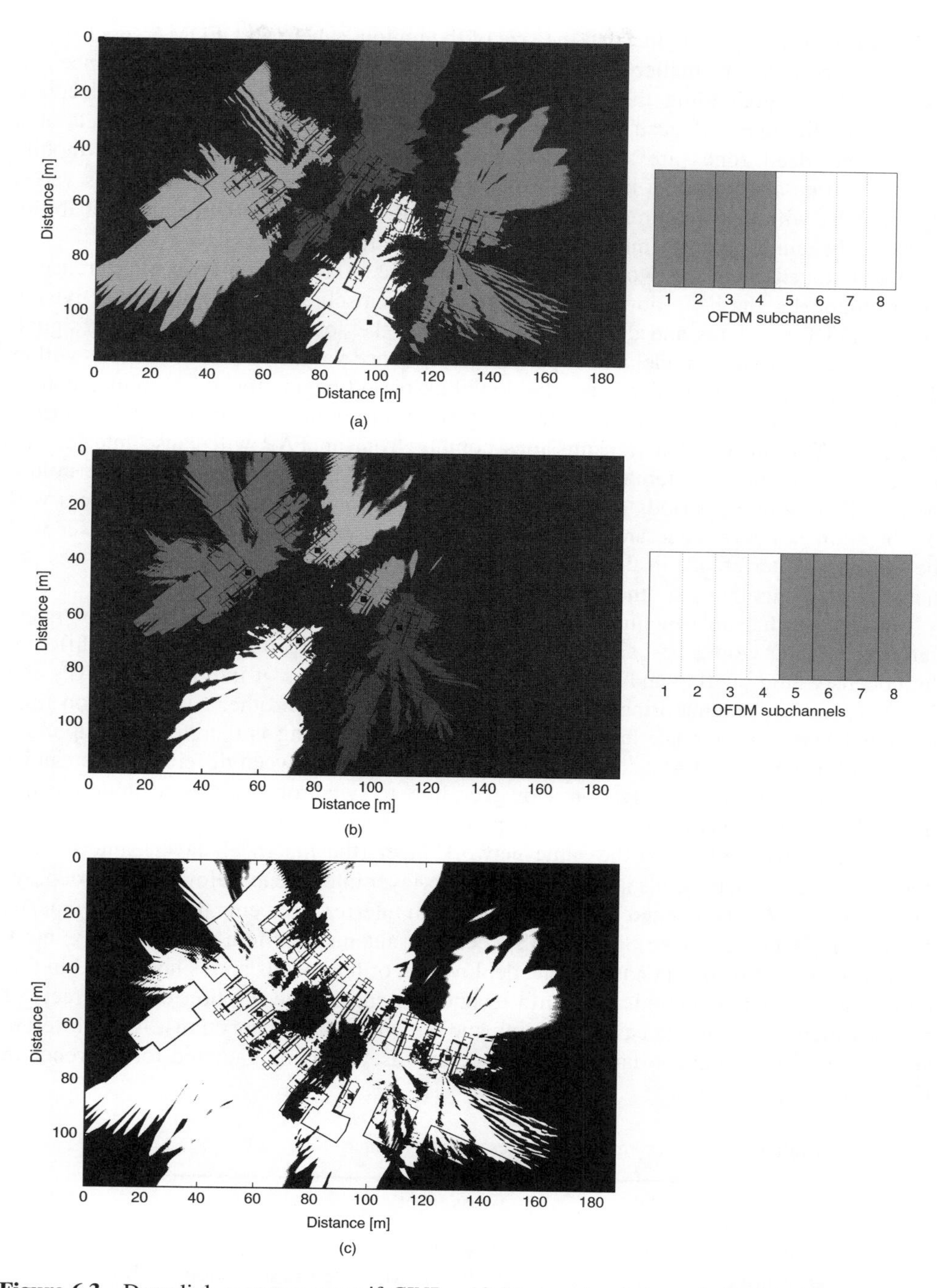

Figure 6.3 Downlink coverage areas if $CINR > 10\,\text{dB}$ at each subchannel. (a) Subchannels 1 to 4 occupied. (b) Subchannels 5 to 8 occupied. (c) Joint coverage areas. In the dark areas, all OFDMA subchannels suffer $CINR < 10\,\text{dB}$

It is clear from Figure 6.3(c) that the size of the regions where all OFDMA subchannels are interfered is much smaller than the dead zones for individual subchannels. Hence and as a result of subdividing the spectrum usage into subchannels, dead zones decrease in size. Furthermore, if each femtocell makes use of less than four subchannels at a given time, dead zones are additionally diluted. In OFDMA systems such as mobile WiMAX, time also plays an important role because the allocation of subchannels varies dynamically with QoS requirements, depending on the scheduling. Hence, the probability of two subchannels being simultaneously occupied is further reduced.

In a femtocell, there are two types of source capable of creating co-layer interference to other femtocells: the FAP (downlink) and the users themselves (uplink). In Time Division Duplex (TDD) systems and depending on which is the source of the interfering signal, the approaches used to model and cope with interference vary. If all the femtocells within the same area are synchronized (i.e. the downlink period begins for them simultaneously in time), the aggressors* at an interfered femtocell user are the neighbouring FAPs during downlink. This means that transmissions coming from an FAP will cause interference to UEs of neighbouring femtocells in downlink only. The same applies for the uplink period. If the uplink periods of close femtocells are synchronized, femtocell users will be the sources interference and hence, transmissions coming from a femtocell user will be sensed as interference in the uplink of neighbouring FAPs. Table 6.1 summarizes the network elements that play the role of aggressors in the case of interference.

In the case that no synchronization existed between femtocells, the source of interference in TDD would be undetermined. The uplink and downlink periods of different femtocells would overlap and introduce heterogeneous sources of interference (FAPs and UEs). This way, neighbouring femtocells would overrun each other's transmission time slots and make interference harder to control. Accurate timing is therefore an important feature of TDD-based FAPs so that clock synchronization between different FAPs can be assumed for interference mitigation. However, how to synchronize FAPs accurately is not a trivial task and several solutions are explained in Section 9.1.

Since all MBS belong to the same network layer (the macrocell layer), interference between different macrocells is also a cause of co-layer interference. However, the deployment of macrocells is planned by the operator, with interference being dealt with by means of planning schemes (Base Station (BS) location, antenna azimuth/tilt, frequency, etc.). This problem is thus independent of the deployment of femtocells and is not treated in this book. The following sections contain a technology-specific description of the effects of co-layer interference in the cases of uplink and downlink data connections. Also, different approaches to overcome co-layer interference are discussed and gathered for reference in Figure 6.4.

Table 6.1 The aggressors in cases of interference

	Co-layer	Cross-layer
Uplink	Femtocell users	Any UE
Downlink	FAPs	FAPs and Macrocell Base Stations (MBS)

* The sources of interference are also called *aggressors* under 3GPP terminology.

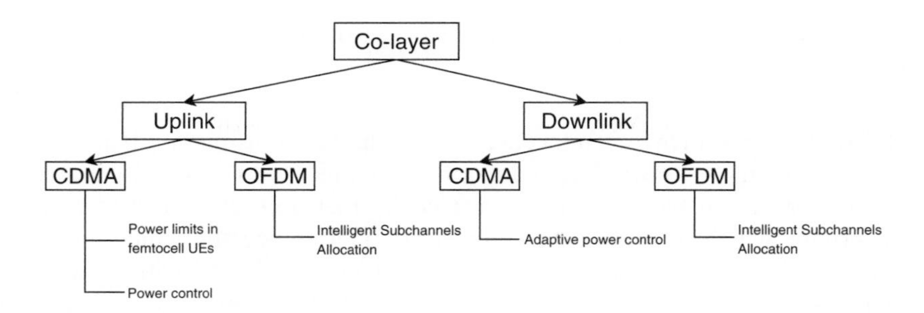

Figure 6.4 Approaches to cope with co-layer interference

Uplink

In uplink co-layer interference, femtocell UEs are the aggressors or sources of interference. On the other hand, neighbouring FAPs are the victim systems. This problem is explained next under different RF technology assumptions.

Uplink CDMA At a given femtocell f_a, the most inmediate interferers are those UEs of neighbouring femtocells, being f_a typically surrounded by several interferers and isolated from them only by means of the outer house walls. The main problem in neighbouring CDMA uplinks is the rise in the noise level due to the spatially distributed nature of the interferers. If neighbouring UEs transmit with high power as requested by their own femtocells, the noise level of nearby femtocells will be degraded and the coverage area of the victim femtocell will be reduced. To cope with this issue and avoid the performance of close 3G femtocells deteriorating, the 3GPP† recommends in [3] the use of interference management techniques. This can help to avoid the interference caused by femtocells with each other.

As an example of interference management and since the density of femtocells can be extremely high in given areas, it is suggested that FAPs impose power limits on their subscribed UEs. This way, the noise rise in the uplink of neighbouring femtocells can be controlled. Furthermore, the users' transmit power must be controlled by the FAP and not by the UE itself. This applies to 3G systems such as Universal Mobile Telecommunication System (UMTS) and High-Speed Uplink Packet Access (HSUPA) and the method for choosing the transmit power level is as follows: In a given femtocell f_a, the FAP scans the transmission band and gathers information about the received power from nearby femtocell UEs. This can be done, for instance, by sensing performed at the FAP itself. The UEs maximum transmit power can then be independently chosen by femtocell f_a taking into account the received power from the other femtocells, i.e. the UEs power will be set so as to receive at the FAP the desired *CINR* for a given service.

In the case of dedicated channel deployments there is no macrocell signal in the same band to provide a reference for the noise level. It is thus especially important in this case that FAPs are capable of performing an effective and reliable sensing of the uplink

† 3GPP is the third Generation Partnership Project, an alliance of telecommunications entities responsible for the standardization of 3G networks.

noise level. To do this, the design of the receiver sensitivity of FAPs must be carefully observed.

Uplink OFDMA In OFDMA femtocells, sensing the full transmission band for nearby emissions might be unnecessary. Depending on the uplink QoS required by a given user, only some OFDMA subchannels will be needed by that user. It is therefore up to the FAP to determine which subchannels are subject to interference and which ones are not. For instance in Figure 6.5 a situation is shown in which user 2 falls within the coverage area of both its femtocell (f_2) and that of a neighbouring femtocell (f_1). In this situation, the subchannels used by user 2 to connect to f_2 will be sensed as being interfered and hence unusable by f_1. Furthermore, femtocell f_1 also has a user of its own (user 1). However, since user 1 does not lie within the coverage area of user 2 there is no way for user 1 to know which uplink subchannels are being interfered. The responsibility for the subchannels allocation of user 1 must thus reside within the FAP.

The case of uplink interference is typically more severe than that of downlink. This is because downlink interference only affects the interfered user, while the interfered uplink subchannels become unusable to all users of the femtocell. Contrary to CDMA, where transmissions occupy the whole licensed band, in the case of OFDMA the availability of subchannels to restrict the emission makes it possible to reduce the interference also in the frequency domain, thus providing higher chances of avoiding interference. Some strategies for the avoidance of interference in OFDMA femtocells are given in Section 6.4.2.

Downlink

In downlink co-layer interference, the FAPs are the aggressors or sources of interference, while UEs of neighbouring femtocells are the victims. A description of this problem follows next.

Downlink CDMA Co-channel downlink interference is also one of the main sources of impairment for femtocells. Since femtocells will be deployed in close positions relative

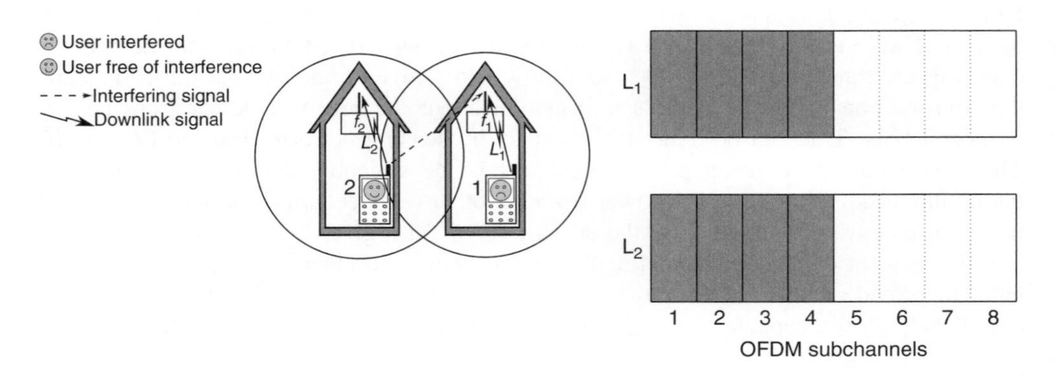

Figure 6.5 Co-layer uplink interference in an OFDMA femtocell network

to each other, they are very likely to interfere with one another by means of power leaks from windows, doors and poorly isolated walls. As previously shown in Figure 6.2, the signals of several femtocells within the same area contribute to raising the interference. Hence, in CDMA systems the noise level increases and creates dead zones where downlink connectivity becomes impossible (cell breathing).

To avoid femtocells causing interference in the downlink to UEs of nearby femtocells (both indoors and outdoors), the first recommendation of 3GPP [3] is that FAPs handle very carefully their transmit power settings by using adaptive power control techniques. This is necessary especially in CSG femtocells as the UEs are not necessarily served by the strongest FAP but by the one to which they are subscribed. As in the uplink case, the downlink transmit power can be decided by each FAP, based on the received power from neighbouring femtocells. Since the appearance of dead zones is mainly due to the addition of several femtocell downlink signals occuring simultaneously, time hopping techniques have been proposed to avoid neighbouring femtocells transmitting at the same time. These are explained in more detail in Section 6.4.1.

Downlink OFDMA A case of co-layer downlink interference occurs when a given femtocell user is located in an area within the FAP premises, where the signal coming from its own femtocell is not high enough compared with the interference coming from surrounding femtocells. This is equivalent to the problem of dead zones shown in Figure 6.2, which as explained in the previous section, can completely disrupt downlink connections in CDMA femtocells.

However in OFDMA systems, the allocation of the subchannels at each femtocell plays a decisive role in the final impact of the interference. Dead zones in OFDMA femtocells depend on the spectrum occupancy at a given location. Two femtocell users could be at the same geographic position and only one of them would suffer interference from surrounding femtocells, i.e. the dead zone would only affect some OFDMA subchannels. To illustrate this, in Figure 6.6 users 2 and 3 want to receive downlink data from their respective femtocell (f_2 in this case). Femtocell f_2 thus allocates subchannels 1 to 4 and 5 to 8 to users 2 and 3 respectively. In this example, both users are located in a dead zone due to interference coming from the neighbouring femtocell f_1. However, f_1 only

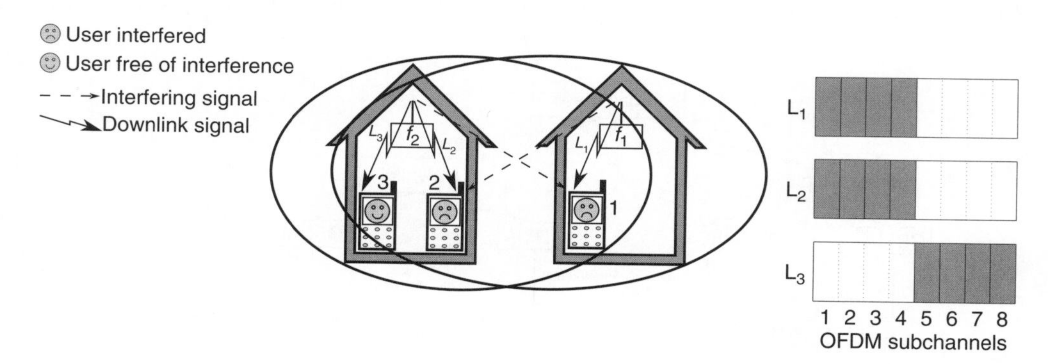

Figure 6.6 Co-layer downlink interference in an OFDMA femtocell network

requires subchannels 1 to 4 to send information to its current user (user 1). Therefore, only the users of subchannels 1 to 4 will be interfered with, with user 3 able to communicate succesfully. Although not necessarily the case, user 1 in this example is also located in a dead zone caused by interference coming from f_2 and also suffers interference due to the occupation of subchannels 1 to 4.

The allocation of the frequency resources is thus extremly important in OFDMA femtocells. Furthermore, the time domain also provides an additional dimension for the management of the subchannels. Resource allocation is thus one of the key technologies for the proper functioning of OFDMA femtocells. These topics are covered in depth in Chapter 8.

6.2.2 Cross-Layer Interference

In two-layer networks, an interfering signal is assumed to produce cross-layer interference if the aggressor and the victim systems belong to different layers of the network. For example, the distortion caused by an emitting FAP (member of the femtocell layer) at the downlink of one or several macrocells (members of the macrocell layer) is a clear case of cross-layer interference. Likewise, it can also be considered as cross-layer interference if the distortion is caused by a macrocell user (member of the macrocell layer) at the uplink of a nearby FAP (member of the femtocell layer). Cross-layer interference is a problem especially in CDMA co-channel deployed two-layer networks, due to the fact that both femtocells and macrocells use the same frequency band. Besides, and due to power control, sudden high transmitting powers can cause the appearance of dead zones, reducing thus the feasibility of these networks. The main problems caused by the presence of cross-layer interference are presented in Figure 6.7 and explained in the following sections. Also to illustrate this phenomenon, Figure 6.8 shows a residential scenario with a low penetration of femtocells (around 14% of the households). It is possible here to see how the coverage area of isolated femtocells invades neighbouring houses, having an even higher power levels than the macrocell itself. If the femtocells were to be deployed in the same band as the macrocell layer (co-channel deployment) and the users inside the houses pretended to use the macrocell, they would suffer severe interference from their neighbours and would not be able to connect. Furthermore in Figure 6.8(b), where coverage areas with $CINR > 10\,\text{dB}$ are shown, dead zones not meeting this requisite are also pictured. It is

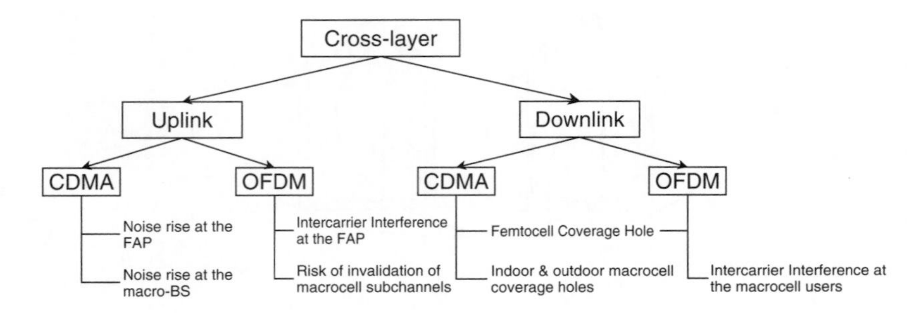

Figure 6.7 Main problems caused by cross-layer interference

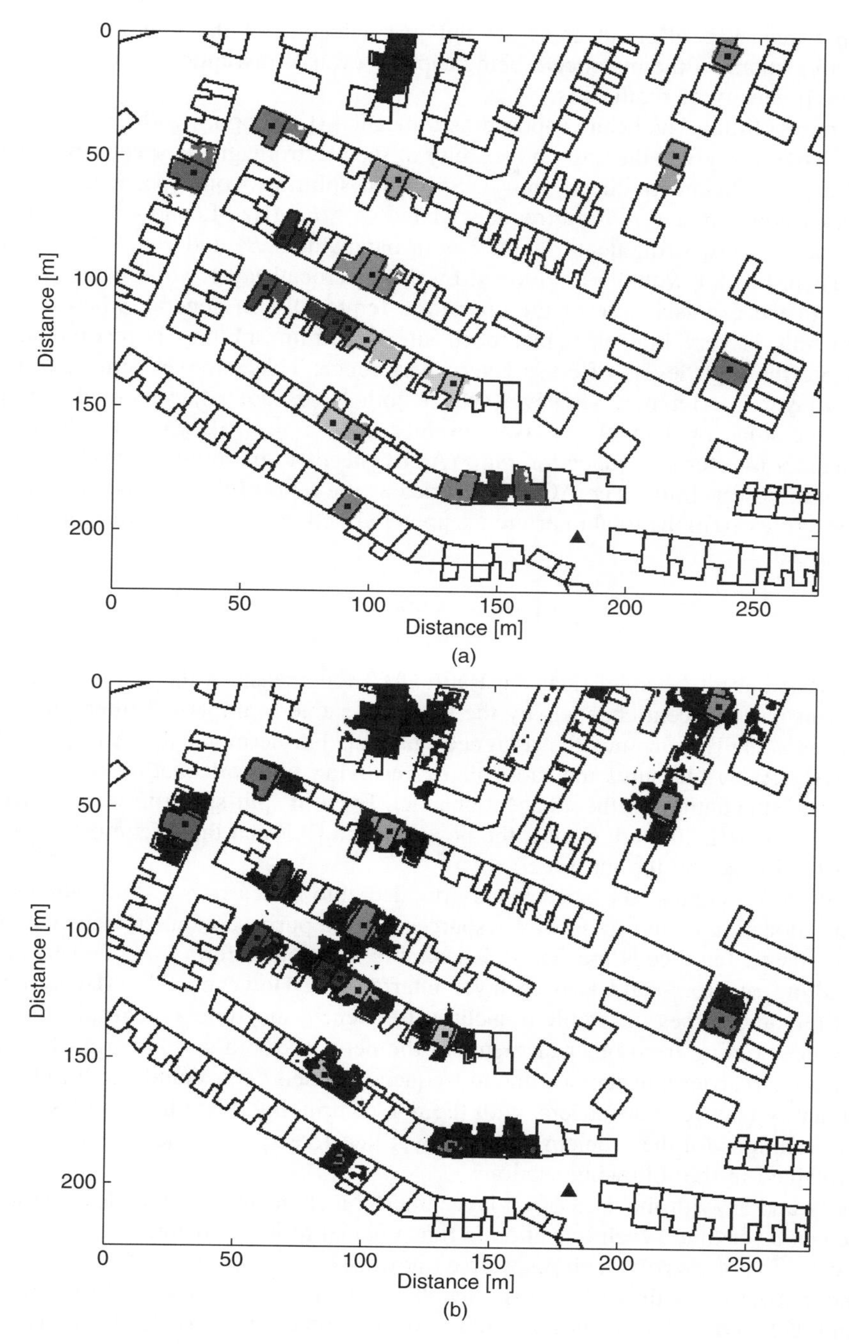

Figure 6.8 Downlink coverage areas of femtocells in a residential area covered by a co-channel deployed macrocell. The dark squares indicate the FAPs, the triangle the MBS and the background white colour represents the macrocell coverage area. (a) Coverage areas if $CINR > 0\,\text{dB}$. (b) Coverage areas if $CINR > 10\,\text{dB}$

then clear that such blackout areas are not just a feature of the femtocell layer. Dead zones will also affect the macrocell, being capable even of disrupting its service in streets with a high density of femtocells.

Spectrum splitting has been proposed as a means [4] of coping with cross-layer interference. However, given the cost and scarcity of the electromagnetic spectrum, this would lead to a less efficient frequency reuse. Spectrum splitting would occur as follows: an operator owning the licence for a frequency band with B MHz of bandwidth would divide such a band into two fragments 1 and 2 with bandwidths B_1 and $B_2 = B - B_1$ respectively. Then, band 1 would be exploited by the macrocell layer while band 2 would be assigned to the exclusive use of the operator's femtocells. In general, when the density of femtocells is very high at a macrocell site, spectrum splitting is recommended [5] to mitigate the high levels of cross-layer interference. This removes almost completely all cross-layer interference. However, when both bands are adjacent in the frequency domain, the adjacent channel can also introduce interference [6], which is why the Adjacent Channel Interference Rejection ratio (ACIR) needs to be minimized when designing the transmit power limits. The ACIR is defined as the ratio of the total power transmitted from the aggressor to the total interference power affecting the victim. It is mathematically expressed as:

$$ACIR = \frac{1}{\frac{1}{ACLR} + \frac{1}{ACS}} \tag{6.1}$$

where the Adjacent Channel Leakage Ratio (ACLR)[‡] measures the ratio of the average power sent into adjacent channels by the transmitter due to imperfect filters, to the average power actually sent into the assigned channel. Furthermore, the Adjacent Channel Selectivity (ACS) measures the ratio of the receiving filter attenuation on the assigned band to the attenuation on the adjacent channel. Even in split-spectrum deployments, the achievable ACIR is limited. Hence, the power of the FAP and the UEs must be regulated to limit the impact on the macrocell.

In OFDMA systems, the spectrum is divided into subcarriers, which allows for an efficient allocation of the frequency resources for the purpose of interference avoidance. OFDMA-based femtocells are hence a desirable solution, offering higher versatility for the handling of cross- and also co-layer interference. However, OFDMA systems can also suffer other types of problem such as frequency and time synchronization issues. Interference coming from other elements of the network (both co- and cross-layer) could introduce intercarrier interference due to frequency offsets. This could result in the loss of orthogonality between subcarriers, with the risk of bringing down the whole system. The existing solutions for the problem of cross-layer interference are summarized in Figure 6.9 and explained in the following sections.

Finally and as with the case of co-layer interference, it must be recalled that network-wide cross-layer time synchronization is fundamental to guaranteeing that femtocells and macrocells do not override their respective transmission periods. One of the main purposes of synchronization is thus to ease the handling of interference issues. In the following, link and technology dependent approaches to the problem of cross-layer interference are presented. These are further summarized in Figure 6.9.

[‡] The ACLR terminology applies only to 3G systems but it is equivalent to the concept of Adjacent Channel Power Ratio (ACPR).

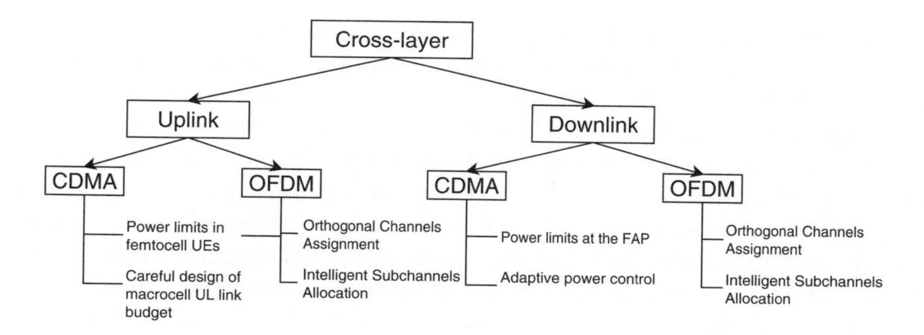

Figure 6.9 Approaches to coping with cross-layer interference

Uplink

Uplink CDMA Cross-layer uplink interference takes place under two different circumstances. In the first case, the femtocell users take the role of the aggressor, with the macro-NodeB being the victim of the interference (interference at the macro-NodeB). When the femtocells are CSG, transmissions coming from femtocell UEs are responsible for the noise increase in the macrocell layer. Therefore, operators need to impose power limits on femtocell UEs in order to guarantee a proper performance of the macrocell. This means that power control algorithms running in the FAPs should not be allowed to request any arbitrary power increase from users in order to keep a low noise level at the macrocell. However, lower transmit powers from femtocell UEs make CDMA FAPs more sensitive to interference coming from nearby users of the macrocell. Therefore the FAP must reach a compromise [3] between reducing the interference caused to the macrocell and protecting itself from the macrocell users. Regarding the power settings for power control, these can be easily decided by the FAP by measuring the pilot power coming from the macro-NodeB in downlink.

Another approach proposed by Qualcomm [5] is to limit the maximum number of femtocell UEs per macrocell site. However, this can only be done if the number of femtocells per macrocell is known and can be controlled. Given the fact that femtocells will emerge opportunistically, this seems rather improbable.

On the other hand in open-access femtocells, macrocell users are allowed to connect either to femtocells or to macrocells, depending on which one provides, at a given instant, the better quality of service. This guarantees that the UEs uplink connections use the least amount of power, hence reducing noise and uplink interference to both femtocells and macrocells. Furthermore, given that interference coming from the macro-NodeB will restrict the size of the femtocells, femto UEs will need to be very close to their FAP for succesful communication. It is thus improbable that they will transmit with high power in the proximities of a macro-NodeB and cause extreme interference. In addition, the noise rise caused by the macrocell's own users will typically be higher than that caused by femtocell UEs.

The second case of uplink interference occurs when users of the macrocell (the aggressors) transmit with high powers in the vicinity of femtocells (the victim). This is a case of interference at the FAP. Building walls typically provide enough isolation for the FAP from the outdoor macrocell users. However, when femtocells are located in areas of poor macrocell coverage, it is more than likely that passing macrocell users transmit with higher powers to reach the macrocell. The Rise-over-Thermal (RoT) noise at the FAP will, in this case, increase. Since the macro-NodeB is not aware of the precise location of its users and whether or not they interfere with any femtocell, the macrocell uplink budget must be designed considering the worst case scenario of interference at a femtocell. This way, an appropriate upper limit can be set to the transmit power of macrocell UEs.

For example, let us assume that each macrocell user m able to cause uplink interference to a close FAP f has a transmit power p_m^M (expressed in watts) where M denotes that the user belongs to the macrocell and m is the user ID. The overall path loss between macrocell user m and FAP f is $l_{m,f}$ and the noise level at the FAP is given by n_f in watts. The total interference plus noise level in_f at FAP f is thus:

$$in_f = n_f + \sum_{m=1}^{N_M} \frac{p_m^M}{l_{m,f}} \tag{6.2}$$

The value of $l_{m,f}$, depends mainly on the system's frequency, the house construction materials and the distance between the macrocell UE and the FAP. The number of users, however, will vary with time and will be different depending on the density of users of the macrocell. The in_f values are depicted in Figure 6.10 at an arbitrary FAP.

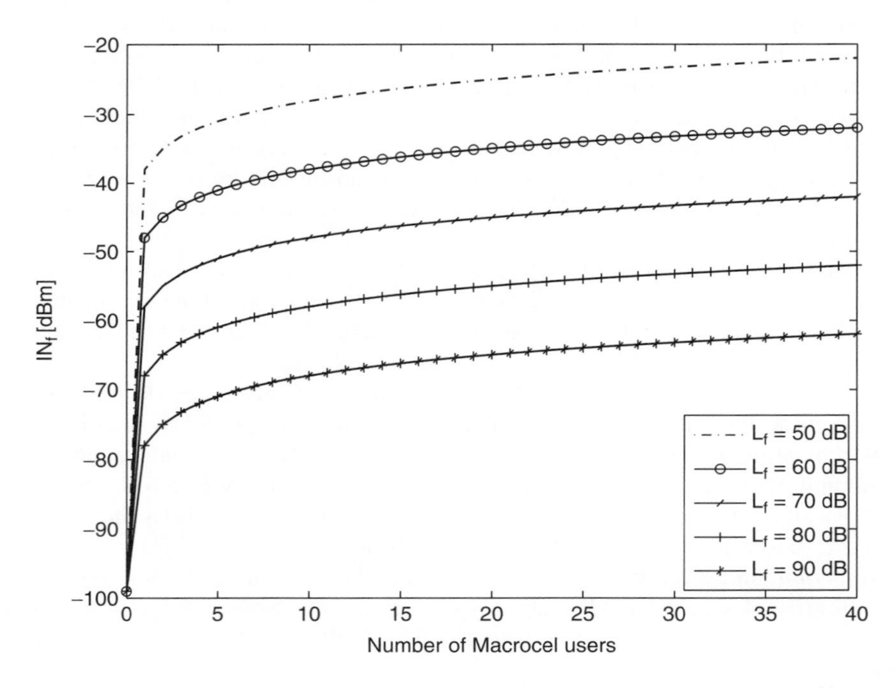

Figure 6.10 Interference-plus-noise level at an FAP when all macrocell users have $P^M = 12\,\text{dBm}$ and the noise level is $N_f = -99\,\text{dBm}$. L_f expresses the outdoor-to-indoor signal attenuation from the macrocell user to the FAP

Depending on the technology to be used (UMTS, HSUPA, ...) graphs like this one help to determine the maximum levels of interference depending on the maximum transmit power of macrocell users.

HSUPA simulations [7] have shown that, if there were a macrocell UE within 15 m of the FAP, the femtocell user would need to transmit with at least 15 dBm and be within 5 m of the FAP in order to enjoy a throughput of 2.8 Mbps. However, it is unlikely that interfering macrocell users will ever be that close to the FAPs. This is mainly because of the presence of dead zones for macrocell users in the proximity of femtocells. If a macrocell user were that close to a dead zone, the macrocell would try its best to hand it off to another CDMA frequency or drop the connection. In general, it can be said that the downlink deadzones guarantee a minimum separation between macrocell users and FAPs.

Uplink OFDMA In the case of dedicated channel OFDMA femtocells, it is possible to divide the licensed band exactly at the frontier between two OFDMA subchannels. This way, an integer number of subchannels is assigned to each network layer. Cross-layer interference is in this case practically non-existent, with adjacent channel interference being the only source of impairment. However, when the femtocells are co-channel deployed, two types of interference can occur in the uplink. As defined for the CDMA case, these are interference at the FAP and interference at the MBS.

When femtocells are located near the macrocell edge, nearby macrocell UEs might be asked by the macrocell to increase their transmit power due to the high distance from the MBS. If the proper OFDMA subchannels are not used, this could cause uplink interference with the FAP. As illustrated in Figure 6.11(a), macrocell user 2 is transmitting with high power in the same subchannels as femtocell user 1. The isolation provided by the house walls might be insufficient in this case, causing the uplink connection of user 1 to drop. In this case, an appropriate allocation of the OFDMA subchannels relaxes the power restrictions imposed on macrocell users in CDMA femtocells. It is thus important that femtocells (especially those located in areas of low macrocell coverage such as the cell edge) plan their uplink subchannels taking into account the spectral occupancy as illustrated in Figure 6.11(a) by links L_3 and L_4.

The other type of uplink interference can occur when femtocells are located too close to the MBS (interference at the MBS). This is illustrated in Figure 6.11(b) where femtocell user 1 is required by its femtocell to transmit with high power.[§] If this transmission is received at the MBS, the subchannels occupied by femtocell user 1 become useless for the rest of the macrocell UEs. As with the CDMA case and in order to guarantee the prevalence of the macrocell resources, the power increase requested by FAPs to its users must be bounded. However, in the OFDMA case, this can be done dynamically and is dependent on the occupancy of the OFDMA subchannels at a given time instant.

Downlink

Downlink CDMA In CSG femtocells deployed in the same band as the macrocell (co-channel deployment), signals coming from the FAPs will cause downlink interference to

[§] Femtocell users are also subject to power control. High power transmissions might be required if, for example, the user is in a remote location within the premises such as a cellar or a garden.

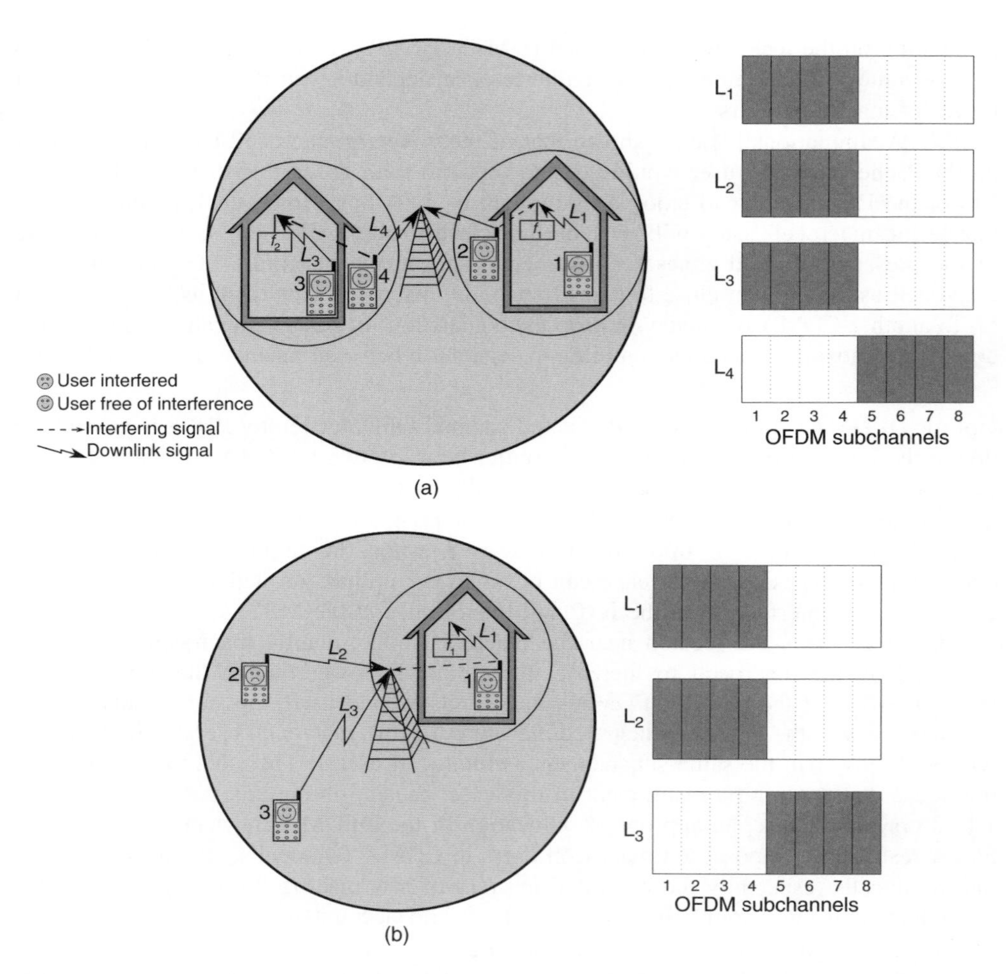

Figure 6.11 Cross-layer uplink interference in an OFDMA co-channel two-layer network. (a) Interference at the femtocell. (b) Interference at the macrocell

nearby macrocell UEs. If this interfering signal is strong enough, dead zones will appear in the macrocell. The femtocell coverage areas of Figure 6.8 can be considered as dead zones from a macrocell perspective because macrocell users located in those regions will suffer extremely high noise levels and will not be able to get any service. Note that the scenario presented in Figure 6.8 is optimistic from the point of view of outdoor macrocell coverage. This is because windows and doors have been eliminated and therefore femtocells do not leak power outdoors. However, it can be seen that the outer walls are still not enough to efficiently isolate adjacent houses and contain the femtocell signal within a premises. Therefore dead zones also appear indoors in neighbouring homes. This becomes even more severe when the FAP is located near windows or doors.

The first solution proposed by 3GPP to cope with the interference caused by FAPs to macrocell users in UMTS and HSDPA is the limitation of the FAP maximum transmit power. However in co-channel deployments, the power limit will need to be adapted depending on the circumstances. It is generally agreed that a fixed maximum transmit power is not feasible in co-channel deployments due to the risk of creating uncontrolled interference. An adaptive power control approach for the Primary Common Pilot Channel (PCPICH) is thus more suitable and FAPs will need to support a wide transmit power dynamic range, which could increase their cost. The algorithms that control the FAP transmit power are left free to implementation by the operator. Another proposed solution for cases of extreme femtocell density is to deploy open access femtocells. This way, operators would supply CSG or open access femtocells to their clients depending on the FAP density of the area where they plan to install it. However, the impact on the macrocell of a femtocell layer composed of CSG and open access femtocells still requires further research.

When the femtocell is located far from the macro-NodeB, due to high levels of indoor femtocell coverage and assuming reasonable isolation from the macrocell signal (femtocell UE far from windows and near the FAP), the interference caused by the macro-NodeB to users of the femtocells is minimum. However in cases where the femtocell is very close to the macro-NodeB, the FAP coverage area is reduced to a few metres due to interference from the macrocell. In these situations the quality of the femtocell signal is strong only when the UE is in the very proximities of the FAP. For instance, a femtocell UE located next to a window is more likely to connect directly to the close macrocell than to its own femtocell. However, these situations do not represent the general case. The Femto Forum [7] indicates that throughputs of 14.4 Mbps can be achieved with HSDPA femtocells when the user is located at least 250 metres from the closest microcell or 1000 metres from the closest macrocell.

Downlink OFDMA As explained above, cross-channel interference can be neglected in the case of dedicated channel femtocells. However, when OFDMA femtocells are co-channel deployed, the case is fundamentally different from the CDMA case.

For example, in Figure 6.12, user 1 is at home receiving data from their femtocell (f_1) through link L_1. In order to guarantee quality of service, f_1 allocates subchannels 1 to 4 to downlink L_1. Meanwhile, macrocell user 3 is walking down the street and passes by the front door of user 1, thus falling inside the coverage area of f_1. If this had happened in the case of a CDMA two-layer network, user 3 would have immediately experienced a sudden rise in the noise level with undetermined effects for the communication. However, in this case, user 3 receives data from a macrocell by means of link L_3, which has been allocated subchannels 5 to 8 by the macrocell. Since these subchannels are not being used by femtocell f_1, user 3 is free of interference and the communication will not be affected. However, this is not always the case. Figure 6.12 also presents a case in which macrocell user 4 is allocated OFDMA subchannels which are being simultaneously used by a nearby femtocell (f_2 in this case). User 4 will be in this case heavily interfered.

By examining Figure 6.12, one could conclude that femtocell user 2 would also suffer interference because of the occupation of subchannels 1 to 4 by the macrocell. However, the house walls typically attenuate the macrocell signal sufficiently and the proximity to the FAP outcomes a reasonably high *CINR* for user 2. Downlink cross-layer interference is thus not extreme for femtocell users.

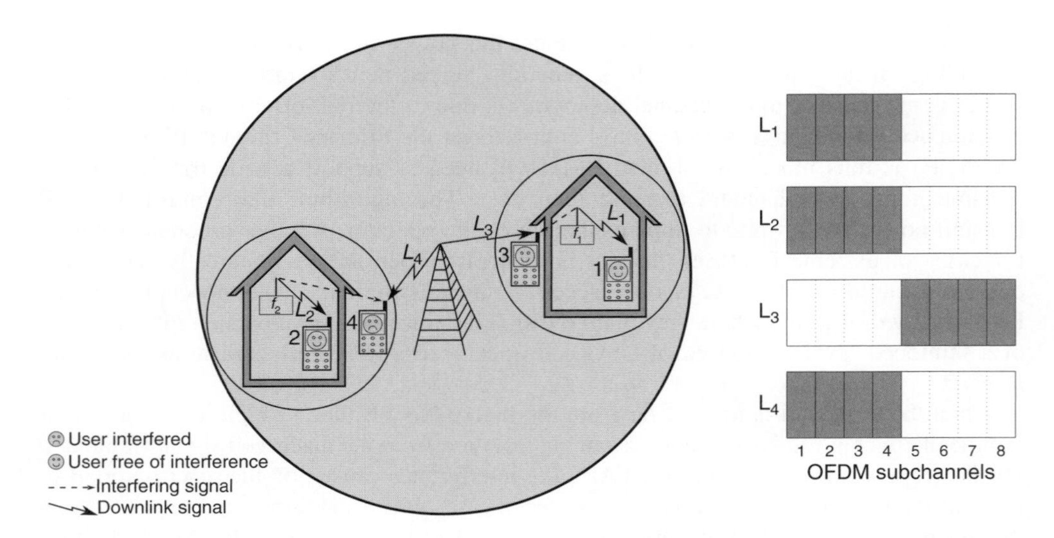

Figure 6.12 Cross-layer downlink interference in an OFDMA co-channel two-layer network

6.2.3 The Near–Far Problem

Power control algorithms are an essential part of today's cellular networks. For instance in CDMA macrocell networks and depending on the service recquired by a mobile user, a Signal to Interference plus Noise Ratio (*SINR*) target is defined at the Node-B in the uplink. If the *SINR* at the macrocell does not meet the target, the connection will be terminated. In order to reach this threshold, there exist several procedures during which the Node-B asks the UE to increase its transmit power. This is known as power control. However, if the power increase of all UEs is not regulated in a centralized manner, it could happen that a mobile close to the Node-B increases its power too much and eclipses the signal of a much further mobile. This phenomenon is called the near–far effect. In order to solve this problem, the Node-B tries to control all the UEs power by avoiding excessive variations in the received power among all the mobiles, i.e. close transmitters will emit with less power than those much further away.

Figure 6.13 illustrates graphically the Near–Far problem. Let us assume that the UE close to the Node-B transmits with power[¶] P_1 and that due to propagation effects, the signal suffers an attenuation of L_1 dB. Similarly, the UE located further from the Node-B transmits with power P_2 and suffers an attenuation of L_2. Since the second transmitter is located further away from the MBS, it is assumed that $L_2 > L_1$. The received powers P_{r1} and P_{r2} from each transmitter at the Node-B in the UL are thus:

$$P_{r1} = P_1 - L_1 \tag{6.3}$$

$$P_{r2} = P_2 - L_2 \tag{6.4}$$

¶ The notation used here assumes the use of lower case for magnitudes expressed in natural units (watts, volts, ...) and upper case for magnitudes expressed in a logarithmic scale (dB, dBm, dBu, ...).

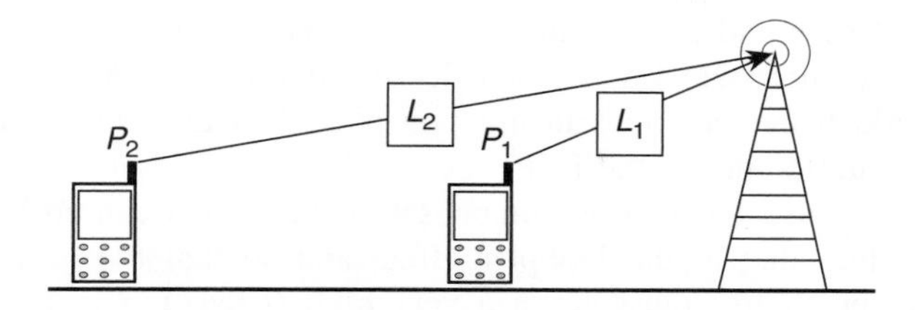

Figure 6.13 Scenario where power control is needed and the near–far problem appears

Assuming a noise level n, the interference I at the reception side of each link is obtained as follows:

$$I_1 = 10 \cdot \log_{10}(p_{r2} + n) \tag{6.5}$$

$$I_2 = 10 \cdot \log_{10}(p_{r1} + n) \tag{6.6}$$

And hence, the *SINR* of each links is computed as:

$$SINR_1 = P_{r1} - I_1 = P_1 - L_1 - 10 \cdot \log_{10}(p_{r2} + n) \tag{6.7}$$

$$SINR_2 = P_{r2} - I_2 = P_2 - L_2 - 10 \cdot \log_{10}(p_{r1} + n) \tag{6.8}$$

In the absence of power control, it is assumed that both UEs emit with the same power ($P_1 = P_2$) and request the same service (i.e. their target *SINRs* are equal). Hence, given that $L_2 > L_1$ it is easy to see from Equations (6.7) and (6.8) that the second UE is heavily interfered by the closest one ($SINR_2 < SINR_1$). This is illustrated in Figure 6.14, where the difference in *SINR* values is made evident. However, if a power control algorithm is used, it could occur that the closest UE is asked to reduce its power so that $P_2 = P_1 + L_2 - L_1$, in which case both *SINRs* would be equal and interference would not block out the second UE.

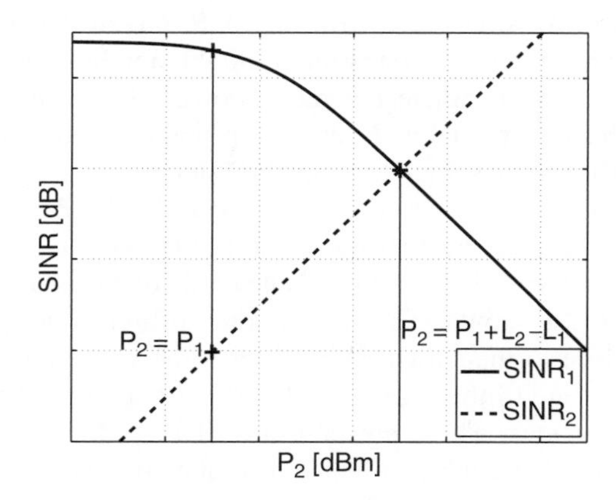

Figure 6.14 Received *SINRs* for different P_2 values in the scenario of Figure 6.13

Similarly, the MBS could also decide to ask the second user to increase its transmit power in order to compensate for the path losses. If this second user were close to a femtocell, this sudden increase in transmit power could cause high interference in the femtocell uplink (interference at the FAP).

Power control can also be seen as compensation for losses caused by fading. In situations of fast fading due to the Doppler effect and so that the user mobility is not compromised, power control must act at a very high speed. For instance in GSM networks, UEs are subject to transmit power changes almost twice every second. In UMTS systems, this occurs at a rate of about 1.5 KHz. The user terminals must therefore have the ability of adapting to the Node-B recquirements quickly and have a large dynamic range of transmission power. This is a common cause for the increase in costs of these devices.

In order to avoid the Near–Far problem within their coverage area, femtocells are also subject to power control. However, in order to avoid causing uplink interference with the macrocell, the access to the medium in femtocells must be handled carefully. This is because, assuming that femtocells are synchronized with the base station [8], femtocell UEs asked to burst up their powers can become a source of interference in the UL of nearby Node-Bs (interference at the MBS). Since the sensitivity threshold of femtocell receivers is lower than that of macrocells, the power increase requested for a femtocell user should not be as large as for macrocell users. Based on this, an RF engineer could assume that femtocell UEs would not transmit enough power for a strong uplink interference to appear at the macrocell. This will, however, strongly depend on the specific location of the femtocell users. Typically, the closer to the macrocell, the more serious the interference will be. The urban and suburban distribution of femtocells is thus a consideration of the highest importance in the design of femtocell-aware macrocell networks.

As pointed out in previous sections, the main solution to the interference problems caused by power control mechanisms is the limitation of the maximum power that femtocell and macrocell UEs can transmit. Therefore, any operator interested in deploying a femtocell layer must first balance very carefully the maximum power that the FAPs and the MBS will ask from their users in given scenarios. Furthermore, power control in femtocells is subject to a problem that is not present in macrocells. Since UEs can be located at very close to the FAP, it can happen that the femtocell requests a very low transmission power from the UE. Given that UEs have a minimum achievable transmission power, the transmitter might not be able to cope with power control requirements, and still radiate too much power. This limits the effectiveness of power-control algorithms in reducing Near–Far problems and requires other solutions like the attenuation of the user's signal at the FAP side.

With respect to the downlink, it was previously seen that FAP power levels are adapted to reduce co-layer interference. Furthermore, another reason given [9] for the limitation of the FAP power levels (typically to values beneath 20 dBm) is that higher values might interfere with the femtocell's own UEs. This is due to high input power levels suffered by UEs that are too close to their own FAPs. Typical power requirements at the antenna connector in 3G mobiles are of the order of −25 dBm, being the behaviour of the terminal for unspecified higher values. Thus, power limits at the FAP also help to increase the coupling loss in the UEs downlink and hence to reduce outage of the FAP own users.

Femtocells were originally designed for solving indoor coverage problems. Therefore households located in the proximities of Node-Bs or base stations and hence receiving good coverage are less likely to install a femtocell and cause cross-layer UL interference. However, it is expected that the operator's marketing strategies will offer different types of advantages [10] to users of femtocells. The aquisition and installation of femtocells will thus be driven not only by the presence of indoor dead zones but also by economic reasons. Femtocells will therefore emerge opportunistically at any random location within the macrocell, even at locations vey close to the Node-B, and they will interfere with uplink macrocell connections due to the power control algorithms. One solution for avoiding this problem [11] is to have the operator divide the licensed spectrum into two fragments (spectrum splitting). This way, the emissions of femtocell users will be restricted to the frequency band allocated to the femtocell layer and will not cause interference to the macrocell. However, Chandrasekhar [12] indicates that this reduces spectrum efficiency and frequency reuse in WCDMA networks. He proposes the use of interference avoidance to cope with the Near–Far problem and claims that such techniques are capable of providing an increase of up to seven times in the density of femtocells compared with the split-spectrum approach.

6.3 Interference Cancellation

The term Interference Cancellation (IC) refers to any method used to minimize the effects of interference in receiver systems. The interest in these techniques in femtocell networks comes from the unavoidable presence of co-channel interference and the need for receiving systems that can operate in the presence of higher interference levels. In principle, any method that allows a receiver to operate with higher levels of co-channel interference can be considered an interference cancellation technique [13]. The sources of co-channel interference can be FAPs and MBS as well as femtocell and macrocell users. This is important because the interfering source will determine which one is the optimal process for interference cancellation.

Note that the terms *co-channel interference* and *co-layer interference* should not be confused. Co-channel refers to the fact that the interfering signal shares the frequency band with the desired signal. On the other hand, co-layer denotes that the layer of the aggressor is the same as that of the victim system (see Section 6.2.1). Although in femtocell networks co-layer interference is typically co-channel (adjacent channel interference is not that extreme), the opposite is not necessarily true.

Most interference cancellation techniques make assumptions about characteristics of the interfering signal such as, for example, the angle of arrival. However, these techniques typically require the use of antenna arrays at the receiving system in order to cancel out the interference. Since femtocells are aimed at improving the mobile market coverage, and the use of multiple receiving antennas in mobile terminals is limited, Single Antenna Interference Cancellation (SAIC) techniques are preferred in order to cope with downlink interference. Therefore in the following sections, a distinction is made between IC techniques that are more suitable for uplink and for downlink connections.

6.3.1 *Uplink Techniques*

Contrary to the techniques to be implemented on mobile terminals, the following IC techniques are not limited by the hardware they run on. They are therefore better suited for implementation on MBS as well as FAPs and hence, for uplink interference cancellation. Furthermore in CDMA cases, if the FAP were aware of the codes used in the umbrella macrocell, this information could be exploited to cancel out cross-layer interference by means of Parallel Interference Cancellation (PIC) [14]. The classification presented here follows the one introduced by Ofcom (Office of Communications) of the United Kingdom [13].

Filter Based

The main objective of IC methods based on filtering is to provide a filter that attenuates parts of the spectrum of the input signal that are highly affected by interference, i.e. spectrum regions with a very low *SINR*. Similarly, the regions of the spectrum where the *SINR* is higher will be amplified. In order for this to work, the filters need to be adaptive and change over time depending on the interference conditions at a given instant. For example, Figure 6.15 shows the structure of a FIR filter of N coefficients. The filter is thus of order N and its coefficients are defined as $a = (a_0, \ldots, a_{N-1})$. For the purpose of IC, the a_i coefficients will be adjusted *on the run* using algorithms like the least mean squares.

There are several ways in which this adaptation can be done. For instance, Forward Linear Prediction (FLP) has been used as a means of predicting the interference values at a given instant from the previous signal values. This is possible since interference is considered to be deterministic and hence easier to predict than the expected signal. Frequency localized interference can be then easily removed [15]. The FLP process can be formulated as:

$$\hat{x}(k) = -\sum_{i=1}^{N} a_i x(k-i) \tag{6.9}$$

with being $\hat{x}(k)$ the predictor of sample $x(k)$ based on the previous N received samples. Then the predicted interference value is subtracted from the received sample to cancel the interference, so the improved received signal value will be:

$$x_{IC}(k) = x(k) - \hat{x}(k) \tag{6.10}$$

The desired signal is obviously slightly distorted so the accuracy of this approach might not be valid for low interference levels. These filters can be implemented by hardware or software. Either way, this would require the inclusion of IC filters in the mobile terminal,

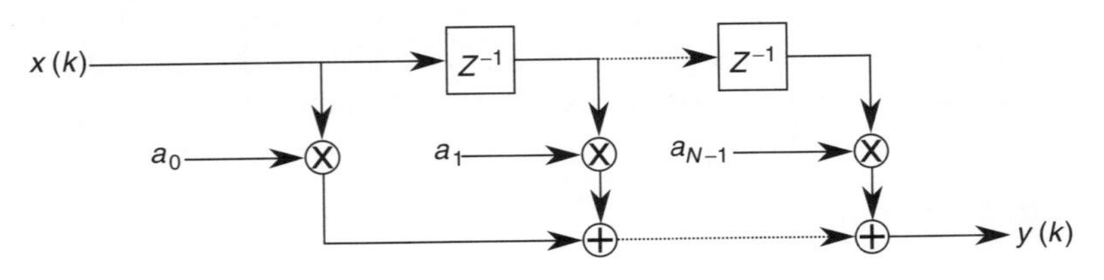

Figure 6.15 Finite Impulse Response (FIR) filter

in the FAP or in the MBS. Therefore, these techniques are both applicable in uplink and downlink.

Multiuser Detection

Multiuser detection techniques are based on the fact that each user has a certain signature waveform. Since they are based on the reception of signals coming from a large number of users, these techniques are exclusively used in uplink connections at the FAP or at the base station. For example, in CDMA systems, each user makes use of a different spreading code, thus generating a specific set of signals. Such a technique relies strongly on the orthogonality of the waveforms from different users. Furthermore, since the joint waveforms of all the users need to be considered, these techniques are also known as 'joint detection'.

In this context, the optimum detector would be the one that selects the most probable signal depending on the observed one and the channel statistics. To do this for a number of users K, the optimum detector would compute the correlations of the 2^K possible signal vectors and choose the one with the higher correlation. Since the complexity of such a detector is too high, sub-optimal detectors like the parallel Minimum Mean Square Error (MMSE) detector [16] have therefore been designed. Other proposed approaches to multiuser IC is Successive Interference Cancellation (SIC) [17], where the strongest user signal is detected, recreated and then subtracted from the received signal. This way, an improved *SINR* is achieved and the process is repeated untill all the users have been detected. However, users with equal power might not be detected correctly.

Compared with macrocell networks, the number of interfering users in FAPs is much lower. This relaxes the conditions that IC algorithms should meet in order to achieve reliable levels of performance. For example, in the case of OFDM systems with a low number of interfering users ($K \approx 4$) and four receiving antennas, a turbo-coded MMSE-SIC algorithm with State Insertion (SI) has proved quite efficient [18]. MMSE-based PIC approaches perform slightly less well on exchange for a less complex implementation of the antenna weight calculation. However, specific research on the applicability of these techniques on hybrid femto/macrocell scenarios is still missing.

Cyclostationary

The statistics of a stationary signal are considered to be time invariant. However, the statistics of cyclostationary signals vary over a certain period of time. In the frequency domain this translates into fluctuations of n frequency bands that are statistically dependent. If the centre frequencies of those n bands add up to some nonzero value, the signal is then considered to be cyclostationary of order n.

Since the main cyclic frequencies correspond to the carrier and the data rate, the cyclic frequencies of co-channel interference can then be detected by exclusion. This can thus be exploited for the removal of interference in certain frequency bands [19].

Frequency Shift (FRESH) filters combine frequency shifted versions of the input signals and they have also been applied to the removal of interference when the desired signal is used as reference. Since the desired signal is obviously not present at the receiving end, blind adaptation of FRESH filters is thus necessary for practical cases.

Higher Order Statistics

These methods are based on the use of signal statistics other than means (first order statistics) and they have been proved to be quite succesful for the purpose of source separation. The objective is to separate independent signals from their addition to many others, which can be effectively done by using spatial diversity with array antennas. However, this restricts once again the use of this technique just to uplink connections and hence, FAPs and MBS.

As an example of this technique, Blind Source Separation is based on the use of higher order statistics and it consists of separating the signals coming from q users, using the p outputs of an array antenna. It is assumed that $p \geq q$ so that there are more observations than unknown variables. This method is called *blind* because the sources are not known in the reception process. The only requisite is that the different signals must be statistically independent, which is typically true for multiuser systems. The sources x_i with $i \in [1, \ldots, q]$ at the time instant n are represented by:

$$X_n = \begin{pmatrix} x_1 \\ x_2 \\ \ldots \\ x_q \end{pmatrix} \tag{6.11}$$

while the output of the array elements is:

$$Y_n = \begin{pmatrix} y_1 \\ y_2 \\ \ldots \\ y_p \end{pmatrix} \tag{6.12}$$

If M denotes the $p \times q$ channel matrix, and Z_n the Additive White Gaussian Noise (AWGN) vector at time n, then the equation that defines the system is:

$$Y_n = MX_n + Z_n \tag{6.13}$$

The vector of sources X_n is the sought solution but the channel matrix M is also unknown. However, due to the characteristics of the channel, Equation (6.13) is true for a period of $n = 1, 2, \ldots, N$ blocks. Then, higher order statistics of matrix M can be accumulated for periods shorter than N. If the sources remain statistically independent, M can be quickly figured out and hence, the sources vector X_n.

These types of method typically require preprocessing of the received signal components such as passing Y_n through a bank of parallel filters or even whitening Y_n. Such processing capabilities are prohibitive in mobile terminals and could also increase the cost of FAPs. Other IC methods might then be desirable for femtocell access points.

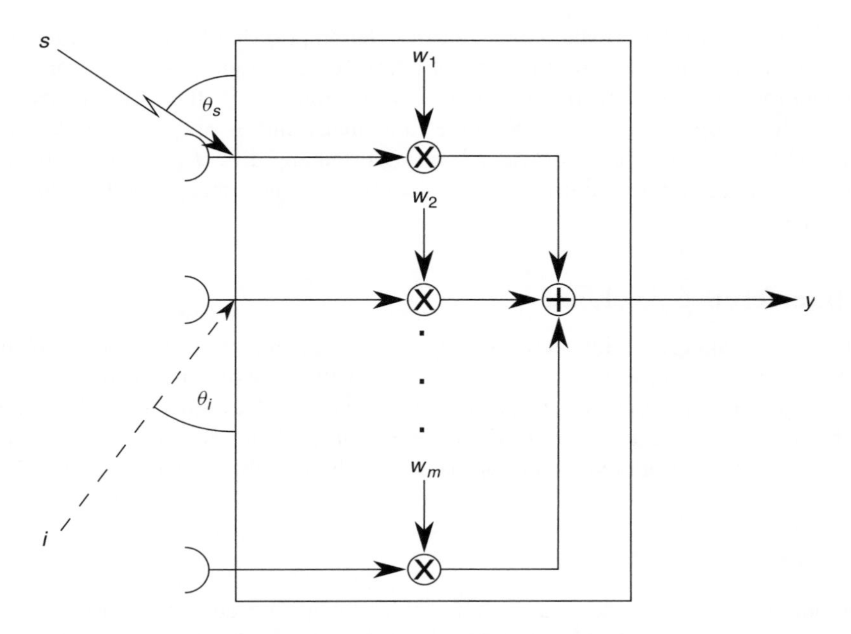

Figure 6.16 Beamforming receiving system

Spatial Processing

In femtocells, there might be cases where the aggresor and victim systems are spatially separated. The FAP could, for instance, receive a strong uplink signal from a CSG femtocell subscriber within the house, while an interfering signal from a macrocell user located outside arrives through a window. It is situations like this one that make techniques such as beamforming and null steering interesting for FAP designs. The main idea is to assign different weights to the elements of the receiving antenna array so as to form beams in given directions. This way, the direction of the interfering signal i can be attenuated, while the desired signal s is amplified. Figure 6.16 illustrates the architecture of a beamforming system. An FAP equipped with this system would estimate the optimum set of weights $w_1, \ldots, w_m$ so that the *SINR* of the ouput signal y is maximized.

Other spatial processing techniques such as Multiple Input Multiple Output (MIMO) can also be applied in femtocells for the separation of co-channel interference in the uplink. The interfering signals are produced by nearby macrocell users and by neighbouring femtocell users. There are thus as many interfering input signals as interfering users in the surroundings of the FAP premises. Then, spatial diversity is exploited at the receiving end by means of, for instance, an Optimum Combining (OC) [20] method.

6.3.2 Downlink Techniques

Downlink interference cancellation techniques are related to the implementation of IC techniques on mobile terminals. As in the uplink case, the objective is to mitigate the effect of co-channel interference. However, mobiles impose hardware limitations such as the number of antennas and circuitry so not every IC technique is suitable for downlink implementation. Single Antenna Interference Cancellation (SAIC) [21] seems to be a promising technology based on an adaptive filter that requires only one single receiving antenna. *SINR* gains of up to 15 dB have been reported and different strategies have come up in the last years, some requiring training sequences and others based exclusively on blind estimation.

6.4 Interference Avoidance

The main disadvantage of interference cancellation techniques is that they are usually expensive to implement, thus increasing the cost of the network. Even in WCDMA networks, where they are supposed to perform best, the tendency now is to drop their use [22]. This is mainly due to errors in the cancellation process and therefore, interference avoidance is being considered as an approach with higher chances of success [23].

6.4.1 CDMA

Time-hopping has been proposed as a feasible means of reducing cross-layer uplink interference. In the case of 3G networks, time-hopping can be implemented using TH-CDMA technology. The basic idea is not to transmit over the whole spectrum all the time but only during short periods, and staying idle during the remaining period. The subdivision of the transmission period is illustrated in Figure 6.17. This approach assumes that there is no communication between femtocells and macrocells. There is thus, in principle, no procedure for the synchronization of the transmission slots of different cells. Therefore, the moment of transmission is chosen independently in the two layers of the network. If, for instance, the period of a CDMA transmission is T and the time-hopping scheme divides T into N_{hop} hopping slots. Each femtocell UE can then choose for transmission one of the hopping slots of length T/N_{hop} according to a random noise-like function or in some coordinated way within its femtocell. According to [12] this approach reduces the interference between different femtocells (co-layer interference) and between femtocells and macrocells (cross-layer interference) by a factor of N_{hop}.

If there are too many UEs in the femtocell, a scheme where all the users of the same femtocell transmit in the same time slot is recommended as a good approach for reducing

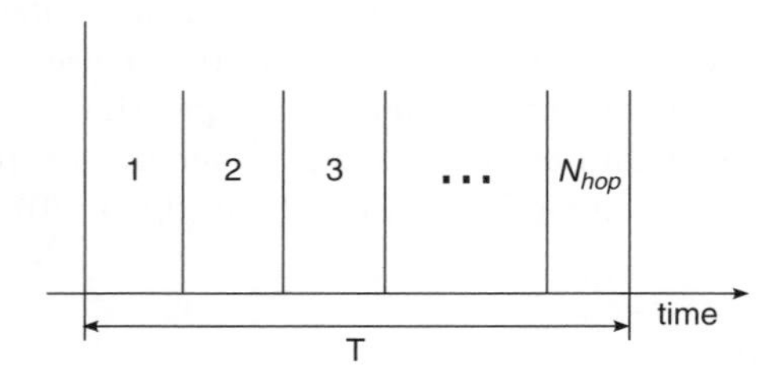

Figure 6.17 Division of the CDMA transmission period T into N_{hop} time slots for TH-CDMA interference avoidance

the outage probability in the uplink of CDMA Base Transceiver Stations (BTSs). This scheme is called joint hopping and the transmission slot is independently chosen by each FAP. With this scheme, femtocell UEs do not disturb each other within their own femtocell when transmitting in the same time slot, thanks to the averaging of aggregate interference in CDMA systems. Furthermore, and since the selection of the transmission slot is independent between femtocells, co-layer interference from neighbouring femtocell UEs is also decreased by a factor N_{hop}. In the same way, this also reduces cross-layer uplink interference at the FAP coming from macrocell users. However, if there is only one user per femtocell, both joint and independent hopping schemes have a similar performance.

Figure 6.18 illustrates an example of a potential slot assignment using joint hopping TH-CDMA. It is then clear how with this configuration, there is a risk of co-layer interference in the uplink only between femtocells 2 and 6, as well as between femtocell 3 and the macrocell M. However since femtocells 2 and 6 are far apart, it is probable that they will not even interfere with each other, thus exploiting spatial diversity. Meanwhile, the transmissions from the users of femtocell 3 could be unsuccessful due to cross-layer interference from nearby macrocell users. However, femtocell 3 is not near the macrocell edge so macrocell users in that area are probably transmitting with low power and will thus not interfere the FAP either. The assignment of the transmission slot changes at least every CDMA period T. Therefore, if a given assignment causes outage to a given user, it is unlikely that such a user will be interfered with again during the next period.

On the other hand, in non-CDMA systems like OFDM, a time hopping scheme different from joint hopping is recommended. This is because due to the lack of CDMA interference averaging, one single macrocell interferer is enough to produce outage to femtocell users. Therefore, a random access transmission scheme is preferred. This approach is effective because it is improbable that two femtocell users independently decide to use the same time slot.

The use of antennas with N_{sec} sectors in FAPs is also suggested in [12] as a way of reducing the interference caused by close macrocell UEs in the uplink. Such a configuration would reduce cross-layer interference by a factor of N_{sec}, which is especially severe to femtocells located near the edge of the macrocell since they would receive interference from several macrocells. Chandrasekhar [12] proposed configuring these femtocells so as to recquire higher reception powers in the uplink compared with inner FAPs, However, equipping femtocell access points with sectorial antennas would imply a price increase.

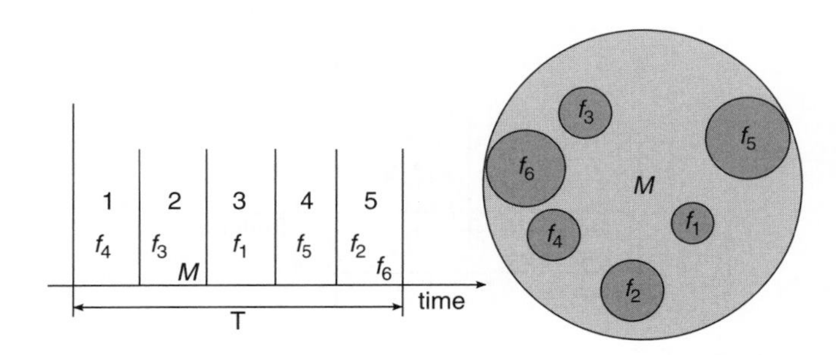

Figure 6.18 Random transmission slot assignment throughout joint hopping in TH-CDMA. In this example the scenario contains one macrocell, $N_f = 6$ femtocells and $N_{hop} = 5$

At the time of the writing of this book, FAPs already exist on the market [24] that use omnidirectional antennas so it is unsure if FAPs equipped with sectorial antennas will succeed.

Another interference avoidance technique proposed in [12] is to make use of a femtocell exclusion region, which consists of silencing femtocells that are too close to a macrocell. This guard zone [25] or interference range can also be defined as the minimum distance that an interferer can be placed from a receiver in order to cause negligible distortion in the communication. The exclusion region is used in conjunction with a layer selection handover policy, which assumes that femtocells will be configured as open access. This policy allows the macrocell UEs located within the femtocell coverage area to perform a handover to the femtocell. However, given the outcome of several customer surveys, femtocell manufacturers such as Motorola [26] seem to be more keen on producing CSG femtocells.

6.4.2 OFDMA

Compared with CDMA, OFDMA systems have the advantage of providing two dimensions (time and frequency) for the management of radio resources. They therefore provide a much higher versatility for the design of interference avoidance techniques [27]. In the case of femtocells and given their short range of coverage, the sensing of nearby channels is simpler to implement than in macrocell networks. For instance, it would be possible to use measurement reports from users in order to choose intelligently the most appropriate subchannel and time slot to perform the transmission. The present section will cover the latest findings and proposals for the allocation of resources in femtocell networks with the objective of minimizing the effects of interference.

In order to reduce cross-layer interference in two-layer networks, some companies [28] have opted for a split spectrum approach (see Figure 6.19). This assumes the division of the available spectrum between the two layers, i.e. all of the operator's macrocells will use one set of subchannels, while the femtocell layer will use the remaining ones. This way, co-channel cross-layer interference is eliminated. However, from a frequency reuse perspective, each layer only accesses a fragment of the whole licensed spectrum, placing the efficiency of this approach under discussion.

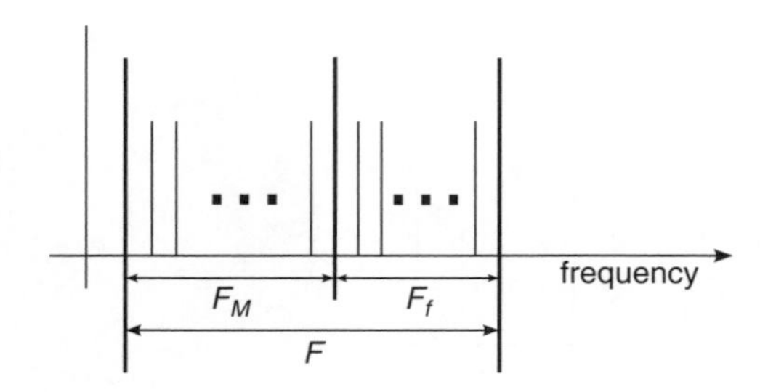

Figure 6.19 Assignment of F frequency subchannels in a two-layer OFDMA network. F_M subchannels will be used by the macrocells, while F_f subchannels will be assigned to the femtocell network layer

Nevertheless and based on such an orthogonal division of the frequency resources, Chandrasekhar proposes [29] a spectrum allocation strategy so that the spatial frequency reuse is maximized. The spatial reuse of the spectrum is typically quantized (see [30]) by means of the Area Spectral Efficiency (ASE), which is measured in b/s/Hz/m^2 and expresses the average throughput per frequency subchannel and unit of area. Let's define the ASE of the macrocell and femtocell layer as ASE_M and ASE_f respectively. The objective of this approach is to obtain the spectrum division that maximizes the network's global ASE under the following assumptions:

1. The access method to the femtocells is based on a Closed Subscriber Group (CSG), i.e. only licensed users (typically the owners) are allowed to use each femtocell.
2. There is instantaneous feedback of the Channel State Information (CSI).
3. The power is uniformly distributed across subchannels by the FAPs and macrocell Base Station.

In this method, it is assumed that a total of $F = F_M + F_f$ subchannels is available, with F_M allocated to the macrocell layer and F_f to the femtocells. Therefore, the portion of spectrum assigned to the macrocell layer is defined as $\rho = F_M/F$ and one important planning decision should be a proper dimensioning of the parameter ρ. For instance in [29], ρ is chosen dynamically depending on the specific QoS requirement of one layer with respect to the other. This is done by means of a parameter η, which represents the ratio of expected throughput per user in one layer to the overall expected throughput per user. In other words, it expresses which layer expects a higher throughput.

Since femtocells emerge opportunistically, a protocol for the access to the OFDMA subchannels should be decentralized. Although theoretically possible, it is not feasible for an operator to control remotely the different features of their femtocells, so distributed approaches are necessary (see Chapter 8). Following Chandrasekhar's strategy, the different femtocells will thus access the medium by choosing a random set of the subchannels allocated to their layer.[‖] For instance, if each femtocell accesses k subchannels of the F_f allocated to the femtocell layer, then $\rho_f = k/F_f$ can be defined as the portion of allocated spectrum accessed by each femtocell. This way, the whole network ASE can be defined as:

$$ASE = \rho ASE_M + (1 - \rho)ASE_f \tag{6.14}$$

It has also been shown that under the previous assumptions, the different area spectral efficiencies can be estimated using:

$$ASE_M = \frac{T_M}{|\mathcal{H}|} \tag{6.15}$$

$$ASE_f = \frac{N_f \rho_f T_f}{|\mathcal{H}|} \tag{6.16}$$

where T_M is the long term throughput in each subchannel of the macrocell layer, $|\mathcal{H}|$ the coverage area of a macrocell providing service to a region denoted $\mathcal{H}$, N_f the average

[‖] Due to its similarities with the protocol used in ALOHAnet networks for access to the medium, the authors of [29] have named this approach *F-ALOHA*.

number of femtocells per macrocell site, and T_f the expected femtocell throughput in each frequency subchannel. The objective is thus to find out the optimal network configuration such that the area spectral efficiencies are maximized.

From the point of view of the macrocell, [29] analyses and presents two scheduling policies for the allocation of subchannels to the network elements: Round Robin (RR) and Proportional Fair (PF). In the RR scheme, the users of the macrocell are allocated transmission time slots of the same length and always in the same order. Therefore, when the last user has used its first time slot, the transmission right is passed on to the first user and the cycle starts over again. In the proportional fair scheme, the users are scheduled for transmission depending on their throughput needs. Users with a higher transmission rate relative to their mean transmission rate will thus be selected first. Numerical simulations and analytical results show that the proportional fair approach doubles the subchannel throughput T_M in the macrocell layer. Therefore, a PF access relaxes the spectrum needs of the macrocell, when it comes to achieving a given data rate.

From a femtocell standpoint, the problem to be solved consists of finding the optimum ρ_f that reduces the co-layer interference and therefore maximizes the ASE_f. It has been found that, due to raising interference from other femtocells, ASE_f reaches a maximum value $ASE_{f,max}$ at a certain femtocell density. The conclusion is that the ASE_f per subchannel is upper bounded. A direct consequence of this is that, for increasing femtocell densities, the average femtocell throughput** over the whole coverage area $\mathcal{H}$ will grow linearly with $(1 - \rho)$. Further results show that in low femtocell density areas ($N_f \approx 10$) the best approach is to allow the femtocells to access all F_f allocated subchannels, i.e. $\rho_f = 1$. However, at higher femtocell densities, ASE_f reaches $ASE_{f,max}$ at lower values of ρ_f, which need to be determined by simulation.

6.5 Interference Management with UMTS

In December 2008, Femtoforum released a document [7], in which the impact of UMTS femtocells on the macrocell layer is evaluated. Several simulation-based results are presented. However, to avoid providing results that outperform real networks, extreme cases have been chosen in order to study both co-channel and adjacent channel interference. These are summarized in Table 6.2 and are explained in the following sections.

Table 6.2 Interference scenarios analysed by Femtoforum [7]

	Co-channel interference	Adjacent channel interference
Cross-layer	A B C D	G H I J
Co-layer	E F	

** The network-wide femtocell throughput is calculated from Equation (6.16) as $|\mathcal{H}| \cdot W \cdot F \cdot (1 - \rho) \cdot ASE_f$ where W is the bandwidth of each subchannel.

6.5.1 Co-Channel Interference

Scenario A: Macrocell Downlink Interference with the Femtocell User In this scenario an FAP is located close to a window so that it is directly visible from a macrocell. It was verified that if the distance between the UE and the macrocell Node-B is more than 1000 m, an High Speed Downlink Packet Access (HSDPA) throughput of 14.4 Mbps can be reached. However, for shorter distances it is required that FAPs should be closer to the UE. In urban environments with microcells, the same throughput can be reached if the distance between the user and the microcell is less than 250 m. According to statistical data provided by the operators, and taking into account the probability of a femtocell being located at a short distance from the macrocell, it is estimated that the probability of a UE receiving −48 dBm signal from the microcell is 0.01%, which means that the user can still receive service within a few metres of the FAP. Moreover, for users located too close to the microcell it could even be possible to handover to the microcell.

Scenario B: Macrocell Uplink Interference with the Femtocell User An scenario with poor macrocell radio coverage is considered. The FAP is in CSG mode and an outdoors user UE_1 establishes a connection to the macrocell. Another user UE_2, which is a subscriber of the femtocell, performs a call through the FAP, and is interfered with by the uplink of UE_1. It has been thus verified that the femtocell user has enough power to sustain a voice call. For High Speed Uplink Packet Access (HSUPA) the user is required to move closer to the FAP (5 m) in order to reduce the impact of cross-layer interference.

Scenario C: Femtocell Downlink Interference with the Macrocell User A macrocell user is connected to the macrocell network at the cell edge. This user is located in the same room as an FAP to which no access is allowed. The macrocell user is thus interfered in the downlink by the fully loaded FAP. The conclusion is that the use of adaptive power control is necessary to maintain capacity, which can benefit from an increase of up to 16%.

Scenario D: Femtocell Uplink Interference with the Macrocell User A femtocell user is located next to a window that is directly visible to a macrocell located at an approximate distance of 30 m. This femtocell user is at the edge of the range of the femtocell, and is thus transmitting at full power. The analysis with a femtocell user with 1.5 Mbps HSUPA data service has shown that a noise rise of approximately 1.3 dB occurs. However, it is also noted that a macrocell user operating at the same position and on the same service is expected to cause a larger noise rise than the femtocell user. A solution to reduce the uplink interference caused by the femtocell is to allow a femtocell user located at the edge of the femtocell to handover to the macrocell.

Scenario E: Femtocell Downlink Interference with Nearby Femtocell Users In this scenario, two apartments with FAPs (AP_1 and AP_2) are adjacent to each other. The subscriber of AP_2 visits his neighbour's apartment and is on the edge of coverage to his femtocell (AP_2). The owner of $AP1$ establishes a call requiring full power, and that is why the throughput of the femtocell in the downlink is affected by the downlink of neighbouring femtocells. It is concluded that compensation via adaptive power control is required but there will always be situations where a FAP in one apartment will cause dead zones for its neighbour.

Scenario F: Femtocell Uplink Interference with Nearby Femtocell Users Two apartments with FAPs (AP_1 and AP_2) are adjacent to each other. The subscriber of AP_2 visits his neighbour's apartment and is on the edge of coverage of his own femtocell. Then, a user of AP_2 establishes a call through AP_2 while located close to AP_1. It has been verified that the closer the aggressor (e.g. the user of AP_2) is to the victim FAP (e.g. AP_1), the greater the victim's range reduction is. However, if the radiated powers are dynamically optimized, this range reduction can be mitigated.

6.5.2 Adjacent Channel Interference

Scenario G: Macrocell Downlink Interference with the Femtocell User An FAP is located next to a window and is directly visible to a macrocell (approximately 30 m). The macrocell becomes 50% loaded, while a femtocell user is connected to the FAP at the edge of its range. According to the results the femtocell user will be impacted less than 0.01% of the time by the macrocell.

Scenario H: Macrocell Uplink Interference with the Femtocell User A weak signal is received from the macrocell within the apartment where an FAP is located. A user UE_1 with no access to the femtocell is located next to the FAP and performs a call at full power. Simultaneously, another femtocell user UE_2 has an ongoing call at the edge of femtocell coverage. Two cases have been studied: the first assumes that the macro layer is deployed on the adjacent frequency to the femto layer, while the second case considers a 10 MHz separation between the two carriers. It is concluded that if a minimum separation between the macrocell and the femtocell frequencies is not maintained, the femtocell receiver is not able to decode the signal at the required QoS level.

Scenario I: Femtocell Downlink Interference with the Macrocell User Two users, UE_1 and UE_2, are in an apartment containing an FAP. UE_1 is connected to the femtocell at the edge of coverage while UE_2 is connected to the macrocell at the edge of coverage and located next to the femtocell transmitting at full power. Two services, voice (12.2 kbps) and HSDPA (14.4 Mbps) have been tested. In terms of voice service, the femtocell downlink interference can block the macrocell connection if the macrocell user is located close to the macrocell edge. In terms of HSDPA, the performance of the macrocell user at the macrocell edge is not degraded further by this level of downlink interference.

Scenario J: Femtocell Uplink Interference with the Macrocell User The femtocell user is located next to the window, directly visible to a macrocell. The femtocell user is connected to the femtocell at the edge of its range, and is transmitting at full power. The results recommend fixing a maximum allowed transmission power to femtocell users in order to avoid interference. This value should be between 0 dBm and 5 dBm to satisfy coverage requirements.

6.6 Conclusion

This chapter has defined the different types of interference that arise in two-layer networks. These have been further illustrated using simulation results in realistic scenarios for different types of femtocell (CDMA and OFDMA). Then, the effects of interference and their potential solutions have been explained and classified according to the link direction (uplink and downlink). Besides, interference cancellation techniques have also been presented, as well as the current research trends on interference avoidance in femtocell networks.

Although the problem of interference in two-layer networks has been shown to be complex, it is not unsolvable. As seen throughout this chapter, several solutions have already been proposed for different system configurations, thus making femtocells viable and leaving the door open for future approaches that increase the performance of these networks.

References

[1] S.-F. Su, *The UMTS Air-Interface in RF Engineering: Design and Operation of UMTS Network*. McGraw-Hill Professional, May 2007.

[2] H. Holma and A. Toskala, *WCDMA for UMTS*, 3rd ed. John Wiley & Sons, 2004, ch. 3, p. 51.

[3] 3GPP, '3G Home NodeB Study Item Technical Report,' 3rd Generation Partnership Project – Technical Specification Group Radio Access Networks, Valbonne (France), Tech. Rep. 8.2.0, Sep. 2008.

[4] R. S. Karlsson, 'Radio resource sharing and capacity of some multiple access methods in hierarchical cell structures,' in *Vehicular Technology Conference (VTC)*, vol. 5, Sep. 1999, pp. 2825–2829.

[5] M. Fan, M. Yavuz, S. Nanda, Y. Tokgoz, and F. Meshkati, 'Interference Management in Femto Cell Deployment,' in *3GPP2 Femto Workshop*, Oct. 2007.

[6] Ericsson, 'Simulation results for Home NodeB uplink performance in case of adjacent channel deployment within the block of flats scenario,' 3GPP TSG-RAN Working Group 4 (Radio), Sorrento (Italy), Tech. Rep. R4-080152, Feb. 2008.

[7] 'Interference Management in UMTS Femtocells,' Femto Forum, Dec. 2008.

[8] S. Lee, 'An Enhanced IEEE 1588 Time Synchronization Algorithm for Asymmetric Communication Link using Block Burst Transmission,' *IEEE Communications Letters*, vol. 12, no. 9, pp. 687–689, Sep. 2008.

[9] Ericsson, 'Home NodeB maximum output power from the maximum UE input level point of view,' 3GPP TSG-RAN Working Group 4 (Radio), Sorrento (Italy), Tech. Rep. R4-080155, Feb. 2008.

[10] G. Mansfield, 'Femtocells in the US Market–Business Drivers and Consumer Propositions,' in *FemtoCells Europe 2008*. ATT, Jun. 2008.

[11] *Analysis of uplink and downlink capacities for two-tier cellular system*. Institution of Electrical Engineers, Dec. 1997.

[12] V. Chandrasekhar and J. G. Andrews, 'Uplink Capacity and Interference Avoidance for Two-Tier Femtocell Networks,' *IEEE Transactions on Wireless Communications*, February 2008.

[13] M. Beach, *et al.*, 'A Study into the Application of Interference Cancellation Techniques,' Ofcom, Tech. Rep., Apr. 2006, Report No: 72/06/R/038/U.

[14] D. Mottier and L. Brunel, 'Iterative space–time soft interference cancellation for UMTS-FDD uplink,' *IEEE Transactions on Vehicular Technology*, vol. 52, no. 4, pp. 919–930, Jul. 2003.

[15] J. G. Proakis, 'Interference suppression in spread spectrum systems,' in *International Symposium on Spread Spectrum Techniques and Applications*, vol. 1, Sep. 1996, pp. 259–266.

[16] J. Schodorf and D. Williams, 'A constrained optimization approach to multiuser detection,' *IEEE Transactions on Signal Processing*, vol. 45, no. 1, pp. 258–262, Jan. 1997.

[17] J. Holtzman, 'DS/CDMA successive interference cancellation,' in *IEEE Third International Symposium on Spread Spectrum Techniques and Applications*, vol. 1, Jul. 1994, pp. 69–78.

[18] M. Munster and L. Hanzo, 'Co-channel interference cancellation techniques for antenna arrayassisted multiuser OFDM systems,' in *First International Conference on 3G Mobile Communication Technologies*, Mar. 2000, pp. 256–260.

[19] G. Gelli, L. Paura, and A. Tulino, 'Cyclostationarity-based filtering for narrowband interferencesuppression in direct-sequence spread-spectrum systems,' *IEEE Journal on Selected Areas in Communications*, vol. 16, no. 9, pp. 1747–1755, Dec. 1998.

[20] A. Shah and A. Haimovich, 'Performance analysis of maximal ratio combining and comparison with optimum combining for mobile radio communications with cochannel interference,' *IEEE Transactions on Vehicular Technology*, vol. 49, no. 4, pp. 1454–1463, Jul. 2000.

[21] A. Mostafa, 'Single antenna interference cancellation (SAIC) method in GSM network,' in *IEEE Vehicular Technology Conference (VTC-Fall)*, vol. 5, Sep. 2004, pp. 3748–3752.

[22] W. Webb, *Wireless Communications: The Future*. John Wiley & Sons, 2007, ch. 6, pp. 73–74.

[23] S. P. Weber, J. Andrews, X. Yang, and G. de Veciana, 'Transmission capacity of wireless ad hoc networks with successive interference cancellation,' *IEEE Transactions on Information Theory*, vol. 53, no. 8, pp. 2799–2814, Aug. 2007.

[24] ip.access, 'Oyster 3g: The access point,' http://www.ipaccess.com/femtocells/oyster3G.php, 2007.

[25] A. Hassan and J. Andrews, 'The guard zone in wireless ad hoc networks,' *IEEE Transactions on Wireless Communications*, vol. 6, no. 3, pp. 897–906, Mar. 2007.

[26] M. Latham, 'Consumer attitudes to femtocell enabled in-home services–insights from a European survey,' in *Femtocells Europe 2008*, London, June 2008.

[27] D. López-Pérez, A. Jüttner, and J. Zhang, 'Optimisation methods for dynamic frequency planning in ofdma networks,' in *Networks 2008*, Budapest, Hungary, September 2008.

[28] D. Williams, 'WiMAX Femtocells A Technology On Demand for Cable MSOs,' in *Femtocells Europe 2008*, Jun. 2008.

[29] V. Chandrasekhar and J. G. Andrews, 'Spectrum Allocation in Two-Tier Networks,' *IEEE Transactions on Communications*, 2009. [Online]. Available: http://arxiv.org/abs/0805.1226.

[30] M. S. Alouini and A. J. Goldsmith, 'Area spectral efficiency of cellular mobile radio systems,' *IEEE Transactions on Vehicular Technology*, vol. 48, no. 4, pp. 1047–1066, Jul. 1999.

7

Mobility Management

Hui Song and Jie Zhang

7.1 Introduction

Femtocells are designed to produce extended coverage and enhanced capacity, especially in the indoor environment. Apparently, no specific Mobility Management (MM) for femtocells is necessary since they are expected to work within the existing network standards. However, the real implementation breaks many aspects of the current network assumptions. This is mainly due to the following factors:

- Large number and high density of femtocells
  Considering the large number of small sized femtocells within macrocell coverage, current neighbour cell list (32 maximum) is not large enough to include all the femtocell information. Besides, there are only 512 Primary Scrambling Codes PSC in UMTS and 504 Physical Cell Identities (PCI) in Long Term Evolution (LTE) shared within the network. It may not be sufficient to distinguish the cell identity of all the femtocells as in macrocells. It is unlikely to be scalable to broadcast the cell information of femtocells in the network, which will definitely increase much network signalling overhead.
- Dynamic neighbour cell lists
  The consumer can either add or remove a femtocell. The femtocells can be powered either on or off and their locations can be quite dynamic. The neighbour cell list is not stationary compared with those of the macrocells.
- Variant access methods
  A femtocell can give open, semi-open or fully closed access, which is configurable by the operator through the backhaul or by the user under the supervision of the operator. It is not attractive for the hundreds or thousands of users served by a macrocell doing cell measurements on femtocells that they may not have authorized access to.

Femtocells: Technologies and Deployment Jie Zhang and Guillaume de la Roche

- User/operator preference
  For the UE approaching the coverage of an allowed femtocell, the user prefers to assign the highest priority to the femtocell for a better signal quality and cheaper billing package. The operator will also be happy to reduce the loading from the macrocell. In the case when the femtocells are forbidden, the user prefers to stay with the macrocell to avoid call drop or disconnection from the network. The operator also wants to exclude the mobility towards femtocell to reduce the signalling overhead and user complaints.

Overall, MM is one of the most challenging issues when deploying femtocells. In fact, MM is a very important feature for mobile power consumption and signalling load reduction. Femtocell technology has hit a critical milestone in 3GPP with the approval of specifications for UMTS and LTE. This chapter focuses on the common mobility procedures for femtocells in the ongoing new releases (release 8 and release 9) of UMTS and LTE. A description of mobility procedures for femtocells in the current UMTS network (pre-release 8) is also given.

Section 7.2 gives an introduction of 3rd Generation Partnership Project (3GPP) femtocell specification progress. In this section, ways to read report and specification on MM for femtocells are also given. Section 7.3 describes several types of femtocell identifiers that characterize the features of femtocells from different aspects. Section 7.4 deals with the access control of an UMTS/LTE network in which femtocells are introduced. The different access methods for the femtocells are described. In Section 7.5 some problems of current paging procedure caused by femtocells are addressed, and then some advanced paging procedures that consider the femtocell features are discussed. In Section 7.6 cell selection and reselection methods are described in detail. In Section 7.7, the issues of handover are discussed, and the advantages and disadvantages of the techniques used in femtocell networks for handover are presented.

7.2 Mobility Management for Femtocells in 3GPP

In the past few years, all of the femtocell vendors have made their own efforts with different structures and methods to fix the femtocell into the current network (UMTS, CDMA) or Wireless Interoperability for Microwave Access (WiMAX), etc.). Continuing on this path would result in over-hyped technology solutions that would make it difficult to inter-operate with each other and keep the cost of femtocells at a reasonable level. An industry wide standardization becomes essential to enable the widespread adoption and deployment of femtocells by telecom operators around the world.

3GPP started to work on a femtocell standard (including technical report (TR) and technical specification (TS)) in early 2008. Up-to-now, quite a number of proposals on femtocell architecture, mobility management and network security have been proposed. With the freezing of release 8, some basic functionalities in supporting the MM for femtocells have been achieved. However, to meet the restricted time line of release 8, only basic functionalities of MM for femtocells with closed access (known as CSG cells) are considered in release 8. Things like handover from macrocell to femtocell, handover between femtocells and supporting open and hybrid access method are still ongoing and will be handled in release 9. In this section, 3GPP specifications related to the MM and ways to read this information will be introduced.

3GPP has different Working Groups (WGs) working on different aspects of the radio interface and network architecture. The Technical Specification Group (TSG) Radio Access Network (RAN) is where all the radio air interface specifications are handled and contains six WGs as described in the following list:

- RAN WG1 (RAN1) Radio layer 1 (Physical layer) specification,
- RAN WG2 (RAN2) Radio layer 2 (MAC, RLC, PDCP) and Radio layer 3 RR,
- RAN WG3 (RAN3) Iu, Iub, Iur, S1, X2 and UTRAN/E-UTRAN architecture,
- RAN WG4 (RAN4) Radio performance and protocol aspects RF parameters and BS conformance,
- RAN WG5 (RAN5) Mobile terminal conformance testing,
- RAN AHG1 (RAN_AH1) Ad-hoc group coordinating communications between 3GPP and ITU-R.

There are a couple of specifications related to MM available in both release 8 and release 9. The following UMTS/LTE standards and their corresponding WGs, which involve the mobility procedures with Home NodeB (HNB)/Home eNodeB (HeNB) are listed in detail in Tables 7.1 and 7.2.

Femtocells are quite new in the 3GPP standardization process, although many mobility procedures for femtocells have been addressed. Things like how these procedures actually work and their criteria are still under discussion and further studies are needed. One way to find a more detailed and ongoing work on the specific methodology or parameter in the specification process is to go through the contribution papers (called technical documents, or tdocs) that members of 3GPP upload on their meeting website.

Most of the MM procedures (access control, cell selection/reselection and handover, etc.) are discussed in RAN2 meetings. Paging procedure and access control are also discussed in RAN3. After TSG-RAN Meeting #43 [1], the following mobility procedures are also within the scope of RAN3.

- Enhanced handover scenario:
  - in-bound mobility 3G macro to HNB handover including legacy UE aspects,
  - 3G HNB to 3G HNB handover,
  - in-bound mobility LTE macro to HeNB handover,
  - LTE HeNB to LTE HeNB handover.
- Enhanced access scenarios:
  - open access,
  - hybrid access.

For example, the in-bound handover procedure from macrocell to femtocell is not given in any current specification. By searching through the meeting reports, discussions on hand-in procedure in RAN 2 meeting #65bis and RAN 3 meeting #63bis can be found. There are quite a few related documents (R2-092404, R3-090804, etc.) that are listed in the corresponding meeting report. By looking at these documents, one can understand more about in-bound handover problems that are stated and mechanisms that are under investigation.

Table 7.1 Mobility management in 3GPP standardization

NO		Group	Version	Date	Rel.	Title	Description
3GPP 22.011	TS	SA1	V9.1.0	2009–3	9	Service accessibility	It lists the basic mobility functionality (including access control, cell selection/reselection) for the support of Home NodeB and Home eNodeB.
3GPP 22.220	TS	SA1	V9.0.0	2009–3	9	Service requirements for Home NodeBs and Home eNodeBs	It lists different requirements for both Home NodeB and Home eNodeB.
3GPP 23.830	TR	SA2	V0.3.1	2009–3	9	Architecture aspects of Home NodeB and Home eNodeB	Different requirements and solutions (including mobility procedures) for supporting Home NodeB and Home eNodeB are discussed.
3GPP 25.304	TS	RAN2	V8.5.0	2009–3	8	User Equipment (UE) procedures in idle mode and procedures for cell reselection in connected mode	The mobility procedures with Home NodeBs are listed for UTRA system.
3GPP 25.331	TS	RAN2	V8.6.0	2009–3	8	Radio Resource Control (RRC), protocol specification	Some Home NodeB featured system parameters are introduced to support/optimize the mobility procedures with Home NodeB.
3GPP 36.902	TS	RAN3	V1.0.1	2008–9	8	Self-configuring and self-optimizing network use cases and solutions	

Table 7.2 Mobility management in 3GPP standardization

NO		Group	Version	Date	Rel.	Title	Description
3GPP 25.367	TS	RAN2	V8.1.0	2009–3	8	Mobility procedures for home NodeB, overall description Stage 2	It gives an overall description of the mobility procedures supported in release 8.
3GPP 25.467	TS	RAN3	V8.1.0	2009–3	8	UTRAN architecture for 3G Home NodeB stage 2	It covers specification of the functions for UEs not supporting Closed Subscriber Groups (CSG) (i.e. pre-Rel-8 UEs) and UEs supporting CSGs. It also covers Home NodeB specific requirements for Operations and Maintenance (O&M).
3GPP 25.820	TR	RAN4	V8.2.0	2008–9	8	3G Home NodeB study item technical report	Contains study of items from RAN2, RAN3 and RAN4 point of view.
3GPP 36.300	TS	RAN2	V8.7.0	2009–1	8	Evolved Universal Terrestrial Radio Access (E-UTRA) and Evolved Universal Terrestrial Radio Access Network (E-UTRAN), overall description Stage 2	It defines the user and control interference between HeNB and HeNB-GW. In Appendix F, it discusses mobility and access control requirements associated with Closed Subscriber Group (CSG) Cells.
3GPP 36.304	TS	RAN2	V8.5.0	2009–3	8	User Equipment (UE) procedures in idle mode	The mobility procedures with Home NodeBs are listed for E-UTRA system.
3GPP 36.331	TS	RAN2	V8.5.0	2009–3	8	Radio Resource Control (RRC), protocol specification	Some Home eNodeB featured system parameters are introduced to support/ optimize the mobility procedures with Home eNodeB.

7.3 Femtocell Characterization

As illustrated by problems addressed in Section 7.1, it is very difficult or even infeasible to treat femtocells as normal macrocells. The network cannot afford broadcasting the femtocell information over the network, and the signalling overhead between the macrocell users and femtocells will have a big impact on the performance of the network. The identifiers and mechanisms that can characterize the aspects of the femtocells are necessary in reducing the impact and enhancing the mobility procedures of supporting femtocells.

7.3.1 Distinguish Femtocell from Macrocell

With the help of the identifiers and mechanisms that can distinguish femtocells from macrocells, the whole network can be treated as a two-layer network: macro-network and femto-network as shown in Figure 7.1. A few efforts can then be made to enhance the mobility procedures and reduce the signalling overhead between these two layers. For example, new reselection parameters can be configured between these two layers to prioritize the femtocell users to camp on once they are entering the coverage of the allowed femtocell. Also, these parameters help in preventing the users that are not allowed to use femtocell from scanning or reading the information from the femtocell layer, leading to a longer battery life. The proposed methods that have been discussed in 3GPP are listed below.

Hierarchical Cell Structure

Hierarchical Cell Structure (HCS) can be used as an additional means of better distinguishing between femtocells and macrocells [2]. In GSM and UMTS, HCS is considered as one of the most effective means of solving conflict between continuous coverage and high traffic demand in hot spots. HCS allows network operators to cut their network cells into different categories such as macrocell, microcell and picocell. Each layer can be assigned a different priority (i.e. 0–7). As an naturalized extension, Figure 7.2 shows that

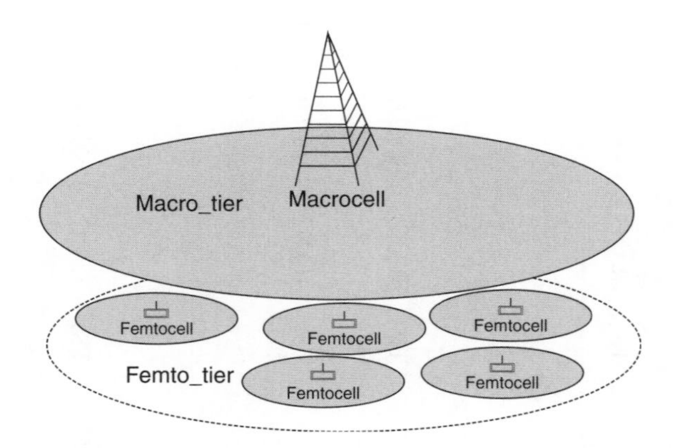

Figure 7.1 Two-layer network

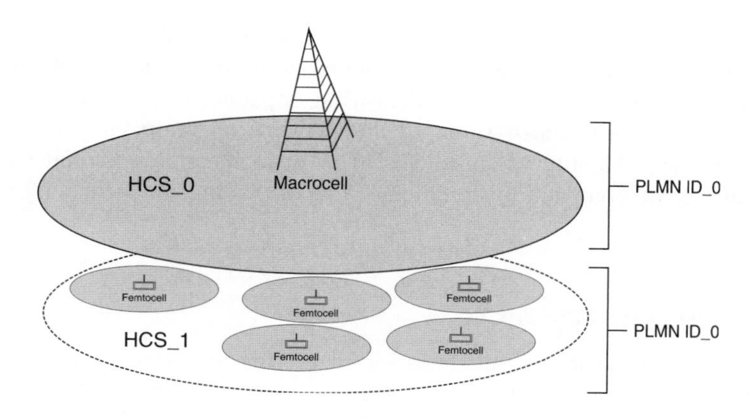

Figure 7.2 Hierarchical cell structure

HCS can be also adopted to allow different cell reselection rules for femtocells. Using HCS does not need any modification of the current technical specification.

Separate Femtocell PLMN ID

Another method of distinguishing femtocells from macrocells is to have a separate femtocell PLMN Identity (PLMN ID) that is distinct from that of the macrocells [2]. In this approach, femtocells will be assigned a different PLMN ID from macrocells to secure femtocell selection and minimize impact on macrocell users as shown in Figure 7.3. UEs that are not allowed to use a femtocell could be configured not to access the femtocell Public Land Mobile Network (PLMN), resulting in better battery performance, and less signalling load towards the Core Network (CN). A separate PLMN ID allows the UE to display the right network identifier, indicating to the user that he has camped on a

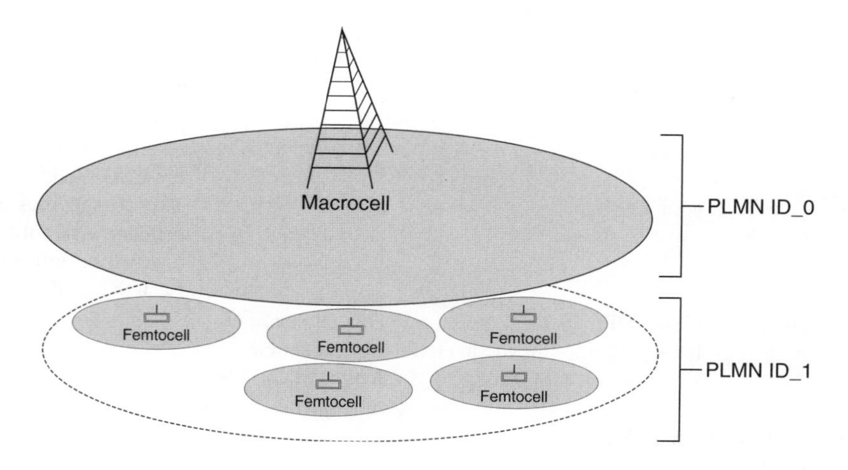

Figure 7.3 Separate femtocell PLMN ID

femtocell. As the straightforward approach for the pre-release 8 network, this method is currently used by many of the femtocell vendors as the main method of distinguishing femtocells from macrocells. However, when using separate femtocell PLMN, additional PLMN IDs are required, which some of the network operators may not have. Also, some old Subscriber Identity Module (SIM)/Universal Integrated Circuit Card (UICC) card may have some compatibility problems with displaying the right femtocell PLMN name.

Reserve Frequency and PSC/PCI

In 3GPP release 8, more specific methods for identifying femtocells have been proposed. To simplify the problem, only CSG cells are considered. As the layer 1 approach, frequency and PSC reservation for UMTS and PCI reservation for LTE have been discussed [3–5]. In UMTS, reserved frequency and PSC have been adopted [6]. In LTE, a set of parameters indicating the reserved PCI range for CSG cells is introduced [7]. Macrocells and CSG cells may broadcast indications of one or more carrier frequencies used for dedicated CSG deployment. This information may be used by the UE with no access to femtocells to avoid unnecessary measurements on that frequency, even when rules would require measurements of this carrier frequency. In UMTS, indications of which carrier frequencies are dedicated to CSG-only deployment may be signalled in System Information Block (SIB)11bis as shown in Table 7.3. Figure 7.4 shows how macrocells and femtocells are separated from each other by assigning dedicated frequencies to femtocells.

Table 7.3 Reserved frequency and PSC in release 8 [6]

Block	Information element/group name	Need	Multi	Type and ref.	Semantics description
SIB3	CSG identity	OP		CSG identity 10.3.2.8	
SIB3	CSG PSC split information	OP			This IE specifies the primary scrambling code reservation information for CSG Cells.
SIB11bis	CSG PSC split information	OP			This IE specifies the primary scrambling code reservation information for CSG Cells.
SIB11bis	Dedicated CSG frequency list	OP	1 to <maxdedicated-CSGFreq>	This IE specifies the frequencies dedicated for CSG cells only.	
SIB11bis	Dedicated CSG frequency	MP			

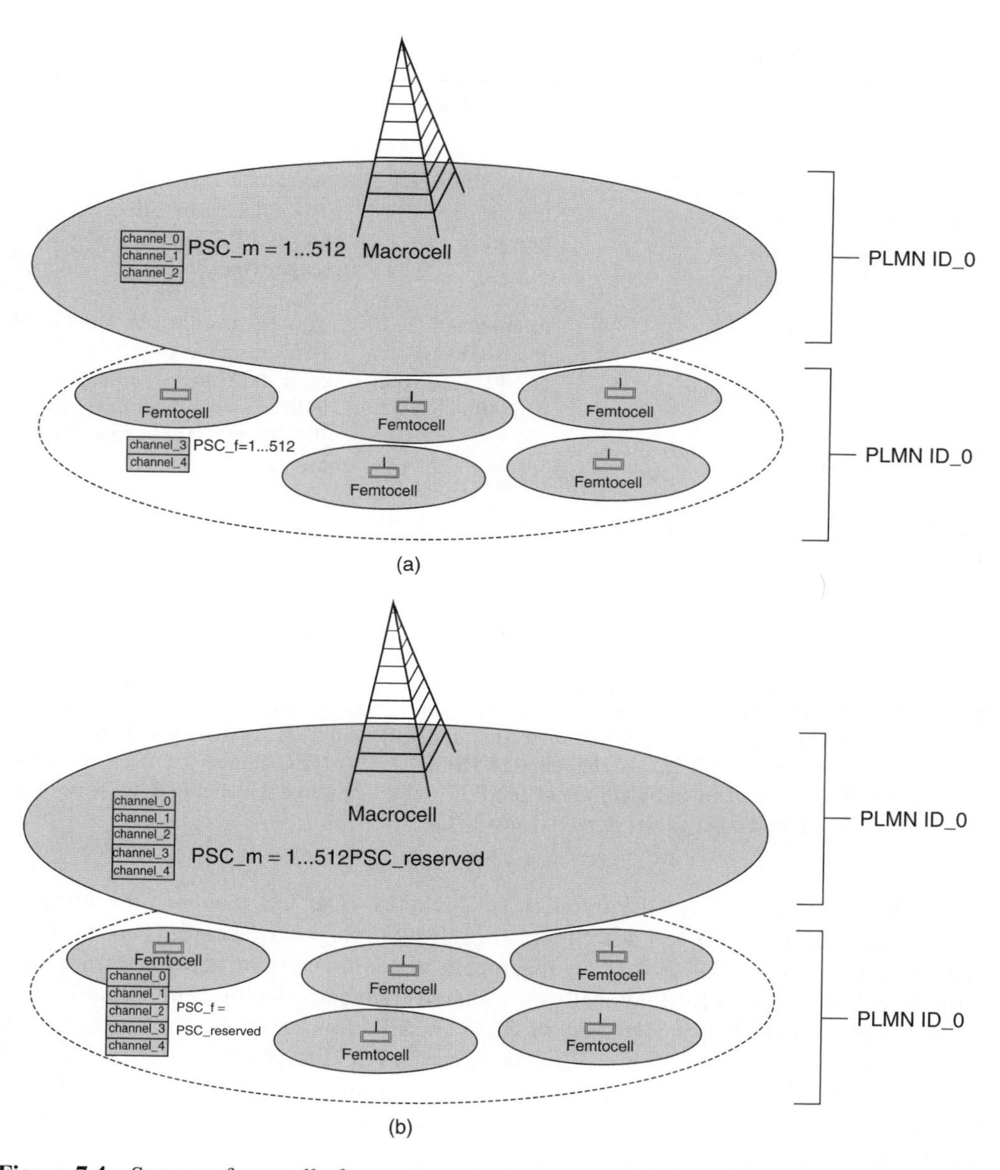

Figure 7.4 Separate femtocells from macrocells by means of dedicated frequencies and reserve PSCs. (a) Dedicated frequency for femtocell. (b) Mixed frequency with reserved PSC for femtocell

With shared carrier frequency deployment, all CSG cells shall broadcast the reserved PSC range in their system information. The non-CSG cells may also optionally broadcast the reserved PSC range. The reserved PSC range is only applicable to the UTRA Absolute Radio Frequency Channel Number (UARFCN) within the PLMN where the UE receives this information. The UE considers the last received reserved PSC range to be valid

Table 7.4 CSG PSC split information in 3GPP release 8 [6]

Information element/group name	Need	Multi	Type and reference	Semantics description
Start PSC	MP		Integer (0.504 by step of 8)	The value of this IE specifies the start PSC of the first PSC range (NOTE 1).
Number of PSCs	MP		Enumerated (5, 10, 15, 20, 30, 40, 50, 64, 80, 120, 160, 256, alltheRest)	This IE specifies the number of PSCs reserved for CSG cells in each PSC range. 'alltheRest' indicates all values from Start PSC to 511. Three spare values are needed.
PSC Range 2 Offset	CV-alltheRest		Integer (8.504 by step of 8)	If this IE is included, the UE shall calculate the second PSC range (NOTE 2). If this IE is not included, the UE shall consider the second PSC range to be not present.

NOTE 1: Let the IE 'Start PSC' = s. and 'Number of PSCs' = n. The complete set of (n) PSC values in range 1 is defined as: s, $((s+1) \bmod 512)$, $((s+2) \bmod 512) \ldots ((s+n-1) \bmod 512)$.
NOTE 2: Let the IEs 'Start PSC' + 'Number of PSCs' − 1 + 'PSC Range 2 Offset' = s, and 'Number of PSCs' = n. The complete set of (n) PSC values in range 2 is defined as: s, $((s+1) \bmod 512)$, $((s+2) \bmod 512) \ldots ((s+n-1) \bmod 512)$.

within the entire PLMN for the duration of 24 hours. The UE may use the reserved PSC information for CSG cell search and (re)selection purposes, according to the UE's implementation. Figure 7.4 shows how macrocells and femtocells are separated from each other by reserving PSCs for femtocells on a shared frequency deployment scenario.

The reserved PSC list is signalled in SIB3 or SIB11bis as shown in Table 7.3. A macrocell will use SIB11bis to broadcast the reserved PSC list, if it broadcasts it. A CSG cell will use SIB3 to broadcast it. A cell should not broadcast the reserved PSC list in both SIB3 and SIB11bis. The detailed parameters and algorithms for calculating the PSC split information are given in Table 7.4.

For LTE, the PCI range information (known as csg-PhysCellIdRange) is signalled in SIB4. This field indicates the set of PCIs reserved for CSG cells on the frequency where this was received. The received csg-PhysCellIdRange applies if less than 24 hours has elapsed since it was received in the same primary PLMN.

CSG Indicator

As a layer 2 approach in 3GPP release 8, CSG Indicator is introduced in [6] and [7] to indicate whether the cell is a CSG cell or not. The information CSG indicator is signalled

Table 7.5 CSG indicator in UMTS release 8 [6]

Block	Information element/group name	Need	Multi	Type and reference	Semantics description
MIB	CSG Indicator	OP		Enumerated (TRUE)	If present, the cell is a CSG cell. Default value is 'FALSE'

Table 7.6 CSG indicator in LTE release 8 [7]

Block	Information element/group name	Need	Multi	Type and reference	Semantics description
SIB1	CSG-Indicator	MP		Enumerated (TRUE)	If set to TRUE the UE is only allowed to access the cell if the CSG identity matches an entry in the allowed CSG cell list that the UE has stored.

in Master Information Block (MIB) and SIB1 in Tables 7.5 and 7.6 for UMTS and LTE respectively.

7.3.2 Find Neighbouring Femtocells

By means of the above parameters or mechanisms, the UE is able to distinguish femtocells from macrocells. During inter-cell mobility, it is very easy for the UE to know the surrounding macrocells using the Neighbour Cell List (NCL). However, it remains a problem for the UE to find the neighbouring femtocells due to the difficulty stated in the introduction. Efforts made to enable UE to find femtocells are listed below.

Neighbour Cell List

For outbound mobility from a femtocell to a macrocell, the NCL in the femtocell is a straightforward way for the UE to be aware of the surrounding macrocells and femtocells. The NCL can be created by the femtocell through self-configuration algorithms implemented by the vendors. The NCL may be updated each time the femtocell is powered on or when the femtocell senses any change of the neighbouring cells.

For inbound mobility from macrocells, it is infeasible to include all of the neighbouring femtocells in the list due to the high volume of femtocells. However, alternative ways have been made by some vendors to support high density femtocells in pre-release 8 network. The main idea is based on the reuse of a certain number of PSCs (i.e. 10). These PSCs can then be programmed into the macrocell's neighbour cell list. When such an NCL is received from the serving macrocell, the UE can then do measurements for these femtocells corresponding to the PSCs as for macrocells. The drawback is that the NCL of macrocells needs to be updated and PSC confusion may happen when the UE detects two nearby femtocells with the same PSC [8].

UE Autonomous Search

In 3GPP release 8, UE autonomous search function is introduced for the UE to find the CSG cells. The UE is required to perform an autonomous search function in order to detect suitable CSG cells [9, 10]. The UE, which cannot access CSG cells, can disable the autonomous search function for CSG cells. If 'Dedicated CSG frequency(ies)' information element (IE) is present, the UE may use the autonomous search function only on these dedicated frequencies and on the other frequencies listed in the system information. To assist the search function on shared carriers, UE may search cells with reserved PSCs/PCIs defined by CSG PSC/PCI Split Information for intra-frequency and inter-frequency measurements and mobility purposes.

The UE autonomous CSG search function is left as a UE implementation issue. In other words, the specification of when and where to trigger the UE autonomous search function is not specified and should be defined by the mobile vendors. Currently, the UE autonomous search function is supposed to find CSG cells only. It will be extended to support searching for femtocells with various access methods (open and hybrid) in release 9.

7.3.3 Distinguish Accessible Femtocell

When using the parameters and mechanisms that can distinguish the femtocell from macrocells, macrocell users can avoid trying to access the femtocell and femtocell users can prioritize the femtocell over macrocells. This works fine for the femtocells of open access type but will not work well with the closed access type. All of the femtocell users will try to camp on the target femtocell regardless of whether they have the authority to access it or not. By using the methods that can distinguish whether the given femtocell is accessible or not, unnecessary signalling overhead can be avoided.

Location Area Identity/Tracking Area Identity

In pre-release 8 UMTS, a set of neighbouring macrocells can share the same Location Area Identity (LAI). However, for femtocells the LAI/Tracking Area Identity (TAI) of neighbours will need to be different for user access control purposes. The LAIs of unauthorized femtocells will be put in the UE's Universal Subscriber Identity Module (USIM) after it receives the Location Area Update (LAU) rejection from these femtocells. The UE will not be trying to camp on the femtocell if the LAI is in the forbidden list. This results in lower power consumption and lower signalling overheads. However, if each femtocell is assigned a unique LAI, the solution will become infeasible for the current CN nodes. Repeating LAI across geographically diverse femtocells or using pseudo LAI has repercussions on network performance [11]. Besides, the LAI conflict is another non-negligible problem. For example, if someone's femtocell has the same LAI as the UE's allowed femtocell and is not allowed on theirs. When the user walks by, the UE will try to camp on that cell and it will receive the LAU rejection (#15 'No Suitable Cells In Location Area' [12]). The UE then record that cell as 'forbidden' and keeps track of the LAI in its 'not permitted list'. When the user reaches his home, he/she might not be able to get onto his/her own femtocell. Another problem is that when a user is passing his/her neighbour's department, the UE may be rejected due to the forbidden LAI. If his own femtocell is on the same frequency layer, the user has to wait 300 s before it can reselect the femtocell as defined in pre-release 8 specification.

Closed Subscription Group ID

One or more CSG cells are identified by a unique numeric identifier called CSG Identity (CSG ID). The CSG ID shall have a fixed value of 27 bits as defined in [13]. The CSG ID is broadcast in SIB3 in UMTS by the CSG cell as shown in Table 7.7 and in SIB1 in LTE as shown in Table 7.8. When the UE is not authorized to access the target femtocell, a new reject cause #25 'Not authorized for this CSG Cell' [12] is used. The release 8 UE will then bar the corresponding CSG ID for a configurable duration instead of the whole frequency. In contrast to pre-release 8 procedures, the UE will neither add the Location Area Code (LAC) to the forbidden LAI list nor bar the entire frequency. Barring only the CSG ID rather than the entire frequency prevents possible service outage when alternative coverage is available on the same frequency. Not forbidding the LAI avoids the difficulties in the femtocell (re)selection due to LAI conflicts.

Table 7.7 CSG identity in UMTS release 8 [6]

Block	Information element/group name	Need	Multi	Type and reference	Semantics description
SIB3	CSG Identity	OP		CSG Identity Bit string(27)	

Table 7.8 CSG identity in LTE release 8 [7]

Block	Information element/group name	Need	Multi	Type and reference	Semantics description
SIB1	csg-Identity	OP		CSG Identity bit string(27)	Identity of the closed subscriber group within the primary PLMN to which the cell belongs. The IE is present in a CSG cell.

7.3.4 *Handle Allowed List*

When a UE trying to camp on a femtocell, either an allowed International Mobile Subscriber Identity (IMSI) list of the femtocell or an allowed femtocell list of the UE is essential in order to check if the UE is allowed to access the target femtocell. In pre-release 8, such a list is organized as femtocell based (allowed IMSI list), which may be handled in femtocell or femtocell-GW. While in release 8, the list is organized as subscriber based (allowed femtocell list) and it is handled in the CN.

In this section, both of the lists and how they are handled in the corresponding network entity will be introduced.

Allowed List in Femtocell/Femtocell-GW

In pre-release 8 UMTS, the allowed list is stored in femtocell or femtocell-GW locally. In this case, each femtocell has its own allowed list that contains the list of IMSIs or Mobile Subscriber Integrated Services Digital Network Numbers (MSISDNs) that are allowed access. The allowed list is managed by either the operator or the owner of the femtocell under the supervision of the operator (i.e. on a secured webpage).

The advantage is that during the UE registration procedure, the unauthorized mobiles will be rejected without sending signalling traffic to the CN as shown in Figure 7.5. This implementation does not need to carry out any modification on the CN to support femtocells.

However, IMSI/MSISDN is generally available only in the Non-Access Stratum (NAS) to protect user identity confidentiality. To store a permanent identifier such as the MSISDN or IMSI locally may become a security concern. In addtion, when managing subscribers for a campus or an enterprise, deployment with many femtocells would introduce substantial overhead and management.

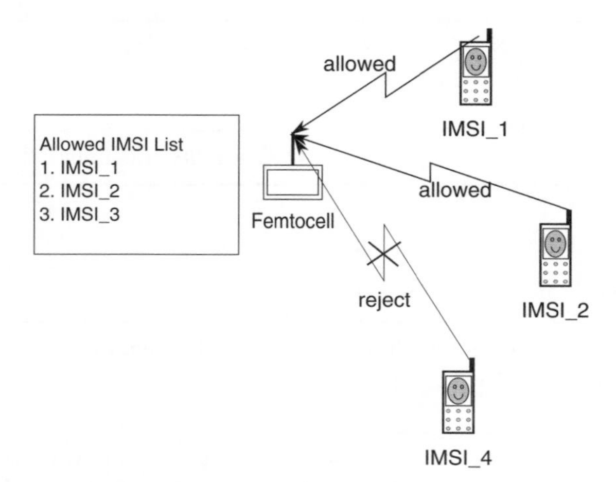

Figure 7.5 Allowed IMSI list stored in femtocell

Allowed List in CN

In 3GPP release 8, an allowed list is provided as a UE's Allowed CSG List (ACL). It lists the corresponding CSG IDs of the femotcells that the UE belongs to. It is agreed that the ACL is stored with user's subscribe information in CN [9, 10]. In the NAS, a CN entity such as Mobile Switching Centre (MSC)/Visitor Location Register (VLR), Serving GPRS Support Node (SGSN) and Mobility Management Entity (MME) keeps the context for the UEs that are currently being served. The UE context may be lost when the UE is deregistered. Thus it is not suitable to store the permanent information such as the UE's ACL. Additionally, the ACL may share CSG IDs across multiple 3GPP access technologies such as Evolved UTRAN (EUTRAN), UMTS Terrestrial Radio Access Network (UTRAN) and GSM EDGE Radio Access Network (GERAN), and should be available to the relevant network elements in these networks. Therefore, the best location for storing the CSG subscription information is at a network element similar to the Home Subscriber Server (HSS). A copy of the ACLs should be stored at the MME/SGSN for the currently serving UEs as shown in Figure 7.6. In addition, for a CSG capable UE, it may keep a copy of its own ACL in the USIM. This list may help to avoid the unnecessary signalling towards an unauthorized femtocell. The usage of the ACL in the idle mode and the corresponding MM procedures are defined in [12–14] and [15].

The synchronization of the ACL in CN and UE may be realized under either an automatic or a manual basis. During the automatic synchronization, while changing the CSG ID of the femtocell or removing a CSG ID from a UE's subscription, the network will update the ACL in CN first and then will send a message to inform the UE to do the ACL update automatically as shown in Figure 7.7.

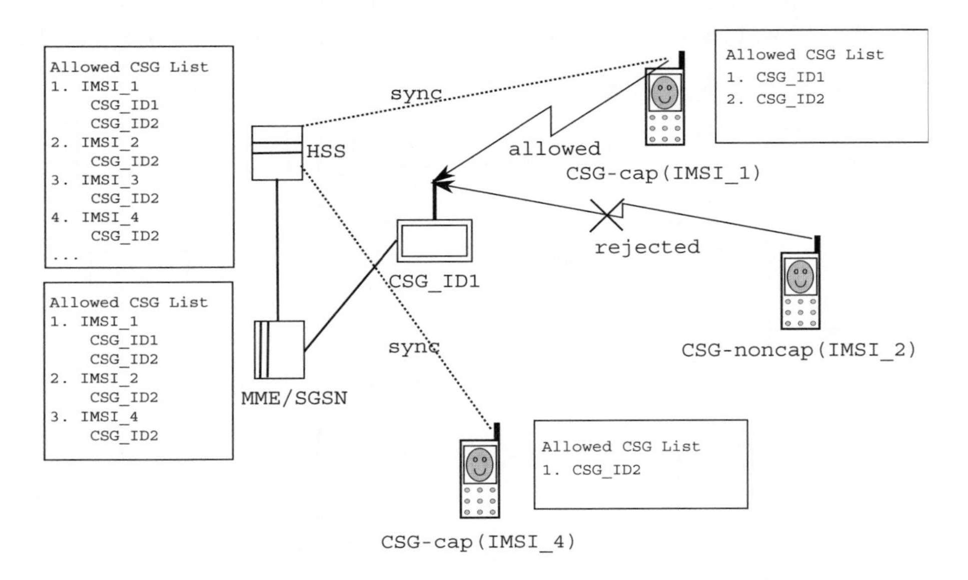

Figure 7.6 Allowed CSG list managed by CN

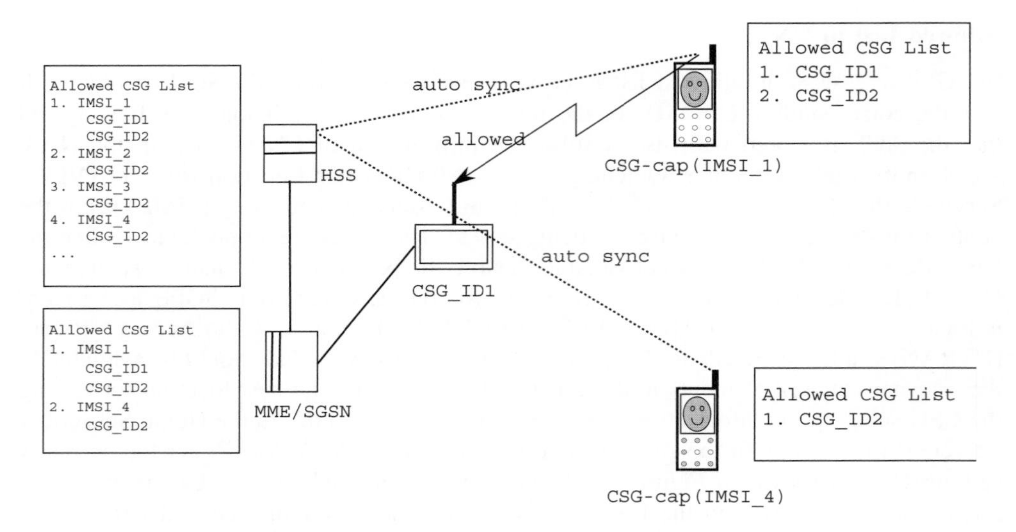

Figure 7.7 Automatic mode of allowed CSG list synchronization

In manual mode, the UE at first will not have a synchronized ACL with CN. When an allowed CSG cell changes its CSG ID and the new CSG ID is not in the UE's ACL, the UE will only select that CSG cell by manual cell selection. It will add the new CSG ID into its ACL after receiving a positive LAU/Tracking Area Update (TAU) feedback. When removing a CSG ID in the subscriber's ACL in CN, without knowing the change, the UE will still try to camp on the femtocell with that CSG ID. After receiving a LAU/TAU rejection, the UE will remove that CSG ID from its own ACL as shown in Figure 7.8.

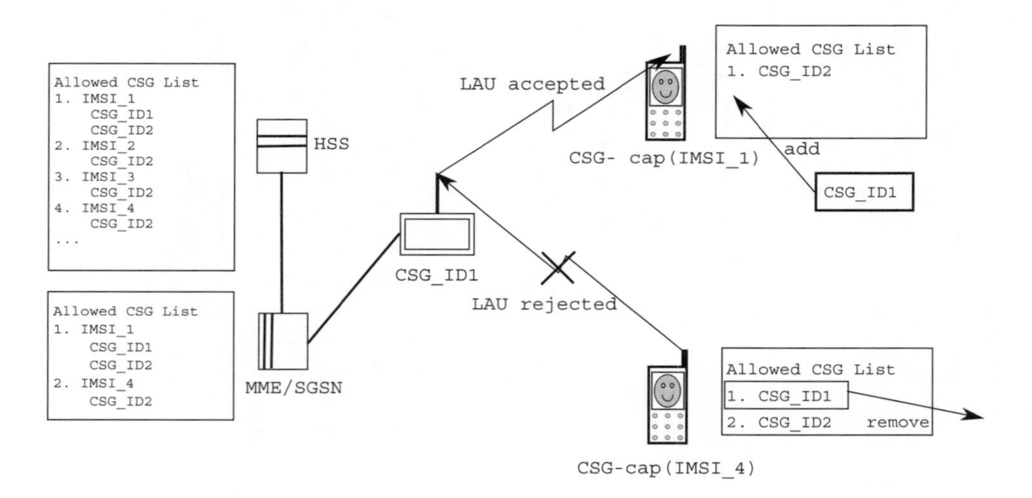

Figure 7.8 Manual mode of allowed CSG list synchronization

7.4 Access Control

In a macrocell network, access control normally happens during LAU/TAU or when UE requests a data transmission service. However, access control needs to be invoked whenever a UE is trying to camp on a femtocell to prevent the unauthorized use of that femtocell. In this section, triggers that enable access control for femtocells whilst remaining compatible with macrocell signalling procedures are presented. Then, the possible locations of access control in the network are discussed. Subsequently, the general access control procedures for femtocell with closed, open and hybrid access modes are introduced.

7.4.1 Access Control Triggers

Location/Tracing Area Update

A common assumption is that access control is triggered by LAU/TAU [2]. In this case, each femtocell is assigned a femtocell specific LAI/TAI different from macrocell. This can ensure that the access control procedure can be invoked when a UE carries out inbound mobility from a macrocell to a femtocell. For a home-use case, each femtocell may be assigned a different LAI/TAI from their neighbour femtocells in order to avoid UE camping on an unauthorized femtocell. For enterprise or metro-zone cases however, the femtocells belong to the same company or campus may share the same LAI/TAI to avoid unnecessary update signalling. The UE that are not allowed in a certain femtocell will then receive a negative response at location registration, meaning that it can not camp on that femtocell normally.

A side-effect of using LAU rejects is that a pre-release 8 UE would ban the whole frequency on which the target femtocell is for 300 s. To reduce time in out-of-service or limited camped state there should always be a frequency available with cells where LAI is allowed (i.e. there should be a non-femtocell frequency layer).

Data Transmission Service

An alternative approach is to allow UEs that are allowed to use a femtocell to roam and camp also on femtocells which they are not allowed [2].

In this approach, the access control would be triggered when data transmission service is requested. The non-allowed UEs will then be redirected or handed over to an available macrocell. Thus, the allocation of LAIs for the femotcell would be able to follow the same route as that for macrocells. A number of femtocells can be configured in the same Location Area (LA). Within such LA, no LAU is needed when the UE is moving from one femtocell to another, leading to lower signalling overheads compared with the LAU approach.

The drawback of this approach is that when the macrocell coverage is not available, it will cause radio connection failures when redirecting the non-allowed UEs to the macrocell.

7.4.2 Access Control Location

Access Control in Femtocell/Femtocell Gateway

In pre-release 8 UTRAN, access control happens in femtocell or femtocell-GW for femtocells while in CN for macrocells. Figure 7.9 shows the access control in a femtocell during UE registration. This method has the least impact on the existing network structure as it does not need any modification on the CN nor any requirement on the UEs. Additionally, during the UE registration procedure, the femtocell can reject an unauthorized user without sending any overhead signalling to CN by the locally stored allowed IMSI list. The main drawback of this approach is that the access method has to be managed and implemented separately.

In release 8 UTRAN, for the support of femtocell for legacy (pre-release 8) UEs, the approach in this scenario is that the femtocell gateway shall perform access control, while the femtocell may optionally perform access control as well [16]. During the UE registration procedure, since the legacy UEs do not understand CSG, they may try to register and camp on any detected femtocells even if they are not allowed to do so. To locate the access control in femtocell/femtocell-GW can significantly minimize such signalling overheads to the CN. Compared with the pre-release 8 UTRAN, Figure 7.10 shows that the femtocell/femtocell-GW needs to fetch the ACL of the registering subscriber from SGSN/MME. Such a procedure may potentially increase the network overheads.

Access Control in UE

In release 8, the UEs that are CSG-aware can do the basic access control to accelerate and enhance the mobility procedures with femtocells. Figure 7.11 shows the access control occurring in a CSG-capable UE during UE registration procedure. Since the CSG-aware

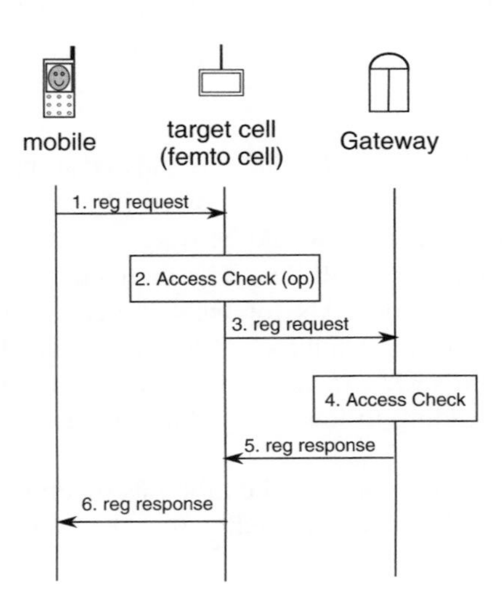

Figure 7.9 Access control in femtocell-GW during UE registration in pre-release 8

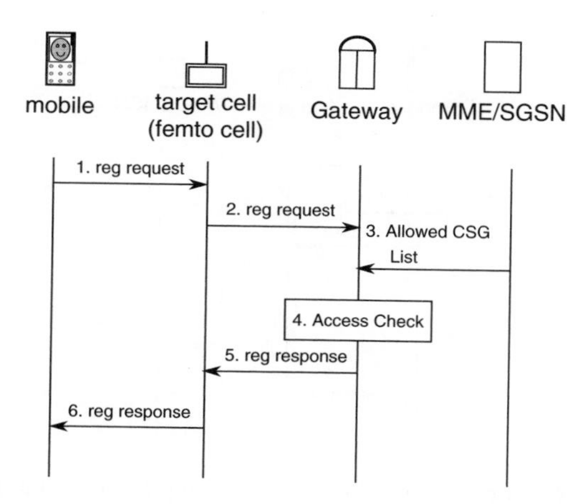

Figure 7.10 Access control in femtocell-GW for non-GSG UE during UE registration in release 8

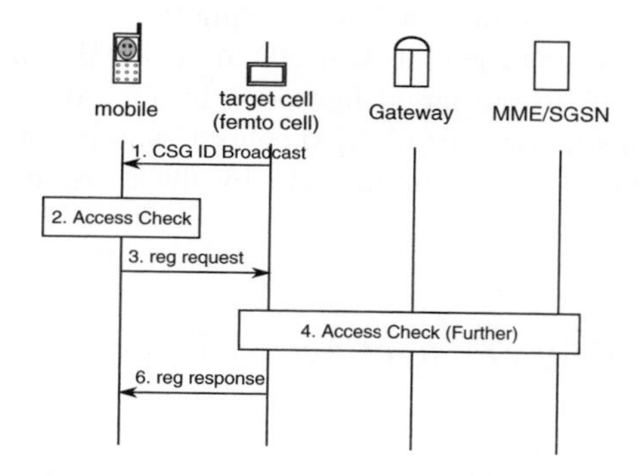

Figure 7.11 Access control in UE during UE registration in release 8

UE can distinguish whether the target femtocell is accessible or not, it can avoid attaching to a CSG cell that is not in its ACL.

Access control in UE requires the UE frequently to decode the SIB of the target cell candidate containing CSG ID, which may impact on the quality of the active service. Additionally, there is a risk that the ACL in the UE may be out of date and the UE may try to access a cell for which the CSG subscription has expired. Since the report from UE is not always trustworchy, the CSG access control cannot be carried out only via the UE.

Access Control in MME/SGSN

In release 8 EUTRAN, MME performs access control for the CSG-capable UE accessing the network through an EUTRAN CSG cell during attaching, detaching, service

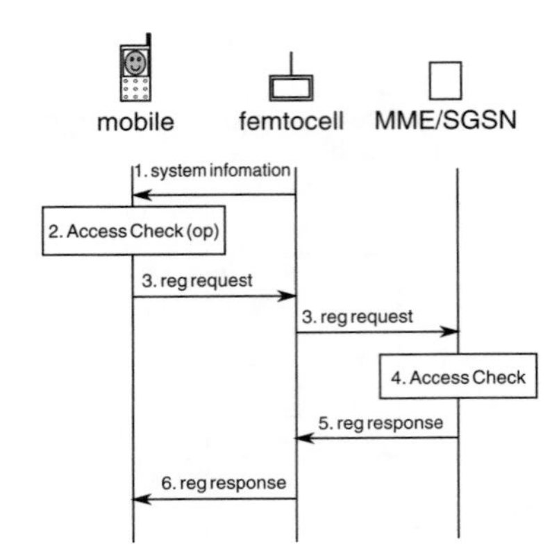

Figure 7.12 Access control in MME/SGSN during UE registration in release 8

request and TAU procedures. The same principle is applied to the access control of UTRAN CSG cells. In this case, MSC/VLR and SGSN perform the access control [17]. Figure 7.12 shows the access control in MME during the UE registration procedure. The femtocell/femtocell-GW needs to send its CSG ID to the MME, and the MME then checks the accessability of the UE to the target femtocell by the ACL. During this procedure, the large number of registration signalling (i.e. user walking along the street and trying to camp on the femtocells) can be avoided by the assistance of the access control in UE.

7.4.3 Access Control for Different Access Types

Closed Access

Closed access mode femtocell is known as the CSG cell in 3GPP and is the only featured access mode in release 8. The CSG cell may be widely used in individual home deployment. In this scenario, the owner of the femtocell does not want to share the femtocell due to the limited source of the backhaul or due to some security concerns. The access control should always be performed whenever a UE is trying to camp on the femtocell. Any UE that is not in the CSG will be rejected by the femtocell.

For a non-CSG capable legacy UE, the femtocell performs an identity request to inquire the UE's IMSI and try to register the UE in the femtocell-GW; femtocell-GW should then perform the access control and may accept or reject the UE for camping on the target femtocell. For a CSG capable UE, the femtocell includes its CSG ID in the initial UE message to CN, and the CN performs UE access control according to the CSG ID.

Open Access

As a new access mode in 3GPP release 9, the open access mode femtocell operates as a normal cell, i.e. non-CSG cell. The operator may deploy an open access mode femtocell to fill some indoor blind spots and some public hot-spots to serve all the users as a macrocell. The mobile network doesn't need to perform any specific UE access control for such a femtocell.

For a non-CSG capable legacy UE, the femtocell doesn't need to perform an identity request to ascertain the UE's IMSI and can use TMSI/PTMSI to register the UE in the femtocell-GW; the femtocell-GW should always accept the request and assign a context ID for the UE.

For a CSG capable UE, since there is no CSG ID for an open access femtocell, the femtocell doesn't include this information in the message, so there is no access control in the CN. Additionally, whether the CN needs to know the UE is camping on an open mode femtocell or not needs further investigation.

Hybrid Access

Information about hybrid/semi-open access can be found in [18]. This access feature will be available in 3GPP release 9. A hybrid mode means that the femtocell can provide a combination of both open and closed access modes at the same time. The hybrid access femtocell is a cell that not only has a CSG ID, but also allows UEs that are not members of that CSG to camp thereon. In this access mode, these UEs may only be authorized a limited QoS service and have lower priority compared with the UEs in the CSG.

For a non-CSG capable legacy UE, though the mobile network allows all UEs to camp and offers services to them whether these UEs are in that CSG or not, the femtocell still needs to distinguish the two types of UE in order to provide their respective QoS services. Thus, the femtocell still needs to perform an identity request in order to obtain the UE's IMSI for access control and service level differentiation purposes. Since the UE access control in a femtocell is optional, femtocell-GW needs to inform the femtocell whether or not the UE is a member of that GSG. If the UE does not belong to the CSG, the femtocell may limit the data rate for the user, and redirect the UE firstly to a macrocell due to the shortage of femtocell resource limitation. In addition, the CN may also need to be informed for different services or charging, etc.

For a CSG-capable UE, the femtocell should also include the hybrid mode information. The CN also needs to inform the femtocell regarding the access control result in order for the femtocell to provide different QoS services and different priorities to different UE respectively.

7.5 Paging Procedure

Due to the high volume and small-sized femtocell deployment, it is well-known that paging messages is a big burden for the femtocell system. Figure 7.13 shows a paging procedure invoked by the UE with IMSI_1. In the normal approach as with a macrocell network, the CN will only distinguish the paging area 1 where the paged IMSI is located. The femtocell-GW will send the paging message to all of the femtocells in paging area 1, which may cause a huge signalling redundancy for the large number of femtocells involved. One of the requirements agreed in [19] is that 'Additional registration and paging load as a result of HNB/HeNB deployment shall be minimized'. This means that paging optimization, namely minimizing the amount of paging messages used to page a UE in femtocells, is a confirmed requirement in 3GPP. In release 8, it was decided that

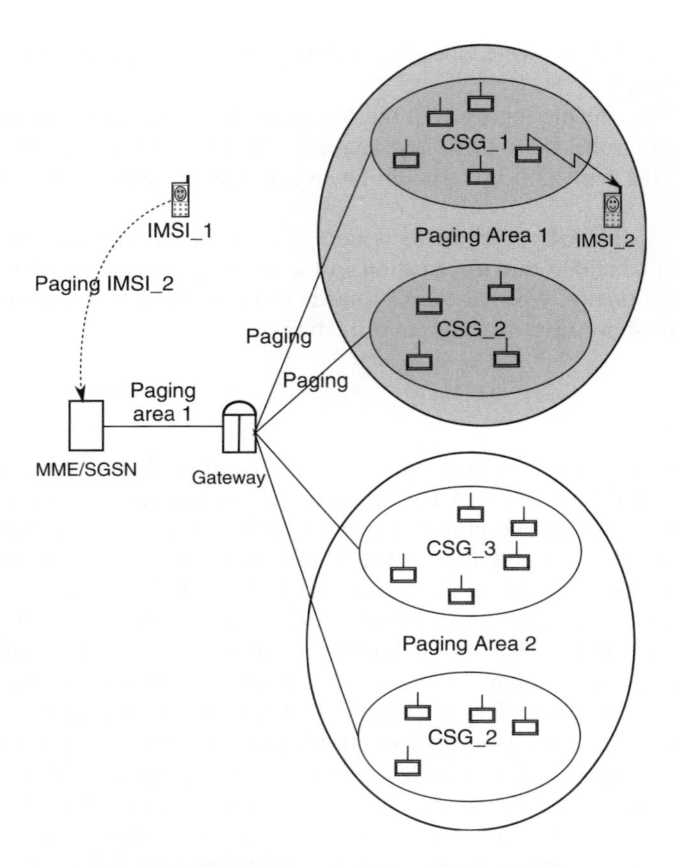

Figure 7.13 Paging procedure with femtocells

MME and/or femtocell-GW can perform paging filtering optionally [20]. The detailed considerations on adapting the paging optimization in UMTS and LTE can be found in [21–24]. In this section, paging optimization in MME/SGSN and in femtocell-GW are both discussed.

7.5.1 Paging Optimization in MME/SGSN

In this method, MME knows the CSG IDs supported by connected femtocells. Since the MME keeps a copy of ACLs for the registered UEs, the ACL of the paged UE is always available to assist in paging optimization. The MME can then use this list to filter the connecting femtocells in the paging area. It will only send a paging message to the femtocells with the CSG ID in the ACL. It should be noted that the ACL is not included in the paging message and will not be sent to the untrusted femtocells.

Figure 7.14(b) shows an example of paging optimization carried out by the MME. In this scenario, the MME receives the paging request from IMSI_1. IMSI_2 is the paged UE, which is located in paging area 1. By knowing the allowed CSG list of IMSI_2

(CSG ID 1 and CSG ID 3), the MME is able to forward the paging message only to the femtocells with CSG ID 1 in the paging area 1.

7.5.2 Paging Optimization in Femtocell-GW

In order to optimize the paging procedure by the femtcell-GW, the Femtocel-GW shall be aware of the CSGs supported by the connected femtocells. This allows the femtocell-GW to identify the appropriate femtocells supporting certain CSGs. The femtocell-GW needs to be informed about the allowed CSG list of the paged UE. This allows the femtocell-GW to understand which of the connected femtocells shall receive the paging message. In order to have a complete paging optimization solution, the allowed CSG list of the paged UE shall be included in the paging message. The ACL of the paged UE needs only to be available at the femtocell-GW, namely it does not have to be forwarded to femtocells or other nodes directly connecting to the MME.

Figure 7.14(a) shows an example of paging optimization performed in the femtocell-GW. The paging message is sent with the allowed CSG list of the paged UE to the femtocell-GW by MME. With the help of the ACL, the femotcell filtering is done by the femtocell-GW. Finally, the paging message is only sent to the femtocells with the allowed CSG ID.

In release 9, since open and hybrid access modes are introduced, all the UEs can always camp on the femtocells with open access and these femtocells do not even have a CSG ID. One user can still have restricted QoS service from a hybrid access femtocell although the CSG ID of the femtocell is not in his ACL. Since the paged UE may also be permitted to camp on these femtocells, the access mode information should be considered when MME and/or femtocell-GW perform paging optimization.

If the femtocell directly connects to an MME, the MME should page the CSGs in the paged UE's ACL in the paged areas as well as the femtocells with open and hybrid access in the paged areas as shown in Figure 7.15(a).

If the femtocell connects to the CN through femtocell-GW and the ACL of the paged UE is sent to the femtocell-GW in the paging message, the femtocell-GW should also page the UE in the open and hybrid femtocells though they may not have CSG ID at all or their CSG IDs are not in the UE's ACL as shown in Figure 7.15(b).

7.6 Cell Selection and Reselection

Cell selection and reselection are two basic mobility procedures in wireless mobile networks. Cell selection takes charge of the selection of a suitable cell to camp on when it is powered on or after having previously lost coverage. Cell reselection however, enables UE to select a new serving cell when it meets the cell reselection criteria. Due to the challenges stated in Section 7.1, cell selection and reselection in the femtocell environment are more complicated than in a macrocell network. In Section 7.3.2, a few methods that enable the UE to find neighbouring femtocells are presented. Femtocell featured cell selection and reselection procedures should prioritize the UE to camp on allowed femtocells whilst avoiding registering on unauthorized femtocells. In this section, a few alternatives in enabling cell selection and reselection with femtocells in pre-release 8 network are

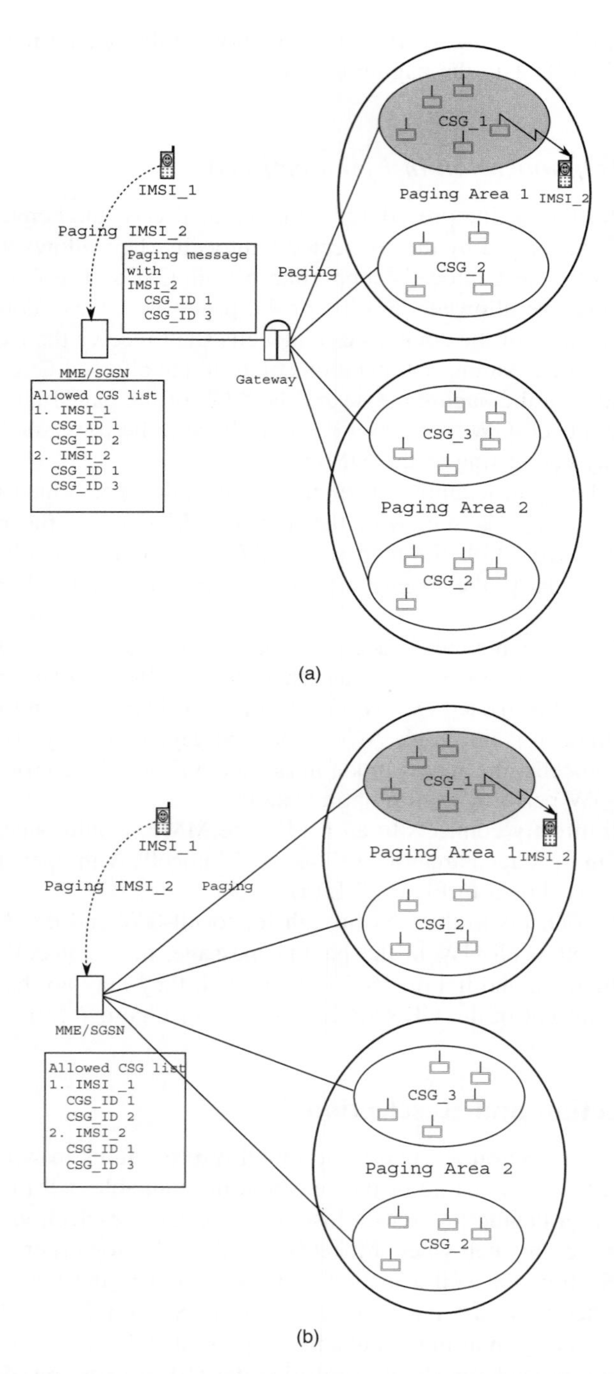

Figure 7.14 Paging optimization procedures in release 8. (a) Paging optimization in femtocell-GW in release 8. (b) Paging optimization in MME in release 8

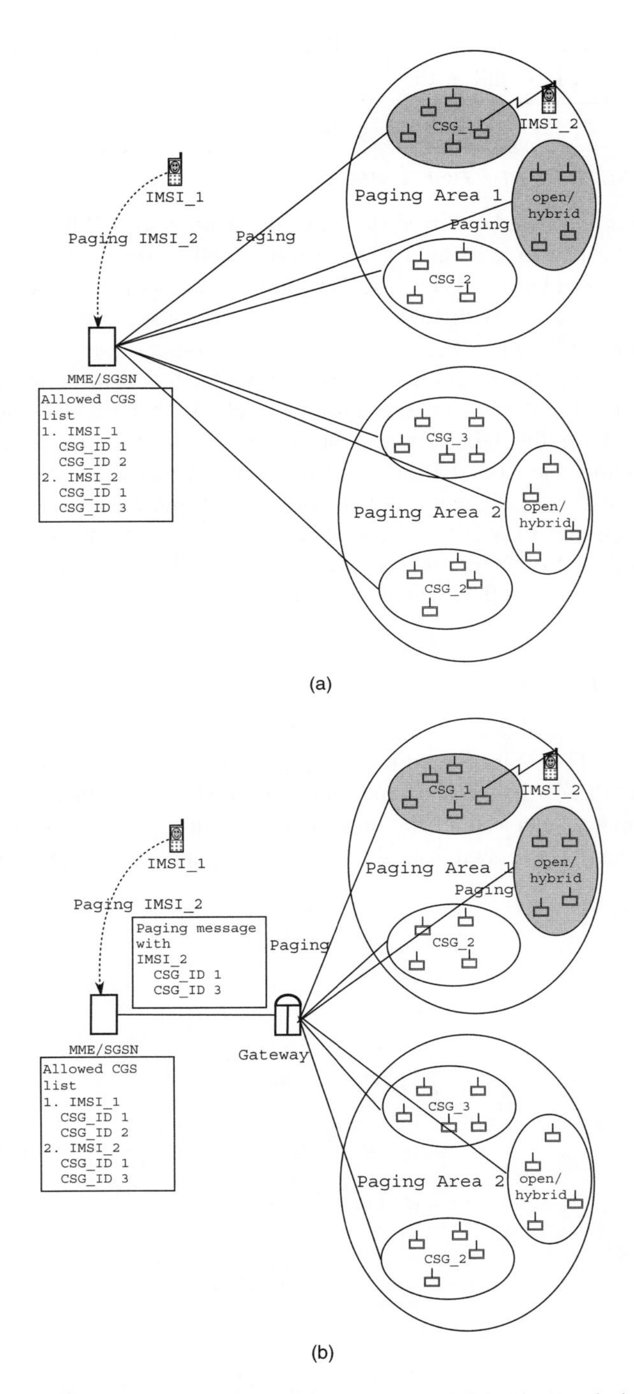

Figure 7.15 Paging optimization procedures in release 9. (a) Paging optimization in MME in release 9. (b) Paging optimization in femtocell-GW in release 9

introduced. The achieved and on-going cell selection and reselection methods in release 8 and 9 are discussed in detail at the end.

7.6.1 Cell Selection in Pre-release 8

Since cell selection is carried out irrespective of the neighbour cell list and will only select the cell with the best received signal as the serving cell, cell selection with femtocells is not a problem in pre-release 8 UTRAN. The policies used in macrocells can be applied in supporting cell selection with femtocells. However, a user may want to prioritize the femotcell over macrocell when the mobile is powered on or loses the coverage from the serving cell. Since Access Stratum (AS) will perform the PLMN search to select the best PLMN before it can pick up the strongest cell of that PLMN and Radio Access Technology (RAT) for the UE, different priorities for macrocell and femtocell can be achieved by assigning separated PLMN ID to them. In this case, the PLMN ID of the femtocell system will be assigned a higher priority than the macrocell layer. It should be noted that an additional PLMN ID is required while some operators may not be able to provide it. According to the report in [2], two main methods that take advantage of the separate PLMN in cell selection with femtocells are listed below.

Manual PLMN Selection

In UTRAN, a user is capable of configuring either manual or automatic network selection mode. Manual PLMN selection enables the user to select the preferred mobile network from the available PLMN list. When separating the PLMN IDs of macrocell and femtocell, the user can manually select either the macrocell or femtocell network (i.e. when the user arrives home, he may try to switch to the femtocell layer) as shown in Figure 7.16. Once

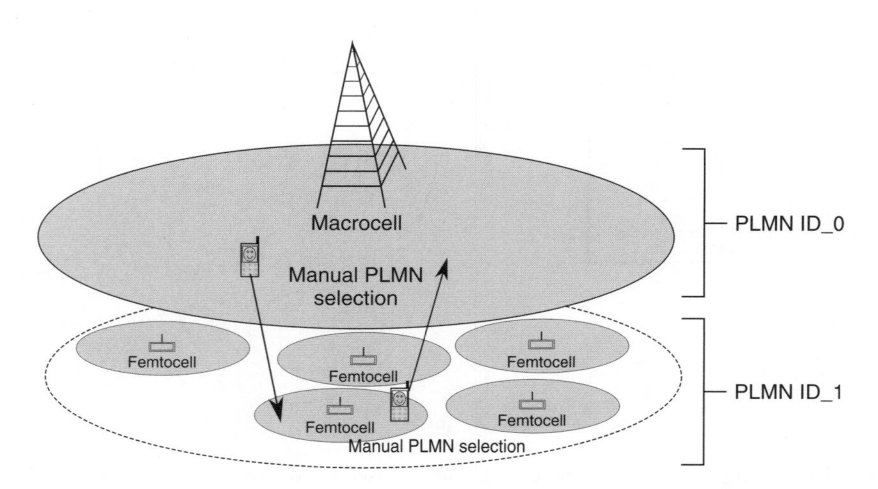

Figure 7.16 Cell selection by manual PLMN selection in pre-release 8

the PLMN is selected the user will stick to that network and no mobility procedure will occur between these two PLMNs.

No mobility procedure between macrocells and femtocells will significantly reduce the signalling load towards the CN and extend the mobile battery life for all of the users. However, manual mode does not allow automatic PLMN selection despite the UE losing coverage. This becomes annoying since the user has to manually select the macrocell layer when he/she goes out.

National Roaming

National roaming is a feature when the mobile is not roaming in its Home PLMN (HPLMN), but on a Visited PLMN (VPLMN) of the same country as the HPLMN. An HPLMN always has a higher priority than a VPLMN. In the automatic PLMN selection mode, the UE can be configured to search periodically for its HPLMN. The range of the HPLMN search timer is from 6 mins up to 8 hours with a default value of 60 mins [25].

One possible femtocell deployment scenario is to consider the femtocell layer as HPLMN and the macrocell layer as VPLMN as shown in Figures 7.17 and 7.18. By setting a proper HPLMN search timer (i.e. 6 mins), the UE can camp on the femtocell as soon as it reaches the coverage of the femtocell and prioritizes it thereafter. However, the UE has to search for the femtocell PLMN periodically even when he is being served by a macrocell, which will drain the mobile battery. Besides, the operator may have set the UTRAN macrocell network as HPLMN, which may significantly impact the operators' service category when switching it to VPLMN.

The other femtocell deployment scenario is to consider the macrocell layer as HPLMN and the femtocell layer as VPLMN as shown in Figure 7.17. The UE will not search for the femtocell PLMN all the time when camped on the macrocell. However, as long as the UE is able to receive signals from the macrocell, it will be very difficult for the UE to select the femtocell even when the user receives a stronger signal from the femtocell.

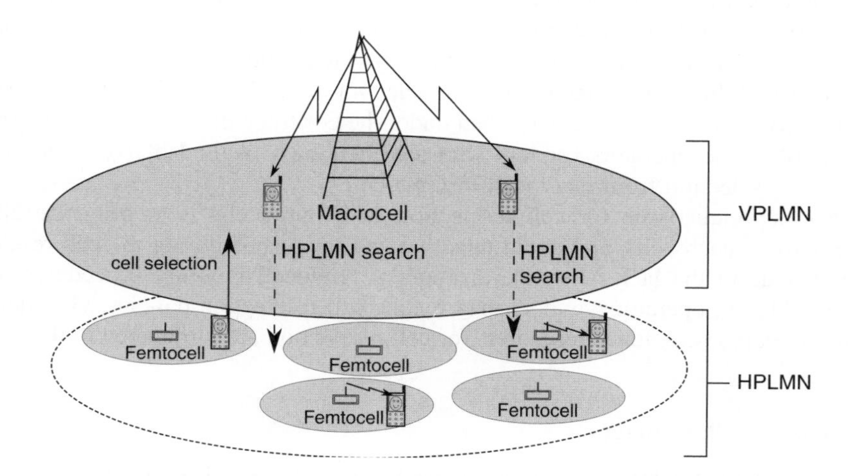

Figure 7.17 Cell selection by national roaming (femtocell as HPLMN) in pre-release 8

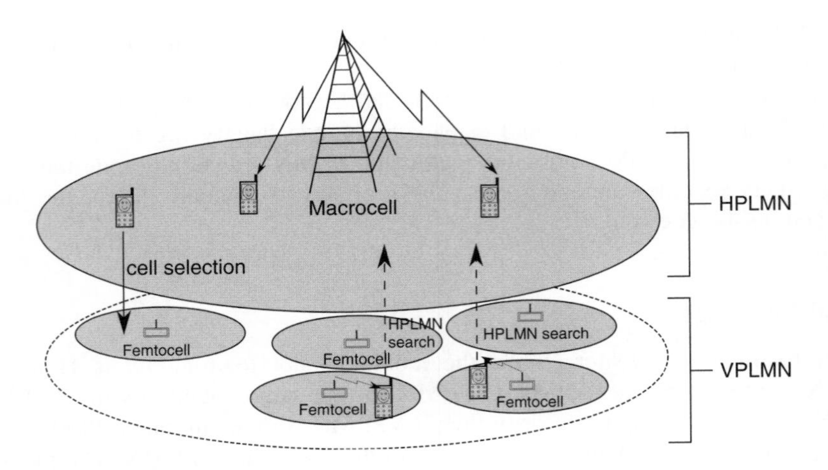

Figure 7.18 Cell selection by national roaming (femtocell as VPLMN) in pre-release 8

In addition, when the UE is camping on the femtocell, it will periodically search for macrocells, which will also drain the battery.

7.6.2 Cell Reselection in Pre-release 8

Cell reselection with femtocells is more complex than cell selection. During cell reselection, the UE needs to carry out cell measurements for intra-frequency, inter-frequency and inter-RAT neighbour cells and rank the cells based on a specified policy. Then the UE reselects the most suitable cell from the current serving cell. While dealing with femotcells in pre-release 8 UTRAN, to provide a reliable and clean neighbour cell list for a macrocell is still a problem as stated in Section 7.3.2. Thus, some operators/vendors do not implement the cell reselection features for the macrocell to femtocell reselection procedure (femtocell to macrocell or femtocell to femtocell reselection is easy, since the NCL of a femtocell will not be large and PSC confusion is unlikely to happen). Macrocell to femtocell mobility only relies on the cell selection procedure, which means users will have difficulty in selecting femtocell when under the coverage of macrocell. Nevertheless, there are still some operators/vendors who use their own method of NCL approach to support cell reselection from macrocell to femtocell.

Another important issue for cell reselection with femtocells is to prioritize allowed femtocells over macrocells, and avoid unauthorized femtocells during the cell reselection procedure. Due to the lack of standardization of femtocell mobility procedures in pre-release 8 UTRAN, operators/vendors may realize cell reselection policies with their own substitution methods. A few alternations for cell reselection are introduced in this section.

Cell Reselection Parameters

Cell reselection parameters are commonly used to adjust the cell reselection strategy in a macrocell network [9, 10]. They can also be used in cell reselection procedures between

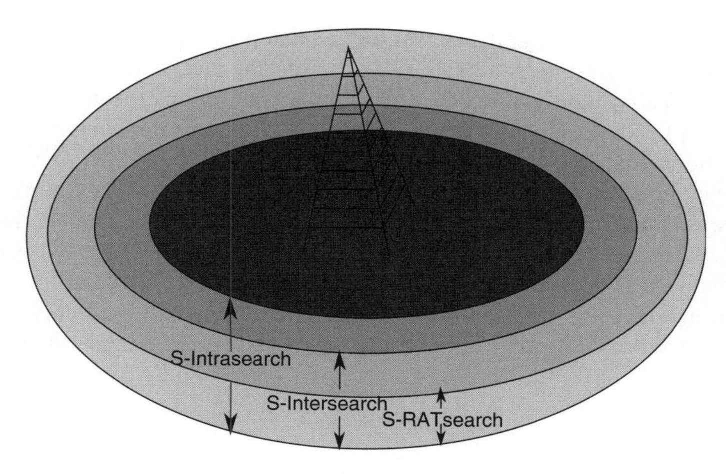

Figure 7.19 Cell search criterion for inter-frequency, inter-frequency and inter-RAT cell searching in cell reselection

femtocell and macrocell. Figure 7.19 shows the searching criterion (S-Criterion) for intra-frequency, inter-frequency and inter-RAT cells. As long as the S-Criterion is fulfilled, the UE will be invoked to do monitoring and measurements of the corresponding neighbouring cells. A lower S-Criteria with a macrocell can help the UE to start searching for femtocells as soon as it reaches the coverage of the the femtocell. A higher S-Criterion for femtocells forces the UEs to stick to the femtocell while they are at home, as shown in Figure 7.20. Since the S-Criterion takes on the whole frequency for all the neighbouring cells, it will impact on the reselection between macrocells as well when in a shared frequency femtocell deployment scenario.

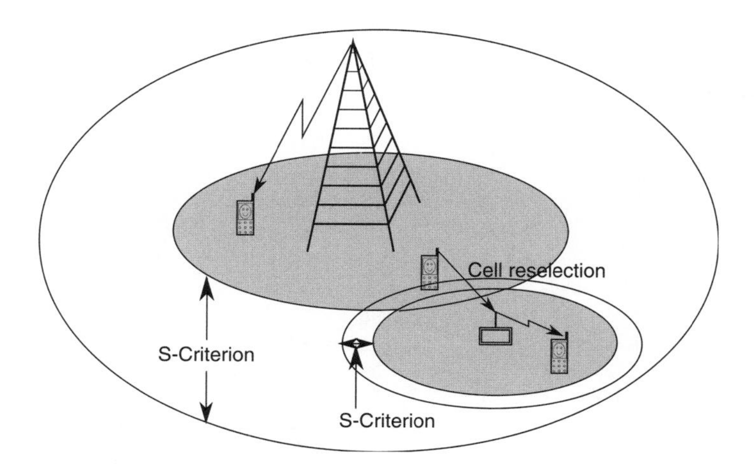

Figure 7.20 Prioritizing femtocell over macrocell by adjusting the cell search criterion in cell reselection

A further consideration in cell reselection is related to the cell ranking criterion (R-Criterion) in which Q-hyst (a hysteresis value for serving cell) and Q-offset (offset values for neighbouring cells) jointly affect the cell ranking. Within macrocell's NCL, a lower Q-offset can be set to femtocells so that they can have a higher rank during cell reselection. Figure 7.21 shows that by setting a negative Q-offset and low Q-hyst, the cell searching point d1 moves backward to d2. While in femtocell, a higher Q-hyst value can help the femtocell keep the users connected within its coverage. Figure 7.22 shows that by setting a positive Q-offset and high Q-hyst, the cell searching point d1 moves forward to d2.

The disadvantage is that all the users in the network will prioritize the femtocell and try to register on it. The users who are not authorized to access will be finally rejected. This will increase the battery consumption of the mobiles and increase the MM signalling load.

Hierarchy Cell Structure

By using HCS, the UEs can perform femtocell measurements even when it is in a good macrocell coverage, and it limits the measurement of either high priority cells or low priority cells that decrease the measurement burden [9]. By setting a proper duration for the HCS penalty timer, the passing UEs on the street can avoid selecting the femtocell,

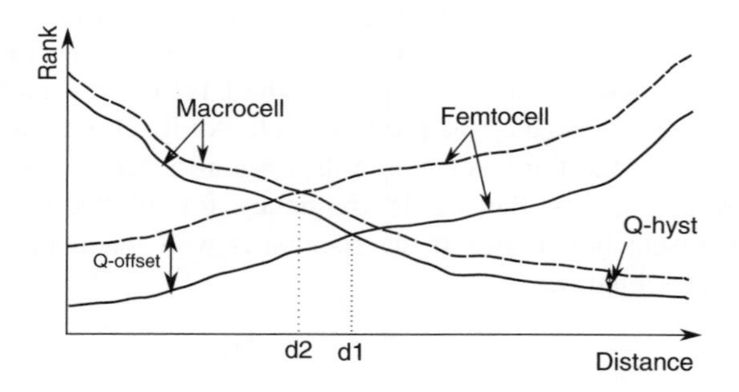

Figure 7.21 Prioritizing femtocell over macrocell in reselection to femtocell by adjusting the cell ranking criterion

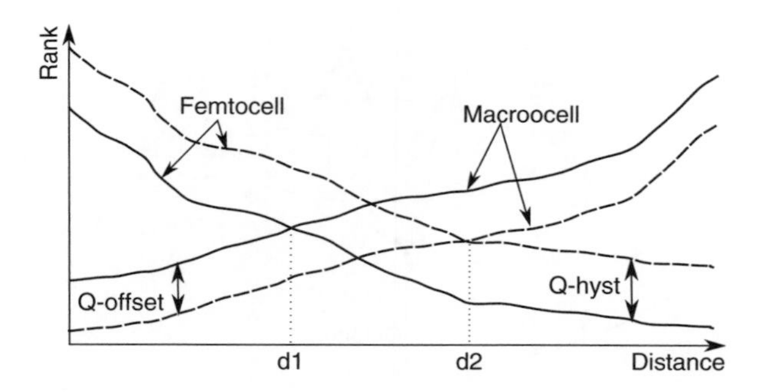

Figure 7.22 Prioritizing femtocell over macrocell in reselection to macrocell by adjusting the cell ranking criterion

and reduce the network signalling overheads. Besides, parameters for the UE with high mobility can be set to stay on macrocells, and only UEs that are relatively stationary or with low speed will select the femtocell layer.

As a mature technology adopted in GSM and UMTS, the biggest advantage of HCS is that it can be used for femtocells in pre-release 8 networks without any modification of the current network. By simply configuring the cell HCS, the UEs that are camped on a macro layer can automatically find a femtocell layer. Once the UE is camped on a femtocell, it will prioritize this cell.

However, when using HCS only, all of the UEs will follow the same cell reselection rules to prioritize femtocells over macrocells and attempt to register thus increasing the loading to the CN. Besides, the NCL is the only method for HCS to find neighbouring femtocells, which remains with quite a few problems in supporting femtocells.

Equivalent PLMN

As introduced in Section 7.6.1, to have separate PLMN IDs of femtocells and macrocells can help to improve the cell selection procedures with femtocells. In order to enable cell reselection between these PLMNs, the Equivalent PLMN (EPLMN) feature can be used [2]. In this approach, the femtocell PLMN is configured as the EPLMNs of the macrocell PLMN. The EPLMN is considered equivalent to the Registered PLMN (RPLMN) regarding PLMN selection, cell selection, cell reselection and handover. The UEs may update their EPLMN lists during LAU. For example, when a UE reaches the overlapping area of a macrocell and an allowed femtocell, the UE shall add the PLMN ID of that femtocell to its EPLMN list. This enables the cell reselection procedures between the macrocell and the femtocell for the UE. Also, while the UE is not allowed to use the femtocell, the femtocell PLMN is removed from its EPLMN list, thus it would never try to camp on that femtocell, thus leading to a longer battery life.

The drawback of the EPLMN method is that the procedures of configuration and update EPLMN list highly depend on the LAU. In addition, the EPLMN feature is only available from release 6, thus old mobiles can not support EPLMN.

7.6.3 Cell Selection in Release 8

In release 8, cell selection follows the strongest cell as the serving cell. By checking the CSG IDs against the UE's ACL during cell selection, the UE can avoid selecting an unauthorized femtocell as the serving cell. In addition, as a new feature for release 8, manual CSG cell selection can be used for the UE to select the serving cell. Figure 7.23 shows an overall idle mode process with femtocells in release 8 [9, 10].

Automatic Cell Selection

In release 8, automatic cell selection procedure for femtocells is similar to that for macrocells. To avoid camping on an unauthorized femtocell, an extra CSG ID check is performed when the target cell is a CSG cell. AS will check the broadcast CSG ID against the ACL

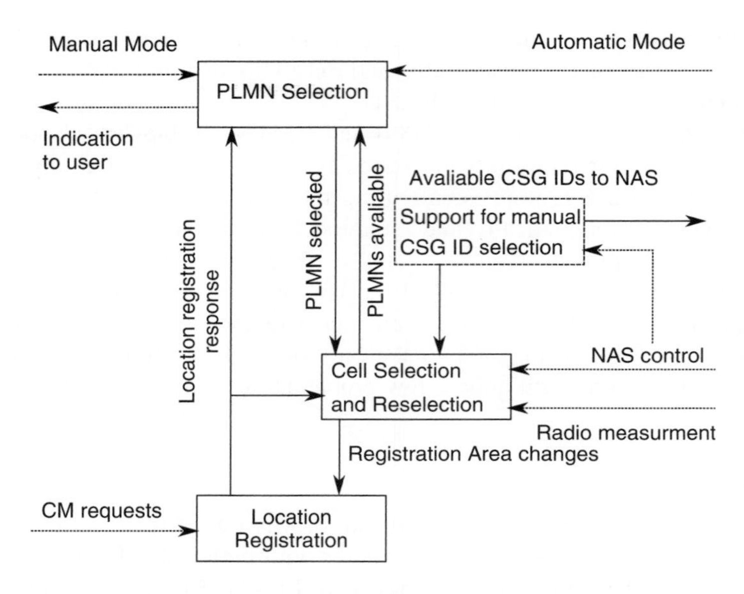

Figure 7.23 Overall idle mode process [10]

provided by NAS whether the CSG cell is suitable for the UE. If the CSG ID of the target cell is in the UE's ACL, the cell is selected for camping on.

In release 9, the support of open and hybrid access femtocells should be further considered. Since femtocells with open access do not have CSG IDs, all the UEs can select those cells without checking CSG ID during cell selection. For the femtocells with hybrid access however, CSG ID checking still needs to be invoked in order to distinguish whether the UE is in its CSG or not. The UEs in the CSG will have full access while others are still able to camp on but may only have limited QoS services from that cell.

Manual Cell Selection

In release 8, a method that supports a user in selecting their serving CSG manually in idle mode is proposed [9, 10]. To avoid interruption to the current UE service, the UE is not allowed to support manual CSG selection in connected mode [26].

As a robust method, the user can trigger CSG manual selection by selecting the femtocell ID from the displayed CSG list. In manual selection, the UE may scan all frequencies in the UMTS Terrestrial Radio Access (UTRA) or Evolved UTRA (EUTRA) band and display a list of found femtocells with femtocell IDs, CSG IDs and corresponding femtocell names if they are broadcasted. The UE may also show indications as to whether or not the found CSG cells are in its ACL. When the user selects an entry in the list, the UE will then perform attach and LAU/TAU procedures with that cell. The UE may normally camp on the chosen cell if it is an allowed CSG cell.

In addition, manual cell selection can help to update the UE's ACL by way of the feedback of the LAU/TAU request to the selected CSG cell. Details can be found in Section 7.3.4.

For the support of manual cell selection in release 9, the UE needs to scan for open and hybrid femtocells as well. The information on femtocells with open and hybrid access will also be displayed on the UE panel. A new field that indicates the access type (closed, open or hybrid) may be included in the listed femtocell information.

7.6.4 Cell Reselection in Release 8

By introducing the CSG ID and UE autonomous search function in release 8, a femtocell-featured cell reselection can now improve the reselection procedures for femtocells. The UE shall use an autonomous search function to detect neighbouring femtocells instead of reading NCL from the serving cell. In addition, the UE can exclude unauthorized femtocells by comparing their CSG IDs with the ACL. In other words, the UE can now carry out cell measurements over suitable femtocells without knowing the NCL from macrocells.

It shall be possible to allow UEs that are allowed to access a given CSG cell, to prioritize their camping towards the CSG cells when in the coverage of the CSG cells. To achieve this, it should be possible either to set the reselection parameters accordingly or other means should allow this [20]. To further improve the cell reselection procedure with femtocells, methods that prioritize femtocells over macrocells during cell reselection are required. Details related to specific mechanisms intended to fulfill this task are still under discussion and may be adopted in 3GPP release 9.

In this section, cell reselection to and from femtocell in intra-frequency, inter-frequency and inter-RAT scenarios are discussed.

Reselection to Femtocell

In release 8, the information on surrounding femtocells is not included in the NCL of the macrocell. An autonomous search function is used to search for CSG cells when the ACL of the UE is not empty, and the UE may disable the autonomous search function while the ACL is empty. If CSG cells are deployed on the reserved frequency(ies), the UE may perform autonomous search only on these dedicated frequencies and on the other frequencies listed in the system information. When the UE has no or an empty ACL, it may ignore cells with PSC/PCI in the stored range 'CSG PSC/PCI Split Information' reserved for CSG cells for intra-frequency and inter-frequency measurements and cell reselections.

When knowing the neighbouring femtocells and macrocells, the UE follows a similar cell ranking procedure to select the cell with highest rank as serving cell. Some femtocell-related policies during cell reselection to femtocell in intra-frequency, inter-frequency and inter-RAT scenarios are listed below.

- Intra-frequency Cell Reselection
  To avoid adding extra interference to the same frequency layer, femtocell prioritization is not supported in intra-frequency cell reselection. For intra-frequency reselection from a macrocell to an allowed CSG cell, the UE follows the same cell ranking rules as those defined for macrocells in [9].

The main concern of intra-frequency reselection is how to design a policy when the best cell is a CSG cell whose CSG ID is not in the ACL. To either allow or disallow intra-frequency cell reselection in this situation, an 'intra-frequency cell reselection indicator' IE is presented in [27] and [28]. When the indicator is set to TRUE, then it would allow UEs to camp on another cell although a non-allowed CSG is the best ranked cell, i.e., causing interference between the non-allowed CSG. However, if the indicator is set to FALSE, then the UE would consider the entire frequency to be barred for some time and select a cell from different frequency or RAT.

Detailed solutions have not been decided so far in 3GPP. For example, in [28], a new intra-frequency reselection threshold is introduced so that the network can have further control over UE's CSG cell reselection. Thus, the UE is allowed to stay on the frequency only if the difference between the strength of the best ranked non-allowed cell and allowed cell (i.e. suitable) is less than the threshold.

- Inter-frequency Cell Reselection

For inter-frequency cell reselection, the priority-based scheme can be used. The priority of CSG cells should be by default higher than macrocells. During cell reselection, instead of ranking all of the measured cells as in macrocells, the UE firstly distinguishes the CSG cells from macrocells (i.e. by reserved PSC/PCI split information). If the tested cell is a CSG cell, the UE will then read the CSG ID of that cell and check it against the ACL. If the CSG cell is allowed to be accessed then the UE will perform cell ranking only on the frequency where the CSG cell is. If the CSG cell remains the highest ranked cell on that frequency, the UE shall reselect this cell irrespective of the cell reselection rules applicable for the cell where the UE is currently camped. If multiple suitable CSG cells are detected on different frequencies and these are the strongest cells on their frequencies, then the UE shall reselect any one of them [9, 10]. If no such CSG cell is detected, a normal cell reselection as defined for macrocells should apply.

The above policies can be used when a cell searching criterion (S-Criterion) is met. However, when the UE finds CSG cells on another frequency, it may prefer to reselect the allowed CSG cell even when the signal strength/quality of the serving macrocell is good. The UE needs a threshold for CSG cells in order to perform a cell reselection evaluation process. Details of the mechanisms still need to be decided in 3GPP. In the literature [29–31], in addition to the normal S-Criterion (S-Intrasearch, S-Intersearch and S-RATsearch) defined in macrocells, a new S-CSGCell criterion (ThreshCSGcell) and a time interval (TreselectionCSG) are proposed in order to enable cell reselection to CSG cells. A new offset (Q-offsetCSG) for CSG cells in cell ranking are also introduced. When the UE detects that the S-CSGCell of a CSG cell is greater than ThreshCSGcell during a time interval TreselectionCSG as shown in Figure 7.24, the UE will compare the CSG ID of that cell with the ACL. If the cell is allowed to access, the UE will perform cell ranking upon the corresponding cell offset on that frequency. The UE will reselect that cell if it remains the highest ranked cell on that frequency.

Moreover, in [32], a proposal to have UE specific settings in order to treat different users in different deployment scenarios is discussed. For this purpose, some cell reselection parameters may be sent to the UE via dedicated signallings.

- Inter-RAT Cell Reselection
  Inter-RAT reselection to an allowed CSG cell is supported when the UE is camped on another RAT. The UE requirements are defined in the specifications of the concerned RAT.

Reselection from Femtocell

Once a femtocell is powered up, it will normally scan the neighbouring macrocells in intra-frequency, inter-frequency or inter-RAT layer. The information on these macrocells will then be automatically configured into the femtocells' NCL by the self-configuration function in the femtocell. The serving femtocell can send neighbouring cell information through its system information block to the UE. Just as in macrocells, the UE can then do corresponding cell measurements for cell reselection purposes without performing an autonomous search function.

- Intra-frequency Cell Reselection
  For intra-frequency cell reselection from a CSG cell to a macrocell, the UE shall apply the same cell reselection rules as defined in macrocells.
- Inter-frequency Cell Reselection
  For inter-frequency cell reselection from a CSG cell to a macrocell, the UE shall consider the frequency of the serving cell to be the highest priority frequency (i.e. higher than the eight network configured values or highest HCS priority) as long as the serving cell remains as the highest ranked cell on that frequency [9, 10]. Otherwise, the normal cell reselection as defined in macrocells should apply.
- Inter-RAT Cell Reselection
  For inter-RAT cell reselection, normal inter-RAT procedures as defined for macrocells in that serving RAT shall apply. For example, for reselection from a UMTS CSG cell to a GSM or LTE macrocell, the UE follows the respective procedures defined in [9].

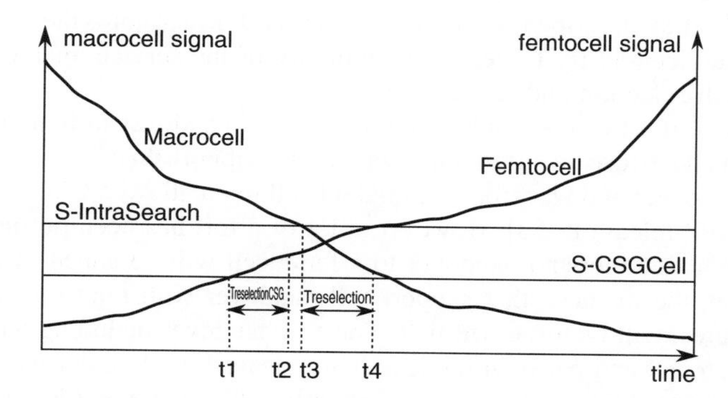

Figure 7.24 New cell search criterion S-CSGCell and time interval TreselectionCSG in cell reselection from macrocell to femtocell

Reselection between Femtocells

A femtocell may also scan for surrounding femtocells in intra-frequency, inter-frequency or inter-RAT layer after it is powered up. Since the neighbouring femtocells are quite dynamic, the femtocell may perform periodical NCL optimization when it is in idle mode. For those femtocells that are listed in the system information, the normal cell measurements can be done by the UE as for macrocells. In any case, to search for CSG cells not listed in the system information of the serving CSG cell, the UE may use the autonomous search function.

- Intra-frequency Cell Reselection
  For reselection between allowed CSG cells in the same frequency layer, the UE follows the same cell ranking rules as those defined for macrocells [9, 10].
- Inter-frequency Cell Reselection
  For inter-frequency reselection between allowed CSG cells, the UE shall consider the frequency of the serving femtocell to be the highest priority frequency when the cell is the highest ranked. If the UE detects a CSG cell on a non-serving frequency, and the cell remains the highest ranked on that frequency, it shall assume that frequency has the same priority as the serving frequency [9, 10]. The UE may reselect the detected CSG cell irrespective of the normal reselection rules of the serving CSG cell. For other cases, the cell reselection rules as defined for macrocells shall apply.
- Inter-RAT Cell Reselection
  If the UE detects one or more suitable CSG cells on another RAT, the UE may reselect one of them according to [10].

7.7 Cell Handover

Cell handover enables the UE to transfer the service seamlessly from its serving cell to the target cell without terminating the service. Since there is no Iur interface for femtocells in UTRA, soft handover is not supported by femtocells. Instead, hard handover is used. In order to support cell handover with femtocells, the issues with NCL, PSC/PCI confusion should be resolved as described in previous sections. Furthermore, the UE should avoid disrupting the service and try to keep the continuity of the service, making it even more complex than cell selection and reselection.

In pre-release 8, most of the vendors did not support handover to femtocells while still other vendors made efforts to find their own way to support this.

In connected mode, handover from a non-CSG cell to an allowed CSG cell is not within the scope of 3GPP release 8 [33]. However, a lot of effort has been put into 3GPP RAN 2 and 3 and the active inbound handover to a femtocell will be added in release 9.

In this section, the methods that support cell handover with femtocells, especially for hand-in procedures, will be discussed first. The cell handover including handover to femtocell, from femtocell and between femtocells in release 8 will be discussed. Meanwhile, it also covers the introduction of proposals currently achieved by 3GPP in order to support inbound handover in release 9.

7.7.1 *Cell Handover in Pre-release 8*

In pre-release 8 UTRA, a UE should initiate the cell measurement of all the neighbouring cells and the serving cell will make the handover decision based on the UE measurement report during cell handover procedure. In order to let the UE recognize the surrounding cells that need to be measured, the information on the neighbour cells needs to be included in the measurement control system information. For this purpose, the NCL is needed.

In this section, the ways to build and configure the NCL for the serving macrocell and femtocell are introduced. Based on such NCL, the procedures during cell handover in pre-release 8 are also discussed.

Cell Handover to Femtocell

In order to address the NCL problem stated in Section 7.3.2, it is crucial to include the femtocell information in the NCL of the macrocell, especially when there is a large number of femtocells deployed under its coverage. Methods to enable inbound handover have been presented [34, 35]. Figure 7.25 shows a possible way of supporting inbound handover from macrocell to femtocell in pre-release 8 UTRAN. The overall description of these procedures is listed below.

- Virtual Neighbour Cell List
  When a macrocell covers a large number of femtocells (i.e. more than 32 in one frequency), it is unlikely to be feasible to configure all the femtocell information into the NCL of the macrocell. To solve this problem in pre-release 8 UTRA, a 'Virtual Neighbour Cell List' is built. In this method, the femtocells share a small

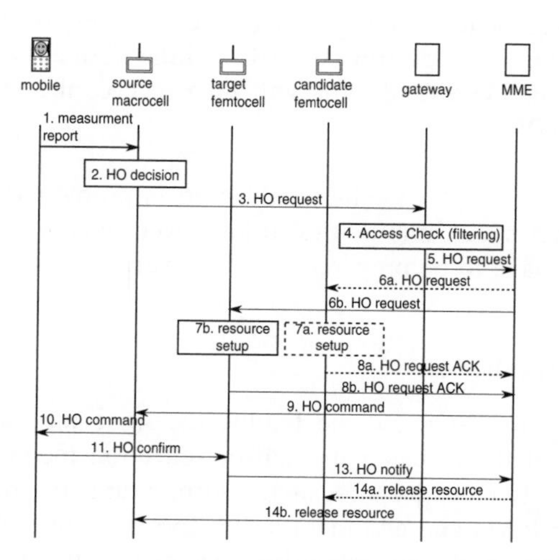

Figure 7.25 A possible way to support inbound handover from macrocell to femtocell in pre-release 8 UTRAN

number of PSCs (i.e. 10) and these PSCs will be reused within the femtocells. These PSCs will then be programmed into the macrocell's NCL. As the PSCs do not represent a certain femtocell and they cannot uniquely distinguish the femtocells, they are called 'Virtual Neighbour Cells'. The serving cell is now able to include this cell list in the measurement control information and the UE can then carry out the cell measurement over these PSCs. It should be noted that the reuse of the PSCs is quite restricted. For example, the number should not be too large in order to fit in the NCL(32 maximum). On the other hand, the number should not be too small in order to ensure that the same PSC will not be assigned to the neighbouring femtocells.

- Identification by Femtocell-GW
  When the 'Virtual Neighbor Cell List' is implemented, the UE is able to report the cell measurements of the femtocells with reserved PSCs to the macrocell. The serving cell can then decide which femtocell (represented by the PSC) the UE may handover to by measurement report. However, because of the reuse of the PSCs and the uncoordinated deployment, the serving cell can no longer identify the candidate femtocells correctly. In this case, the CN will forward the handover request, including the IMSI of the UE, to the femtocell-GW. The femtocell-GW can then compare the UE's IMSI with the allowed IMSI list of each femtocell. If such a cell is found, the femtocell-GW will send the handover request to it and the normal handover procedures will follow.
- Uplink Synchronization
  If the femtocell identified by femtocell-GW is not unique (i.e. the UE may have access to more than one femtocell), the handover request has to be sent to all of them. These femtocells will try to synchronize with the UE by sending a beacon through their downlink synchronization channel. The UE, however, will only receive the synchronization command from the real target femtocell as long as the femtocells with the same PSC are ensured to be separated from each other. The right femtocell will then receive the uplink synchronization message from the UE and the normal handover procedures can follow. The other candidate femtocells will then receive the resource release request from the femtocell-GW.

The overall mechanisms can solve the inbound handover problem in some cases. However, this leads to inefficiencies and ambiguities in handover signalling, as multiple candidate target femtocells may have to be prepared for handover.

Handover from Femtocell

It is easy to configure the NCL for the femtocells, since the number of neighbouring macrocells is very limited. By using the self-optimization method implemented by the femtocell vendors, the femtocell can scan for surrounding macrocells on the operator-defined frequencies or RATs and add them to the corresponding NCL automatically.

With the information of the neighbouring macrocells, the pre-release 8 intra/inter-frequency and inter-RAT handover mechanisms are sufficient to cover femtocell to macrocell active mode mobility.

Cell Handover between Femtocells

Since femtocells normally operate at a very low transmit powers, the number of surrounding femtocells that the serving cell can detect is also very limited. For the femtocells whose cell information are configured in the serving femtocell, the normal handover mechanisms defined for macrocells can be applied.

Although the serving cell can update its NCL by sensing the environment periodically, since the neighbouring femtocells can be quite dynamic, some newly added or powered up femtocells may not be included in the femtocell's NCL. For the handover from a femtocell to a femtocell that is not listed in the system information, the mechanism is not considered in pre-release 8 and possibly could be implemented by a similar solution as introduced in handover from macrocell to femtocell.

7.7.2 Cell Handover in Release 8

By introducing the femtocell-related parameters and functionalities (i.e. CSG ID, reserved PSC/PCI and UE autonomous search function), the network can do a more enhanced and efficient handover compared with pre-release 8 UTRA. For example, a CSG-capable UE can find the surrounding femtocells without asking the serving cell to provide NCL. And a UE is able to filter the non-allowed femtocells and avoid doing cell measurements on these cells.

For handover from femtocell to macrocell or between femtocells, the procedures are straightforward as in pre-release 8. In addition, by using the CSG ID and reserved PSC/PCI, the efficiency of the handover procedures are further improved.

Although the autonomous search function can help the UE to find the neighbouring femtocells without knowing NCL, it will cost time for the UE to scan for femtocells on intra-frequency, inter-frequency or even inter-RAT layers. In Cell_DCH state, handover from a non-CSG cell to an allowed CSG cell is not within the scope of release 8 [33]. The handover mechanisms that can keep the service continuity and low latency are still under discussion in 3GPP and will be supported in release 9.

Handover to Femtocell

During cell handover from a serving macrocell to a femtocell, the UE may trigger the UE autonomous search function to scan for neighbouring femtocells. It may only search on the dedicated frequency(ies), if 'Dedicated CSG frequency(ies)' IE is present. In addition, it can shrink the search space only on the reserved PSCs/PCIs if it knows the PSC/PCI split information. The UE can then carry out cell measurements on these found femtocells and report the measurement results to the serving macrocell. However, it has been shown [8] that due to the reuse of PSCs/PCIs, the problem of PSC/PCI confusion may still occur as in pre-release 8 UTRA. Therefore, there is a need to provide assistance information to the network to identify the correct target femtocell.

Recently, many efforts have been made to solve this problem in both 3GPP RAN 2 and RAN 3 [36–42]. The general idea of these is to enable the UE either to provide the Cell

Global Identity (CGI) or the fingerprint/location information of the femtocells to the serving cell. The serving cell can then identify the femtocell, based on this information. Several types of assistance information and resulting handover procedure are discussed below.

- Reading CGI in System Information
  During the handover procedure, the UE first measures femtocell frequencies by using the normal gap pattern and sends a measurement report with corresponding PSC/PCI to the serving macrocell. Then the macrocell will decide which femtocell the UE may handover to. When the PSC/PCI confusion occurs, the UE is requested to read the system information of the target femtocell to obtain the CGI as shown in Figure 7.26. Generally, there are two ways for the UE to read the system information, listed below.
  - long gap approach
    In this approach, the serving macrocell allocates a long gap to the UE by measurement configuration in order to receive system information of the target femtocell after the first measurement report. The UE can then receive the CGI during the long gap. The CGI will then be sent to the serving macrocell by a second measurement report. After the target femtocell is uniquely identified, the normal handover procedure can be performed.
    If an appropriate long gap is allocated, the UE is able to read the CGI from the target femtocell with ease. However, the additional signallings to reconfigure the gap will increase the complexity of the handover procedures. In addition, for a UE with realtime service (i.e. Voice over IP (VoIP)), the long gap will affect the service continuity.
  - DRX approach
    The serving macrocell can also request the UE to read the system information during the Discontinuous Reception (DRX) periods. To reduce the delay in the handover procedure, it is believed that the UE may autonomously try to read the CGI of the neighbouring femtocells before the serving cell meets the measurement report condition [38, 43].

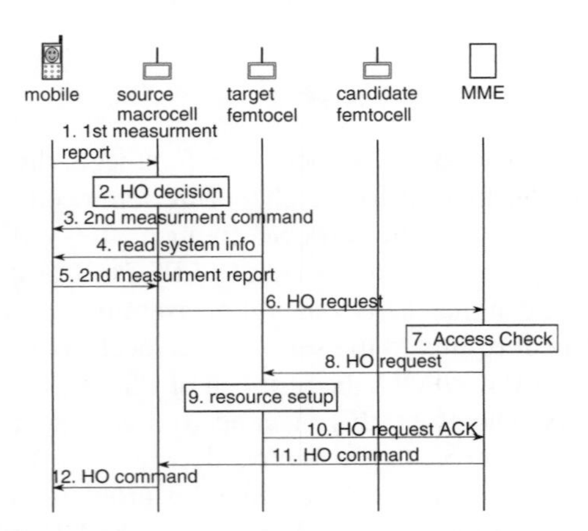

Figure 7.26 Solve PCI/PSC confusion by reading CGI in system information in handover from macrocell to femtocell in release 8

By using the DRX approach, no extra gap reconfiguration is needed and the service continuity is not disturbed. However the DRX period needs to be long enough for the UE to read the CGI in the system information from the target femtocell. For the real time service, like video calls, such a long DRX period is unlikely to be available when the UE is in active mode.

- Using Fingerprint Information
As stated above, methods to identify the femtocell by reading CGI could be inefficient and may increase latency during cell handover. The fingerprint of a cell (which may consist of the cell's cell ID, location and its neighbouring cell information, etc.) is used to characterize that cell uniquely. Such a feature makes it a possible alternative for determining the identity of the femtocell thus reducing the system information reading. In [8], a solution based on fingerprint information is proposed. The solution can be either UE-based or network-based.
  - UE based approach
  With this approach, the UE is able to create the fingerprint for the femtocell once it has been visited. These fingerprints will be stored in association with the UE's ACL in the USIM. When a femtocell is detected or measured, the UE will determine the identity of the femtocell by verifying its fingerprint with the stored information. If the femtocell is recognized, the UE will then include the corresponding cell information (including CGI and CSG ID) in the measurement report and send it to the serving cell.

  This approach has very little impact on the CN. However, the UE needs to be able to create and manage the fingerprints of the femtocells, which are quite dependent on the UE implementation. For the open/hybrid access femtocells in release 9, it may not be possible to put all the fingerprints of these cells in the UE. Furthermore, the effectiveness of this approach is quite sensitive to the accuracy of the fingerprint measurements.
  - CN based approach
  With this approach, the CN will determine the correct target femtocell instead of the UE. This requires that the CN has the fingerprints of all the installed femtocells. Also, the UE needs to send fingerprints with the cell measurement of the detected femtocell to the serving cell. The CN will then verify the identity of a femtocell by its fingerprint.

  This approach requires the network to understand and maintain fingerprints of each femtocell, which may have a big impact on the current CN. Besides, such a large database may not be easy to maintain and manage. As the femtocells can be either added or removed and the locations of the femtocells are not stationary, this adds the additional problem of updating the fingerprint information. As in the previous approach, this also has an issue of location accuracy.

Handover from Femtocell

Femtocells normally hold the information of the surrounding macrocells in their NCL by a self-configuration function. Furthermore, the PSC/PCI confusion is not likely to happen in this situation. The handover procedure from a femtocell to a macrocell is expected to be the same as procedures specified in [6] and [7].

Handover between Femtocells

In release 8, femtocells are usually expected to be deployed in a home-based scenario. A handover between femtocells rarely happens in such a scenario since the owner of the femtocell would not like the neighbour to camp on his/her femtocell. In case the UE has the CSG ID of the target femtocell in its ACL, the CN based Radio Access Network Application Part (RANAP)/S1 is sufficient to handle the handover procedures between these two femtocells.

However, for the enterprise or metro-zone scenario, the femtocells normally belong to the same CSG. A user may walk through one femtocell to another very often and most of the handovers are femtocell-to-femtocell handovers. This may cause a large amount of signalling latency if the CN based RANAP/S1 handover is used.

It was recently agreed in RAN#43 to introduce a few femtocell enhancements for release 9 [1]. One of these enhancements is to improve the femtocell-to-femtocell connected mode handover. A number of mechanisms for supporting enhanced femtocell-to-femtocell handover is proposed for both UTRAN and EUTRAN [44–48]. The general idea is to move most of the handover signallings to femtocell-GW and/or femtocell so that signalling latency to the CN can be reduced. In this section, the femtocell-to-femtocell handover based on CN, femtocell-GW and femtocell are discussed.

- CN Coordinated
  In the CN coordinated scenario, the CN manages the whole handover procedure and all of the handover signalling goes down to the CN as shown in Figure 7.27. The access type (i.e. open, closed and hybrid) and the CSG ID, if the target femtocell has them, will be obtained by the CN and a corresponding access control will be carried out. If the target femtocell is allowed to be accessed, the CN will then route the handover request towards the target femtocell-GW. The femtocell-GW will finally send the RANAP handover request to the target femtocell.

  Modification is unnecessary to support CN coordinated handover between femtocells. In addition, the CN coordinated handover is applicable for use in handover between intra- and inter-GW femtocells. However, the CN will have an extremely heavy load when handling a large number of femtocell-to-femtocell handover requests.
- Femtocell-GW Coordinated
  In the femtocell-GW coordinated scenario, most of the handover message is handled by the femtocell-GW instead of CN, as shown in Figure 7.28. In this case, the femtocell-GW reads the cell information of the target femtocell and performs the access control for the non-CSG UE. For the CSG-capable UE, the access control shall be done by the CN and the result will be sent back to femtocell-GW. If the target femtocell is allowed access, the femtocell-GW will then send the handover request to it. Since the femtocell-GW only has the information of the connected femtocells, it is applicable only to the intra-GW femtocells. If the source and target femtocells belong to a different femtocell-GW, the CN coordinated handover procedure should be invoked instead.

  By handling the handover procedure using the femtocell-GW, the handover latency and the load of CN are reduced. However, new functionalities need to be added to the femtocell-GW so that it is able to read and forward the handover request message.

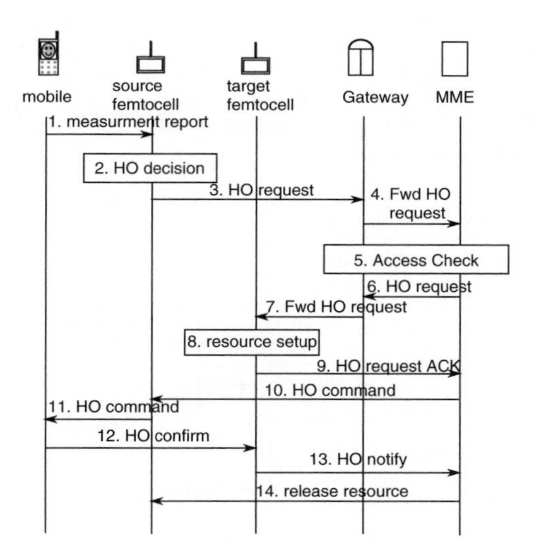

Figure 7.27 CN coordinated cell handover procedures between femtocells in release 9

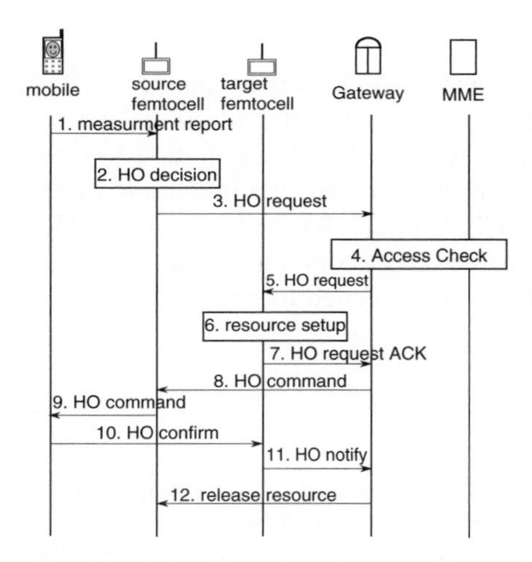

Figure 7.28 Femtocell-GW coordinated cell handover procedures between femtocells in release 9

- Femtocell Coordinated

  In order to enable femtocell coordinated handover, a new interface between femtocells needs to be created. Similarly to the cell handover over X2 in EUTRAN, the handover is directly performed between the source and target femtocells as shown in Figure 7.29. In this case, fewer handover messages will be needed in order to perform handover between femtocells. Functionality of Access Control should be performed in the femtocell-GW for the non-CSG UE and in the CN for the CSG-capable UE. For

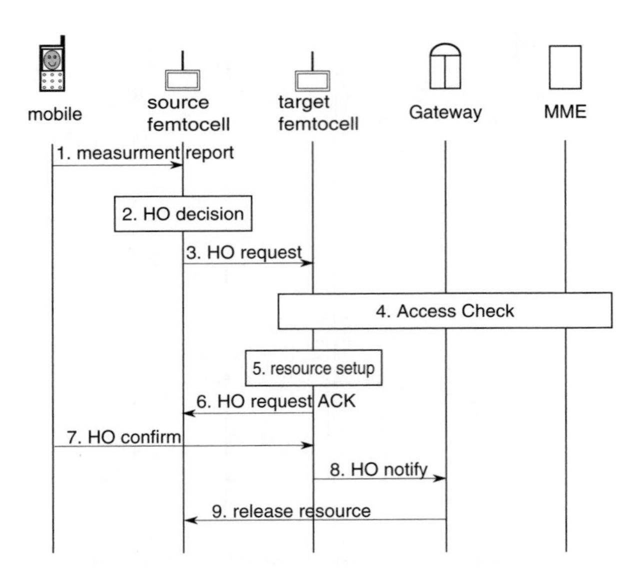

Figure 7.29 Femtocell coordinated cell handover procedures between femtocells in release 9

the special but common intra-CSG handover, the access control can be ignored for the UE's access to the target femtocell. After receiving the handover complete message, the femtocell-GW will trigger the UE deregistration towards source femtocell.

The direct message switching makes the preparation phase very fast. The handover procedure latency and CN load can be further reduced. However, this requires a new interface in order to enable direct communication between femtocells. Besides, modifications of femtocells are also needed to enable them to encode/decode the handover message.

References

[1] Alcatel Lucent, 'Support of Home NB and Home eNB enhancements RAN3 aspects,' 3GPP-TSG RAN, Biarritz, France, Tech. Rep. RP-090349, Mar. 2009.
[2] 3GPP TR 25.820, '3G Home NodeB Study Item Technical Report,' 3GPP-TSG RAN, Valbonne, France, 8.2.0, Sep. 2008.
[3] Nokia Siemens Networks, 'Simple CSG for REL8,' 3GPP-TSG RAN, Shenzhen, China, Tech. Rep. R2-081735, Mar. 2008.
[4] Huawei, 'Consideration on reserving Phy ID for UTRA hNB cell,' 3GPP-TSG RAN, Kansas City, Missouri, USA, Tech. Rep. R2-082282, Mar. 2008.
[5] Qualcomm Europe, 'H(e)NB PCI/PSC and Frequency Identification,' 3GPP-TSG RAN, Prague, Czech Republic, Tech. Rep. R2-085475, Oct. 2008.
[6] 3GPP TS 25.331, 'Radio Resource Control (RRC),' 3GPP-TSG RAN, 8.6.0, Mar. 2009.
[7] 3GPP TS 36.331, 'Radio Resource Control (RRC),' 3GPP-TSG RAN, 8.5.0, Mar. 2009.
[8] Motorola, 'PCID confusion,' 3GPP-TSG RAN, Seoul, Korea, Tech. Rep. R2-092307, Mar. 2009.
[9] 3GPP TS 25.304, 'User Equipment (UE) procedures in idle mode and procedures for cell reselection in connected mode,' 3GPP-TSG RAN, 8.5.0, Mar. 2009.
[10] 3GPP TS 36.304, 'User Equipment (UE) procedures in idle mode,' 3GPP-TSG RAN, 8.5.0, Mar. 2009.
[11] ARICENT, 'Challenge in Deployment of UMTS/HSPA Femtocell,' White Paper, 2008.

[12] 3GPP TS 24.008, 'Mobile radio interface Layer 3 specification; Core network protocols; Stage 3,' 3GPP-TSG CT, 8.5.0, Mar. 2009.
[13] 3GPP TS 23.003, 'Numbering, addressing and identification,' 3GPP-TSG CT, 8.4.0, Mar. 2009.
[14] 3GPP TS 23.122, 'Non-Access-Stratum (NAS) functions related to Mobile Station (MS) in idle mode,' 3GPP-TSG CT, 8.5.0 Release 8, Mar. 2009.
[15] 3GPP TS 24.301, 'Non-Access-Stratum (NAS) protocol for Evolved Packet System (EPS) Stage 3,' 3GPP-TSG CT, 8.1.0, Mar. 2009.
[16] 3GPP RAN, 'Reply LS on access control for CSG cells,' 3GPP-TSG RAN, Prague, Czech Republic, Tech. Rep. R2-086035, Oct. 2008.
[17] 3GPP CT1, 'LS on access control for CSG cells,' 3GPP-TSG RAN, Prague, Czech Republic, Tech. Rep. R2-084948, Aug. 2008.
[18] Vodafone Group, 'Open and semi-open access support for UTRA Home NB,' 3GPP-TSG RAN, Prague, Czech Republic, Tech. Rep. R2-085280, Oct. 2008.
[19] 3GPP TS 22.220, 'Service requirements for Home NodeBs and Home eNodeBs,' 3GPP-TSG SA, 9.0.0, Mar. 2009.
[20] 3GPP TS 36.300, 'Evolved Universal Terrestrial Radio Access (E-UTRA) and Evolved Universal Terrestrial Radio Access Network (E-UTRAN); Overall description; Stage 2,' 3GPP-TSG RAN, 8.8.0, Mar. 2009.
[21] Nokia Siemens Networks, 'Paging Optimisation in LTE CSGs,' 3GPP-TSG RAN, Athens, Greece, Tech. Rep. R2-090317, Feb. 2009.
[22] ZTE, 'Paging Optimisation,' 3GPP-TSG RAN, Athens, Greece, Tech. Rep. R3-090461, Feb. 2009.
[23] Huawei, Airvana, 'Paging Optimization for 3G HNB,' 3GPP-TSG RAN, Seoul, Korea, Tech. Rep. R3-090784, Mar. 2009.
[24] Samsung, 'Paging optimization for hybrid/open HeNB,' 3GPP-TSG RAN, Seoul, Korea, Tech. Rep. R3-090860, Mar. 2009.
[25] 3GPP TS 22.011, 'Service accessibility,' 3GPP-TSG SA, bb, Tech. Rep. 9.1.0, Mar. 2009.
[26] Panasonic, 'CSG Manual Search and Selection for Connected mode UE,' 3GPP-TSG RAN, Prague, Czech Republic, Tech. Rep. R2-086306, Nov. 2008.
[27] 3GPP RAN, 'LS on reselection handling towards non-allowed CSG cell,' 3GPP-TSG RAN, Jeju, Korea, Tech. Rep. R2-084891, Aug. 2008.
[28] Qualcomm Europe, 'Intra-frequency reselection indicator for CSG cells,' 3GPP-TSG RAN, Prague, Czech Republic, Tech. Rep. R2-085383, Oct. 2008.
[29] Qualcomm Europe, 'Parameter for HNB White List Cell Selection,' 3GPP-TSG RAN, Jeju, Korea, Tech. Rep. R2-084346, Aug. 2008.
[30] Sharp, 'Problems associated with using a constant CSG offset when performing reselection ranking,' 3GPP-TSG RAN, Prague, Czech Republic, Tech. Rep. R2-085107, Oct. 2008.
[31] Huawei, 'Inter frequency Cell Reselection from macro cell to CSG,' 3GPP-TSG RAN, Prague, Czech Republic, Tech. Rep. R2-085659, Oct. 2008.
[32] Qualcomm Europe, 'UTRA HNB Idle Mode (Re)selection and UE Access Control,' 3GPP-TSG RAN, Warsaw, Poland, Tech. Rep. R2-083392, Jul. 2008.
[33] 3GPP TS 25.367, 'Mobility Procedures for Home NodeB; Overall Description; Stage 2,' 3GPP-TSG RAN, 8.1.0, Mar. 2009.
[34] Alcatel Lucent, 'Femtoceller,' Technical Seminar, 2009.
[35] ip.access, 'Femtocell handover,' White Paper, 2008.
[36] Panasonic, 'CSG cell handover,' 3GPP-TSG RAN, Sorrento, Italy, Tech. Rep. R2-080884, Feb. 2008.
[37] Huawei, 'Inbound mobility Issues for UEs in RRC Connected,' 3GPP-TSG RAN, Warsaw, Poland, Tech. Rep. R2-083514, Jul. 2008.
[38] Telecom Italia, Qualcomm Europe, Samsung, 'Way forward for handover to HeNB,' 3GPP-TSG RAN, Jeju, Korea, Tech. Rep. R2-084534, Aug. 2008.
[39] Qualcomm Europe, 'Connected mode mobility in the presence of PCI confusion for HeNBs,' 3GPP-TSG RAN, Seoul, Korea, Tech. Rep. R2-092113, Mar. 2009.
[40] InterDigital, 'Inbound handover to CSG and hybrid cells,' 3GPP-TSG RAN, Seoul, Korea, Tech. Rep. R2-092142, Mar. 2009.
[41] Alcatel Lucent, 'Handling of CSG for in-bound Mobility,' 3GPP-TSG RAN, Seoul, Korea, Tech. Rep. R3-090745, Mar. 2009.

[42] Samsung, 'Inbound mobility for HeNB,' 3GPP-TSG RAN, Seoul, Korea, Tech. Rep. R3-090858, Mar. 2009.
[43] Samsung, 'Hand-in to a CSG cell,' 3GPP-TSG RAN, Seoul, Korea, Tech. Rep. R2-092404, Mar. 2009.
[44] ZTE, 'Handover procedure between HNBs,' 3GPP-TSG RAN, Seoul, Korea, Tech. Rep. R3-090759, Mar. 2009.
[45] Huawei, 'Discussion of HeNB to HeNB Handover,' 3GPP-TSG RAN, Seoul, Korea, Tech. Rep. R3-090805, Mar. 2009.
[46] 3GPP, '3G HNB to 3G HNB Handover,' 3GPP-TSG RAN, Seoul, Korea, Tech. Rep. R3-090856, Mar. 2009.
[47] Samsung, 'HNB to HNB Handover Architecture,' 3GPP-TSG RAN, Seoul, Korea, Tech. Rep. R3-090868, Mar. 2009.
[48] NEC, 'HNB to HNB Relocation: possible solutions,' 3GPP-TSG RAN, Seoul, Korea, Tech. Rep. R3-090933, Mar. 2009.

8

Self-Organization

David López Pérez and Jie Zhang

Since the number and position of the femtocells are unknown by the operator, and because they could be moved or switched on/off at any time by the users (individualistic nature), classic network design cannot be applied to configure and optimize a femtocell network. Moreover, and due to the non-technical expertise of the femtocell customers, femtocells must be autonomous units able to integrate themselves into the existing radio access network of the operator, causing the least impact on the existing wireless communication systems. Consequently, femtocells must be plug-and-play devices that exhibit a significant degree of self-organization.

The use of sophisticated self-organization techniques will play a very important role in successfully deploying and managing a large femtocell layer over the existing macrocell network. Self-organization will allow femtocells to:

- integrate themselves into the existing networks with minimal human involvement,
- learn about their radio environment (neighbourhood and interference),
- tune their parameters accordingly.

In this way, femtocells will be able to react to the changing conditions of the network, traffic and channel, and mitigate cross- and co-layer interference. The result of this being the enhancement of the overall performance of the system.

In this chapter, an overview of self-organization in femtocell deployments is presented. In addition, the key self-organizing features that a femtocell must have are analysed in detail. The rest of the chapter presents the following:

- in Section 8.1, the operator's need for Self-Organizing Networks (SONs);
- in Section 8.2, the life cycle of self-organization: measurements, self-configuration, self-optimization and self-healing;
- in Section 8.3, the need for self-organization in femtocell networks;
- in Section 8.4, the booting procedure of a femtocell after power up;

Femtocells: Technologies and Deployment Jie Zhang and Guillaume de la Roche

- in Section 8.5, the different techniques for sensing and learning about the changes on the environment of the femtocell;
- in Section 8.6, femtocell self-configuration and self-optimization techniques.

8.1 Self-Organization

8.1.1 Context

In the early years of wireless systems, when this field was only a research area and the number of users was small compared with the available resources, the efficiency of wireless networks was not an issue, and engineers were only worried about keeping them working. However, over the last two decades, this field has matured and the number of customers and demands of users have dramatically increased. As a result, the planning and optimization of the current networks has become a key factor not only for making these wireless systems functional, but also to increase the revenue of the operators and the satisfaction of the users.

In the 1990s, Global System for Mobile communication (GSM) and Digital Communication System (DCS) [1] networks were deployed to fulfill the user necessities, mostly based on voice services. In these kinds of system, the spectrum is divided into several channels. Multi-user access is achieved by assigning different channels and time slots to different users. At that time, the acquisition of new cell locations and the installation of new cell sites in order to extend the radio coverage were one of the major concerns of network operators. In addition, great efforts were made by Radio Frequency (RF) engineers in the area of automatic frequency planning [2], and antenna azimuth/tilt selection [3] as a way of mitigating interference and enhancing the performance of the network. These tasks, together with the assignment of cell identities, and the set up of neighbouring lists and handover parameters, represented one of the first efforts in the area of what is known nowadays as network planning and optimization. However, the capacity improvements through optimization were not enough to cope with the continuously increasing traffic demands of the customers. Therefore, the development of new technologies that could replace/complement the existing GSM and DCS systems in order to provide new services and higher throughputs was necessary.

At the beginning of this decade (2000), Code Division Multiple Access (CDMA) [4] appeared as one of the most suitable technologies for use in future network deployments. Nowadays, it is used to fulfill the demands of users, based not only on voice but also on data services due to the expectation and popularity of mobile Internet applications. In this kind of system, all the users share the same spectrum, and multi-user access is achieved by assigning different pseudo-random codes with special interference properties to different users [5]. Therefore, radio frequency planning is completely unnecessary, and interference avoidance is handled by means of network planning, power control and these codes. In this third generation (3G) of cellular networks, given the huge number of parameters to configure, and the trade-off between coverage and capacity (cell breathing phenomena [6]), networks will not be able to operate efficiently without the help of advanced network planning and optimization tools. These tools cover a wide spectrum of topics, e.g. base station location, antenna azimuth/tilt selection, power control algorithms, scrambling code assignment. However, the use of these tools require a large human involvement and expertise.

In addition, High Speed Packet Access (HSPA) [8] has emerged in the last few years (2005–2008) as a new way to improve the throughput of the user, allowing high speed connectivity. This new technology moves some Medium Access Control (MAC) capabilities from the Radio Network Controller (RNC) to the NodeB, e.g. fast scheduling, fast Hybrid Automatic Repeat reQuest (HARQ). In this way, these functionalities are closer to the air interface, being the system most responsive to changes in the traffic and channel due to delay minimization. Moreover, new techniques such as Adaptive Modulation and Coding (AMC) and fast power control are supported in the DownLink (DL) and UpLink (UL), respectively. As a result, new challenges in terms of planning and optimization need to be addressed by the network designers in order to successfully combine both networks, CDMA and HSPA. This fact increases the work load and the expenses of the operators.

Furthermore, it is predicted that in the near future the load on the network and the demands of the users will continue growing (Figure 8.1), and new services will be needed, requiring higher levels of Quality of Service (QoS) and data rates. To satisfy these requirements, vendors and operators are working on the development of new technologies such as Wireless Interoperability for Microwave Access (WiMAX) [9] and Long Term Evolution

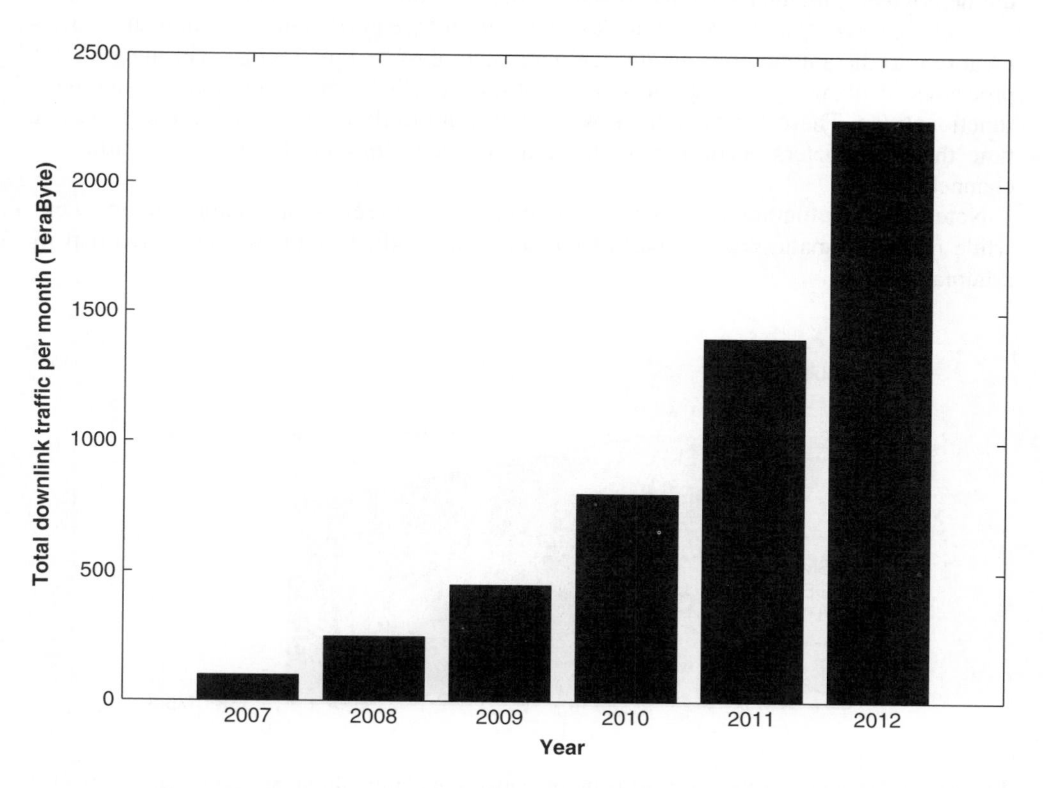

Figure 8.1 Wireless network traffic prediction from 2007 to 2012 [7]

(LTE) [10], which are considered the most suitable options for future cellular system deployments. They are both based on an Orthogonal Frequency Division Multiplexing (OFDM) [11] Physical (PHY) layer, which supports several key features necessary for delivering broadband services at high mobility, e.g. scalable channel bandwidths, high spectral efficiency, multipath robustness. These technologies will realize the possibility of a truly mobile broadband connection, at the expense of increasing again the complexity and cost of the network.

8.1.2 *Definition*

The pressure to be competitive forces network operators to take the reduction of the complexity and cost of the current networks as a key driver for future deployments (Figure 8.2). Due to this fact, the minimization of the operational effort and cost, and the enhancement of the network performance through self-configuration and self-optimization is of great interest. Standardization bodies and groups such as 3rd Generation Partnership Project (3GPP) [12], Next Generation Mobile Networks (NGMN) [13], Socrates Project [14] and Femto Forum [15] have already identified self-organization as being needed for the deployment, maintenance and sustainability of future wireless networks.

A self-organizing network, defined as a network that requires a minimal human involvement due to the autonomous and/or automatic nature of its functioning, will integrate the processes of planning, configuration and optimization in a set of autonomous/automatic functionalities. These functionalities will allow femtocells to scan the air interface, and tune their parameters according to the dynamic behaviour of the network, traffic and channel.

Note that an autonomous activity is one that does not require any human involvement, while in an automatic process, part of the action is handled by the machine, and part by a human being.

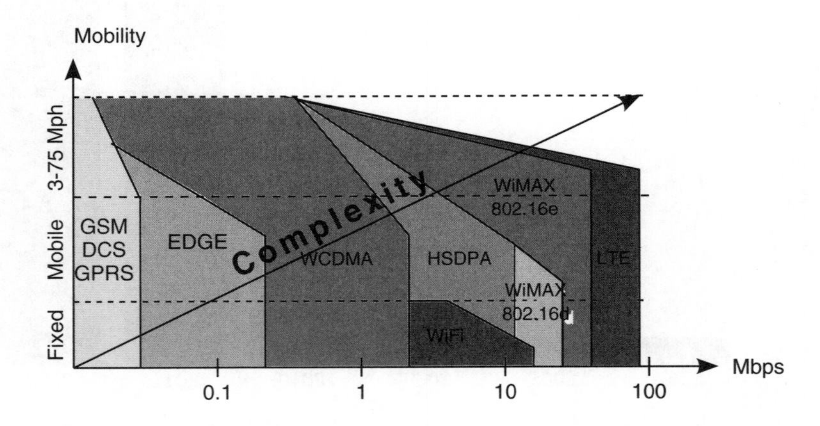

Figure 8.2 Diagram of the trade-off between network mobility, throughput and complexity [16]

8.1.3 Drivers

The need for self-organization in future wireless networks is driven by the following aspects [17]:

- To achieve a substantial reduction in the CAPital EXpenditure (CAPEX) and OPerational EXpenditure (OPEX) of the network by reducing the human involvement.
  Due to the complexity of current wireless networks and given the huge number of parameters to configure, network operators invest large amounts of money in planning the deployment of new cell sites, and optimizing the performance of the operative Base Stations (BSs). To carry out these tasks, operators use different tools, e.g. planning tools, drive test, system statistics, interference trace, and they have to post-process the resulting data. This translates into a large human involvement, and thus cost.
  Moreover, when new technologies such as LTE will be deployed by the network operators, it is expected that in many regions more than three technologies will coexist (2G, 3G and LTE). However, it should not incur an additional operational expenditure.
  The reduction of the human involvement by means of self-organization will allow network operators to remain competitive due to the reduction of the cost. It will enable, for example, the possibility of offering lower prices, improving the user satisfaction.
- To optimize the performance of the network in terms of coverage, capacity or QoS.
  Automated processes such as the configuration of the neighbour list, handover parameters or resource allocation will allow the system to adapt itself rapidly to the changes in the network and the fluctuations in the traffic and channel conditions. Avoiding long periods of time between manual optimization processes, the BSs will be more responsive, and the performance of the overall system will be optimized.
- To allow the deployment of a larger number of small cells.
  Self-organization will allow the final customer to deploy a new type of BS called Home Base Station (HBS) or Femtocell Access Point (FAP), since these devices will be able to integrate themselves into the existing network, avoiding operator involvement. In addition, this device will help the operator to extend the indoor coverage where it is limited or unavailable, and to enhance the user experience due to the short distance between transmitter and receiver.

8.2 Self-Configuration, Self-Optimization and Self-Healing

The nodes of a SON must be self-configuring and self-optimizing units able to integrate themselves into the network of the operator, minimizing the involvement of the RF engineers in planning and optimization tasks. By automating these procedures the nodes of a SON will be more responsive to the changing conditions of the environment, enhancing their performance due to the minimization of the delay between consecutive optimization tasks.

In this section, we are differentiating between four phases in the life cycle of a SON (Figure 8.3): measurements, self-configuration, self-optimization, and self-healing

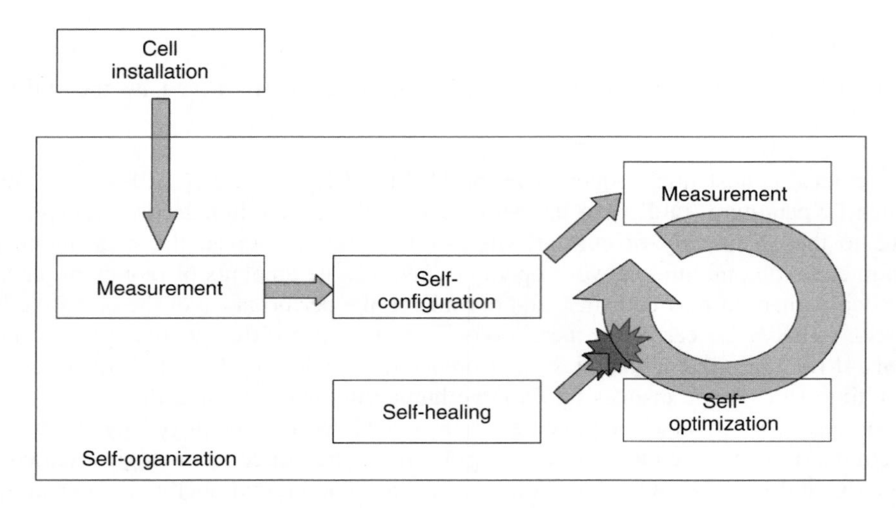

Figure 8.3 Life cycle of self-organizing cell

[18, 19]. Subsequently, an example of the life cycle of a given self-organizing BS is presented for illustration porpuses.

The measurements phase plays a very important role in the life cycle of a SON. BSs collect measurements in order to assess the behaviour of the network, and trigger the adequate actions. These measurements are taken via multiple sources e.g. Operation Support Subsystem (OSS), network counters, measurements coming from neighbouring nodes or user terminals. To provide relevant information to the different configuration and optimization tasks, these raw data, e.g. traffic patterns, user mobility, interference/fading conditions, must be processed. The required format, accuracy and periodicity of the delivered information depends on the specific mechanism to be self-configured or optimized, and these determine the quality of the tuning.

A newly added BSs sets up its software and parameters (neighbouring list, handover configuration, pilot power) using self-configuration. In this way, this BS integrates itself into the network, while aiming to minimize the impact on existing BSs and User Equipments (UEs). Moreover, existing BSs can use the self-configuration phase to react to the introduction of new BSs or new features in the network, e.g. antennas, services, bearers. Before these new upgrades become operative, an initial reconfiguration of a number of algorithms and parameters is generally required.

In the self-optimization phase, the processed measurements are used periodically to adjust the algorithms and parameters of the BS to the changing conditions of the environment. Using the knowledge of the environment, the coverage and capacity of the network can be optimized, filling coverage holes and providing interference mitigation. In case the self-optimization is incapable of meeting the performance requirements, the BS will trigger alarms with accompanying suggestions for human intervention [18], e.g. deploying a new BS.

Self-healing techniques will resolve the loss of coverage and/or capacity due to failures in the network, e.g. damaged BSs. This is done by adjusting the algorithms and parameters in surrounding BSs (cooperation). Once the failure is solved, all parameters are restored to their original configurations.

8.3 Self-Organization in Femtocell Scenarios

8.3.1 Context

An extensive deployment of femtocells is foreseen. According to a recent survey [20], it is estimated that by 2012 there could be 70 million Universal Mobile Telecommunication System (UMTS) femtocells installed in homes, serving more than 150 million customers. Although this number may be overestimated, it gives an idea of the acceptance of femtocells.

Femtocells are low-power base stations initially designed for indoor usage that allow cellular network providers to extend indoor radio coverage where it is limited or unavailable. On the one hand, femtocells provide radio coverage of a certain cellular network standard, e.g. GSM, UMTS, WiMAX or LTE. On the other hand, they are connected to the service provider via a broadband connection, e.g. Digital Subscriber Line (DSL) or optical fibre.

The main differences between femtocells and macrocells are as follows:

- femtocells will be deployed in large numbers compared to the macrocells,
- they can be turned on and off or moved at any time by the customer,
- femtocells are initially designed to provide indoor coverage,
- they are low-cost and low-power devices,
- a femtocell may be used only by a few users (subscribers),
- physical access to the femtocell is unlikely for the operator,
- femtocell access can be restricted to a Closed Subscriber Group (CSG).

Femtocells offer advantages to both customers and network operators:

- Users will enjoy better signal qualities due to the reduced distance between the transmitter and the receiver, the result of this being more reliable communications and higher throughputs, as well as power and battery savings.
- From the operator's perspective, femtocells will extend indoor coverage and enhance system capacity. Femtocells will also help to manage the exponential growth of traffic, thanks to the hand over of indoor traffic to the backhaul link. Moreover, femtocells will reduce the deployment and maintenance cost of the system, since they will be paid for and installed by the customers.

However, these benefits are not easy to accomplish, and a high degree of self-organization is needed to deploy a femtocell layer successfully. For example, the negative effects of cross- and co-layer interference, which could counteract these benefits and downgrade the system performance, must be handled by means of self-organization (power/frequency allocation).

8.3.2 Objectives

The need for self-organization in femtocells deployments is twofold:

- On the one hand, due to the large number and individualistic nature of the femtocells, they cannot be installed and/or maintained by the operators.
- On the other hand, due to the non-technical expertise of the femtocell customers, femtocells cannot be configured or optimized by the customers.

Therefore, the key future of a femtocell must be its plug and play nature for both the operator and the user. Self-organization in femtocell scenarios must provide the following main features among others:

- network and neighbourhood discovery,
- automatic selection of the physical cell identity,
- configuration and optimization of the neighbouring cell list,
- configuration and optimization of the handover parameters,
- configuration and optimization of the RF parameters (power and frequency).

This self-organization can be performed on a regular basis or event triggered [21], and it will allow femtocells to optimize their coverage and capacity, and minimize the probability of ID collision and interference. More detailed information about these features is presented in Section 8.5 and Section 8.6.

8.4 Start-Up Procedure in Femtocells

When a customer buys a femtocell, the network operator provides the customer with the femtocell device and a femtocell ID. This femtocell ID will be used to register and authenticate the femtocell in the network after switching on. Moreover, when the customer buys the femtocell, he/she must provide some information to the operator. For example, the address where the femtocell is going to be installed and the list of femtocell subscribers (registration data). Furthermore, in order to let the customer update the list of subscribers, the operator also gives him/her a secure web site. It is to be noted that, the list of authorized users resides in the core network.

After acquiring the femtocell, the customer only needs to plug the femtocell into a power source and Internet connection to start using it. The customer cannot be assumed to have the knowledge to install or configure the femtocell, hence these processes need to be automatic. Therefore, after power on, the first thing the femtocell does is to connect to the network of the operator through the backhaul connection.

The femtocell is then authenticated and registered into the system as an operative device by the OSS, using the femtocell ID. Afterwards, the femtocell can update its software by downloading the latest available version from the OSS. Note that this software update can also be triggered by the OSS at any time after power on. Subsequently, the femtocell verifies the functioning of this new software, self-testing the installation (Figure 8.4).

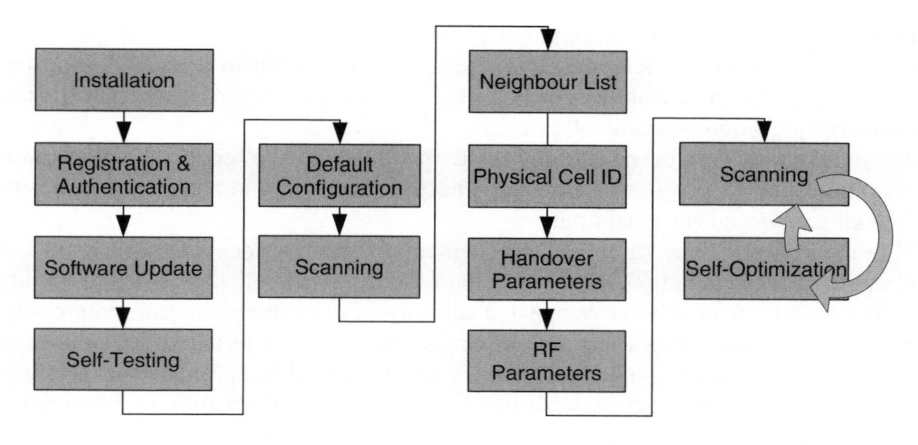

Figure 8.4 Femtocell start up procedure

At this point, the radio parameters of the femtocell must be set to a default configuration. This is done in two steps.

- Fundamental information such as:
  - frequency for DL and UL,
  - scrambling code list, or
  - radio channel bandwidth

  must be provided during the booting procedure by the operator over the backhaul link.
- Network configuration parameters:
  - location, routing and service area code information,
  - neighbouring list,
  - physical cell ID,
  - RF parameters (pilot and maximum data power...)

  can be automatically calculated from information on the macrocell layer provided by the operator (OSS data), and from information on the femtocell layer provided by the users (registration data). These data arrive at the femtocell through the backhaul link.

If the core network does not support this configuration or cannot supply any suggestions, the femtocell will derive these parameters, using data gathered by monitoring the radio channel. However, setting up the femtocell parameters from a blind configuration using only sensing techniques will delay the booting procedure, and might result in an undesirable performance. Therefore, it is advisable that a default configuration is provided by the femtocell firmware or through the backhaul.

The sensing of the radio environment is done by the network listening mode, designed to scan the air interface. By decoding the existing broadcast and control channels, the femtocell synchronizes its internal oscillator and synchronizes the femtocell to the external network. The information derived form the initial sensing is also used to detect new neighbouring macrocells and femtocells. In this way, the default configuration of the femtocell can be set up or reconfigured [22]. For example, the femtocell can add or remove new neighbouring relationships, select/re-select its physical cell ID in order to minimize the collision probability, or tune its handover parameters in order to facilitate the handover procedure towards other cells.

After the femtocell has been self-configured, the life cycle of a femtocell moves towards a self-optimization loop, since the femtocell needs dynamically to adapt its parameters to the changing environment conditions.

Using the network listening mode and other inputs, e.g. broadcast messages, measurement reports, cognitive radio (Section 8.5) the femtocell will collect statistics to optimize its performance dynamically (coverage and capacity). For example, in order to provide an adequate signal quality to its users, and minimize the impact (interference) on other cells, the femtocell will adapt its power and channel usage, as well as optimizing its neighbouring list and handover parameters according to the gathered information (Section 8.6).

8.5 Sensing the Radio Channel

Femtocells must be aware of the presence of neighbouring cells and their power and spectrum allocation in order to maintain the femtocell coverage and avoid interference. Different strategies can be used to achieve this cognitive radio stage, in which the femtocell is able to learn about the structure and behaviour of the network and the channel conditions. The main sources of information for self-organizing algorithms are summarized below.

8.5.1 Network Listening Mode

In this first approach, the sensing capability is implemented in the femtocell device itself. This way, the femtocell will be able to scan the air interface, detect neighbouring cells and tune its network and RF parameters accordingly (Figure 8.5).

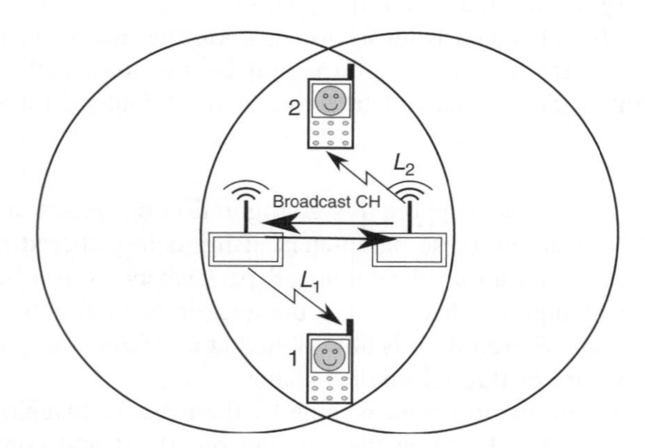

Figure 8.5 Network listening capability. Monitoring the channel, both femtocells will be aware of the presence of the other, and they can coordinate their resource allocation (for example, subchannel assignment in case of OFDMA femtocells) in order to minimize interference

The implementation of a sniffer or network listening capability is essential in order to automate the tasks of cell planning and optimization within a femtocell network. Using this functionality, the femtocell will periodically switch on the sniffer to check network settings, synchronization and interference conditions.

In this case, the femtocell behaves similarly to a user terminal operating in network listening mode. When a UE wants to transmit, it synchronizes with the strongest BS, and decodes the broadcast and control channels before making an attempt to access the network. In a similar way, the femtocell will use the network listening capability to synchronize with the operators network and decode the broadcast and control channels of the neighbouring cells. Moreover, the information collected by the network listening mode can be also used in order to identify surrounding cells and to which operator they belong, distinguish if they are macrocells or femtocells, and estimate the path loss to them [21].

Furthermore, the femtocell can use its own statistics (failed handover ratio, dropped call ratio, blocked call ratio, uplink interference) in order to assess the behaviour of the network, and trigger different configuration and/or optimization actions.

8.5.2 Message Exchange

Femtocells might be able to broadcast information messages that will be received by their neighbouring femtocells, containing interference measurements (receive signal strength) or information about the power and/or scrambling/sub-channel allocation of the broadcaster, e.g. the priority and probability of usage of different RF resources could be broadcast. This way, femtocells will be aware of the present actions and future intentions of neighbouring femtocells, and they can act cooperatively (Figure 8.6).

These messages can be exchanged over existing interfaces or over new ones:

- Femtocells can exchange messages over the femtocell gateway. The source femtocell will send its message to the femtocell gateway, and the femtocell gateway will forward this message to the target femtocell/femtocells.
- Another solution is to establish a new interface between femtocells. This alternative has similarities with the X2 interface defined in LTE to allow communication between eNodeBs, but it has not been extended yet to femtocells [23].

However, these two techniques, network listening mode and message exchange, are limited by the coverage area of the femtocells. For example, if two femtocells are not within range of each other, they will not be able to notice the presence of the other or exchange information (over direct interface) with them. As a result, these femtocells will not be able to coordinate their resource allocation, and users located in the cell edge of these two overlapping femtocells will suffer from physical cell ID collision, inter-cell interference, etc.

This scenario is plotted in Figure 8.7, and it is known as hidden femtocell problem.

8.5.3 Measurement Reports

To solve the hidden femtocell problem, Measurement Reports (MRs) created by the terminals and reported to the femtocells can be used. In this way, a user situated in the cell edge

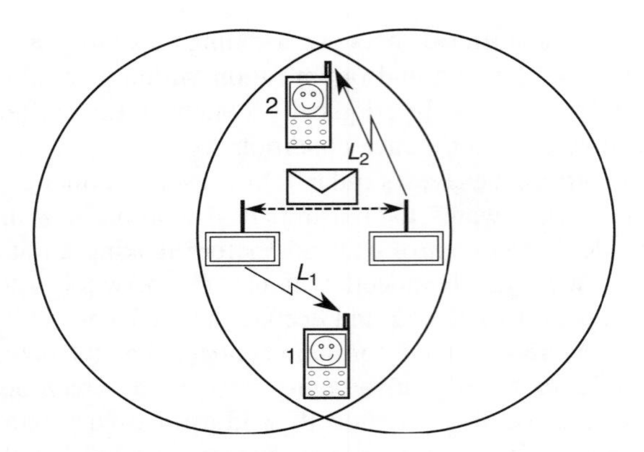

Figure 8.6 Message exchange capability. Exchanging messages, both femtocells will be aware of the actions of the other, and they can coordinate their resource allocation (for example, subchannel assignment in case of OFDMA femtocells) in order to minimize interference

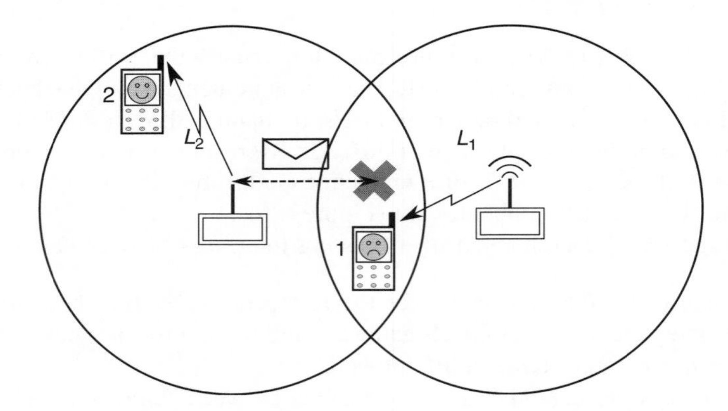

Figure 8.7 Hidden femtocell problem. Since both femtocells are not within range of each other, they can not see each other. As a result, the user situated in the overlapping area suffers from interference due to uncoordinated resource allocation. For example, in case of OFDMA femtocells, users 1 and 2 may be assigned to the same subchannels, resulting in an unsatisfactory performance for user 1 located in the overlapping coverage area of both femtocells

of two overlapping cells can indicate to its serving femtocell the presence and the actions (power and frequency) of other overlapping macrocells and/or femtocells (Figure 8.8).

When a connection is active, the UEs periodically report its signal quality to the femtocell via an MR, which also includes signal measurements from neighbouring cells. If the UE is reporting good signal quality, the femtocell will not take further actions. However, if the signal quality is weak, the femtocell might hand-off the connection to another scrambling/sub-channel/time slot, or initiate a handover to another macrocell or femtocell. Measurements collected through the users attached to the femtocell can indicate, for example, the DL received signal strength of the co-channel and adjacent carrier towards the user terminals.

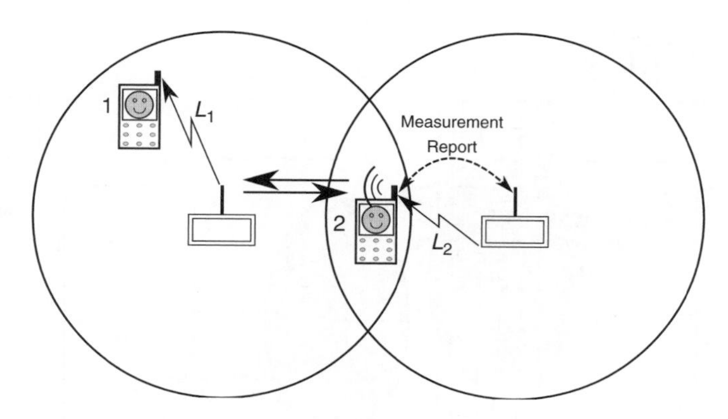

Figure 8.8 Measurement reports. When using measurement reports, the user situated in the overlapping area can alert its serving femtocell of the presence and actions of other hidden femtocells. As a result, the femtocells can coordinate their resource allocation. For example, in case of OFDMA femtocells, knowing that femtocell 1 is using a subset of subchannels for user 1, femtocell 2 will allocate user 2 in other ones. This way, interference is mitigated

MRs are particularly useful because they provide information collected in the environment of the user. Contrary to the information collected by the network listening mode or the broadcast messages, the information provided by the MRs indicates the channel conditions at the position of the user (the one who may suffer from interference).

For example, Figure 8.9 illustrates how the interference conditions of two subscribers of the same femtocell can vary according to their location and environment. Users close to the femtocell will enjoy good signal quality, while users located in opposite rooms, far from their femtocells and close to a neighbouring one will suffer from a large interference.

8.5.4 Cognitive Radio

Since cognitive radios are considered lower priority or secondary users of the spectrum allocated to primary users, it is necessary that these cognitive users do not create interference for potential primary users. Different solutions can be used to underlay, overlay or interweave the secondary user signals with the primary user signals, in such a way that the primary signals are as little influenced as possible by the secondary signals.

In the underlay approach, the secondary users spread their signal over a large bandwidth, minimizing the amount of interference caused to the primary users.

The overlay approach is based on knowledge of the primary user signals by the secondary users. This way, the secondary transmitters can help to relay the primary signal, and the secondary receivers can mitigate interference using dirty paper coding [24].

In the interweave approach, a cognitive radio must be capable of sensing the air interface and opportunistically exploit the unused spectrum by the primary users.

Since underlay techniques are more suitable for spread spectrum technologies, e.g. Ultra Wide Band (UWB), and because it is difficult for the secondary users to obtain the *a priori* knowledge of the primary user signals, interweave techniques are attracting most of the attention in cellular network environments. In a similar way as in the interweave approach,

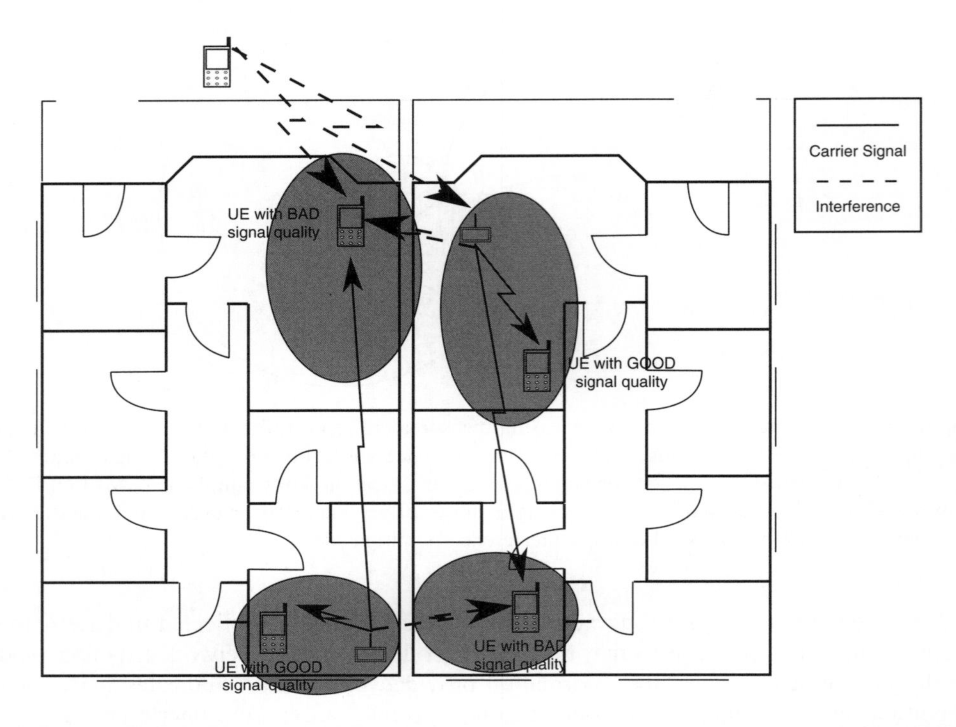

Figure 8.9 User with different channel conditions within the coverage of the same femtocell

femtocells must be able to search the radio channel and estimate which resources are free among the available ones in order to avoid cross-layer and co-layer interference.

Several cognitive radio techniques can be used to detect unused resources across a wide frequency band. For example, match filter, energy or cyclostationary features detection [25]. However, as cognitive radio is outside the scope of this book, they will not be introduced in the following sections.

Nevertheless, let us mention that cognitive radio techniques can be implemented in the femtocell device, but they must not be implemented into femtocell user terminals due to legacy constraints. Femtocells must operate using legacy mobile terminals that do not depend on new user equipment.

Despite having different methods of learning about the air interface, what information should be used and how it should be combined is still an open issue. Moreover, different trade-off has to be taken into account. For example:

- When and for how long a femtocell should collect measurements? Note that a femtocell with only one RF interface cannot collect measurements and transmit or receive simultaneously.

- Measurement reports coming from the user terminals will provide accurate information about the user environment at the expense of raising the overhead information and processing time.

Therefore, further research is needed in this area to understand the benefits and drawbacks of each technique, and to learn how to combine these sources to get the most accurate sensing.

8.6 Self-Configuration and Self-Optimization of Femtocell Parameters

8.6.1 Physical Cell Identity (PCI)

Context

The PCI is normally used to identify a cell for radio purposes, e.g. camping and handover procedures are simplified by explicitly providing a list of PCIs that mobiles must monitor. The PCI of a cell does not need to be a unique network-wide cell identifier. However, this must be unique on a local scale to avoid collision and/or confusion with neighbouring cells (Figure 8.10).

- *Collision* happens when a PCI is not unique in the coverage area of a cell.
- *Confusion* occurs when a PCI is not unique in the neighbourhood of a cell.

Traditionally, the PCI is part of the initial configuration of the cell, and it is set up by the network designers using network planning tools. In this way, collision-free and confusion-free PCI assignments are ensured across the network, thus avoiding possible problems [26].

In GSM/DCS networks, the pair Broadcast Control Channel (BCCH)/Base Station Identity Code (BSIC) is used to identify unequivocally a cell in a geographical area. The BCCH frequency carries crucial control information of the cell, and it should be unique within its

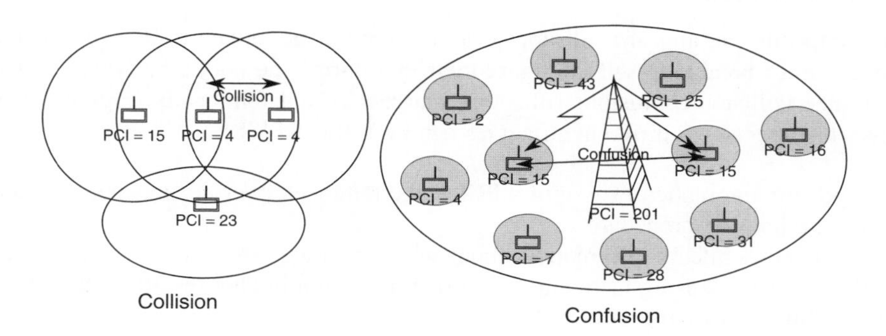

Figure 8.10 Collision and confusion

neighbourhood. However, in proximity to national borders or in areas of tight frequency reuse, it is possible that a mobile will capture more than one BCCH on the same frequency. Then, the BSIC will be used to distinguish the target cell from those other cells transmitting their BCCH on the same frequency. Due to the important role played by the pair BCCH/BSIC, the operator sets up these values using automatic frequency planning tools.

In UMTS and HSPA networks, a set of 512 scrambling codes are reserved to identify unequivocally a cell. The planning process is simpler in respect to that used in GSM/DCS networks, since the available number of scrambling codes is much larger than the number of BCCH frequencies. As a result, no automatic frequency planning is needed, and operators only have to assign different scrambling codes to different cells within neighbourhoods. The assignment of these scrambling codes is done using scrambling code planning tools.

Nevertheless, because there would be too many comparisons to make, terminals could not search for 512 codes without experiencing long delays from power-on to the service availability. To mitigate this delay, the set of 512 codes has been divided into 64 secondary synchronization codes, each of them containing eight primary scrambling codes. During the synchronization procedure, the terminal finds the slot and frame boundary detecting the primary and secondary synchronization codes, respectively. Once the secondary synchronization code has been found, the terminal derives the complete identity of the cell detecting its transmitted primary scrambling code (1 out of 8) [4].

In a similar way, in LTE networks, a set of 510 reference signals is reserved to identify unequivocally a cell. Each reference signal can be divided into two 2-dimensional (frequency and time) sequences, called pseudo-random and orthogonal sequences. There is a total of 170 different pseudo-random sequences, each one corresponding to a cell identity group. Moreover, there are three orthogonal sequences defined within each cell identity group, each one corresponding to a unique cell identity. During the synchronization phase, the terminal derives the specific cell identity from the primary synchronization signal, and the cell-identity group from the secondary one [27].

The Femtocell Case

Since femtocells can appear or disappear at any time or be moved to different locations, femtocells must be able to self-configure their PCI during the booting procedure in order to minimize collision/confusion with other macrocells and femtocells (Figure 8.11). To achieve this target, two different strategies can be followed [28]:

- The operator automatically assigns a PCI to the femtocell over the backhaul connection during the booting procedure.
- The femtocell collects information about what PCIs are being used within its neighbourhood, using sensing techniques, and then it randomly chooses a PCI that does not collide with the existing ones.

Due to the foreseen extensive use of femtocells (thousands of femtocells deployed within the coverage area of several macrocells), the selection of the PCI is not a trivial task, and the reuse of the PCIs in different regions will be unavoidable.

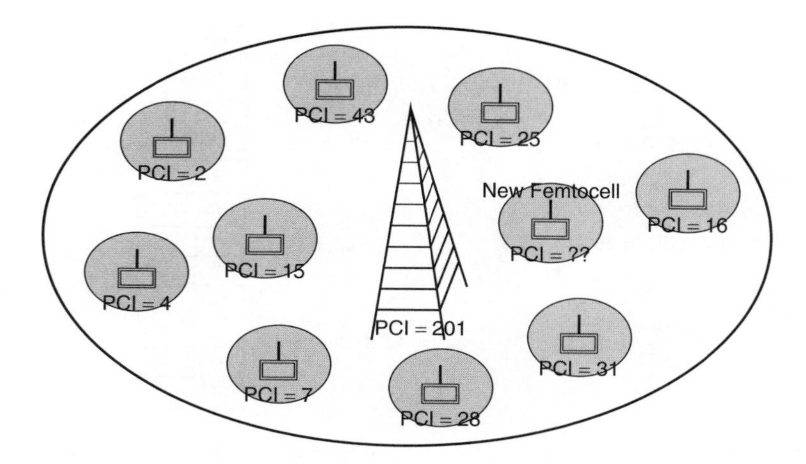

Figure 8.11 Collision and confusion

Considering the handover procedure, for example, if one macrocell has two neighbouring femtocells with the same PCI, it will not be able to distinguish between both (confusion), and the handover is likely to fail, resulting in a dropped call.

To avoid collision/confusion between the macrocell and femtocell layer, the operator needs to identify a set of PCIs for the femtocells that will not be permitted in the macrocells. This kind of planning is necessary to avoid a newly added femtocell from starting to use a PCI that is already being used by a given macrocell. As a result, only a small range of PCIs will be reserved for the use of femtocells [29]. Therefore, in extensive femtocell deployments, PCI reuse among the femtocells that are covered by a single macrocell is unavoidable, causing PCI confusion.

In order to minimize the effect of PCI confusion among femtocells, they must periodically check whether any neighbouring femtocell is using the same PCI, and select the PCI that will result in a better performance [26]. If a femtocell decides to change its PCI, this femtocell should let its connected users know the new PCI before executing any further action. In this way, confusion can be mitigated.

8.6.2 Neighbouring List

Context

In order to select the best serving cell when the terminal is in idle mode, or to aid the handover procedure when the terminal is in active mode, the user handset is continuously making measurements of the received signal strength of neighbouring pilot channels, i.e. BCCH in GSM/DCS, scrambling codes in UMTS. To speed up this procedure and simplify the task of the terminal when monitoring the air interface, the serving cell, in which the terminal is camping, periodically broadcasts the list of pilot channels and cells that it should measure. This list is generally known as the neighbouring cell list.

After receiving this list, the terminal periodically performs the appropriate measurements, and reports back the results to the serving cell (Figure 8.12). Note that the terminal

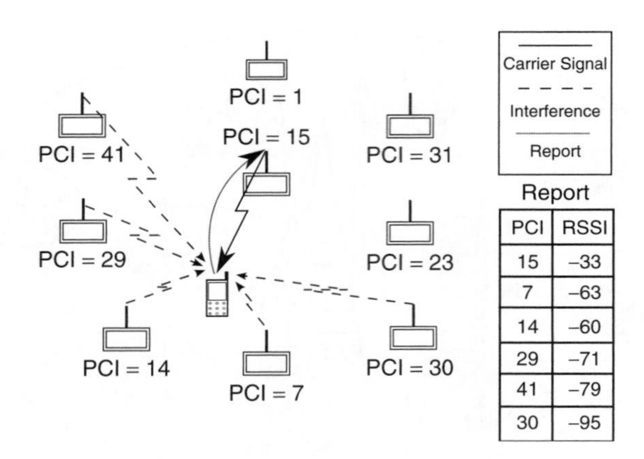

PCI	RSSI
15	−33
7	−63
14	−60
29	−71
41	−79
30	−95

Figure 8.12 Measurements reported by the user terminal to the femtocell based on neighbouring list. RSSI (Received Signal Strength Indicator) [dB]

can report information only about a limited number of neighbouring cells to avoid excessive signalling overhead. For example, in GSM/DCS networks the limit is set to six cells, while in UMTS networks it can vary from six up to 32 cells. If more than the allowed number of neighbouring cells are measured, only the measurements corresponding to the neighbouring cells with the larger received signal strength are reported.

Since a given network, e.g. UMTS, must support intra-system, as well as inter-system handover, the neighbouring list on a given cell can contain neighbouring cells of the same system (UMTS), operating at the same or different frequencies, or cells that might belong to other systems (GSM, LTE, etc.) or even to other operators.

Due to the relevant role played by the neighbouring list in the cell reselection and handover procedure, the configuration and set up of the neighbouring list is an important task in the agenda of network designers. To identify missing or unused neighbouring relationships that should be added or removed to/from the neighbouring list, network operators make use of sophisticated network planning and optimization tools. These tools normally make use of measurements taken from terminals, as well as statics from the network (call drops, handovers). Once the neighbouring cell list is created, it is updated on a regular basis (days or weeks), or upon the identification either of missing/unused neighbours (insertion of a new cell) or trouble situations (handover parameters, call blocks or drops, etc.).

The Femtocell Case

Since femtocells can appear or disappear at any time or be moved to different locations, femtocells must be able to self-optimize their neighbouring list dynamically in order to optimize cell re-selection and handover procedures. Therefore, femtocells must be able dynamically to:

- include new relationships into the list, e.g. if a new cell appears in the neighbourhood;
- remove the inappropriate ones, for example, if there is an unused relationship or if there is a large number of failed handover associated with one of them.

To do this, the femtocell will sense the radio channel searching for neighbouring cells, and will also instruct its connected terminals to do so as part of the normal call procedure. If the femtocell detects, or a user reports, a new cell, the femtocell will include this new cell in its neighbouring list, if it is convenient. It might be helpful for femtocells to inform about changes in their neighbouring relations to the core network in order to share this information with other cells [30], e.g. a femtocell receives its initial neighbouring list through the backhaul in the start up procedure (Section 8.4). Furthermore, different counters can be used to measure when a neighbouring relation is unused or creates too many call drops due to handover.

Moreover, since femtocells must support cell re-selection and handover towards the macrocell layer or other femtocells, the neighbouring list of a given femtocell must consider both macrocells and femtocells. However, the relationship femtocell–macrocell must be treated in a different way to the relationship femtocell–femtocell due to the individualistic nature of the femtocell. For example:

- In certain cases, it might be better not to handover from a macrocell to a femtocell. Imagine a fast macrocell user moving across a large residential area. In this case, it would be better to keep the user in the macrocell than to handover the user across several femtocells or to the macrocell. In this way, the signalling overhead due to handover will be decreased, as well as the probability of dropping a call due to handover failure.
- In other cases, it might be better not to handover from a femtocell to a macrocell. Imagine a femtocell user located indoors but close to an open window. It is possible that the signal strength coming from the macrocell may be larger than the one coming from the subscriber femtocell. However, it may be preferred to keep the user on the femtocell to off-load the macrocell or because the femtocell user could get cheaper calls in the femtocell.

In addition, in macrocells, the neighbouring list has been limited to 32 positions to speed up the measurement and updating. However, in large femtocell deployments this number will be insufficient to handle all femtocells within a macrocell. Therefore, novel structures and algorithms are needed to handle more neighbours and their different nature with a fast response.

In [31], Amirijoo *et al.* present a method for the automatic configuration and optimization of the physical cell identity and neighbouring cell list in LTE networks. The proposed technique is based on measurements reported by the user terminals to detect and resolve PCI conflicts, and update the neighbouring list. In this case, the mobiles are capable of detecting cells that are not included in the neighbouring cell list of the serving cell. Then, the detected cells are included in the neighbouring list of both the serving and the target cell upon previous negotiation. If a PCI conflict is detected, the serving cell reports it to the OSS. The OSS will decide which cell should change PCI in order to resolve the conflict, and will assign a new PCI to the selected cell that is not used within the neighbouring cell list of the neighbours and neighbours' neighbours. This algorithm is fully automated and does not require the involvement of the operators. The algorithm converges to an stable solution, where there are no PCI conflicts and the neighbouring list is complete. However, this algorithm has not been challenged with large femtocell deployments where hundreds of femtocells exist within a macrocell.

8.6.3 Spectrum Allocation

In this section, we overview the two options that operators have when assigning their licensed spectrum to the macrocell and femtocell layers. The idea here is not to self-organize the spectrum allocated to both layers, but to present concepts that are used in the following sections.

The operators can follow two different strategies to assign the licensed spectrum between macrocells and femtocells:

- In an *orthogonal deployment*, a fraction of the spectrum is used by the macrocells, while the other fraction is used by the femtocells.
- In a *co-channel deployment*, both the macrocell and femtocell network reuse the same radio spectrum (reuse factor 1).

In an orthogonal deployment of macrocells and femtocells, cross-layer interference is neglected, and femtocells only need to avoid interference from/to other femtocells. However, as explained in Chapter 6, orthogonal deployments will result in an inefficient usage of the radio spectrum, which is extremely expensive and undesirable for the operators. In the contrary, a co-channel deployment of macrocells and femtocells would enhance the spectral efficiency (bit/s/Hz) at the expense of using more complex interference mitigation techniques.

Orthogonal deployment is supported by companies such as Comcast in the USA, which have acquired spectrum that will be exclusively used by their WiMAX femtocells. However, co-channel deployment seems to be the favourite approach of most of the operators, due to the higher frequency reuse.

In a co-channel deployment of CSG femtocells, the interference conditions are more severe and technically more challenging than in an orthogonal deployment. For example:

- In the DL, a femtocell can jam the communication of a close macrocell user connected to a far macrocell, due to the leakage of power of the femtocell from indoors to outdoors.
- In the UL, a macrocell user connected to a far macrocell transmitting with high power can jam the connections of a close femtocell due to imperfect shield provided by the walls.

This interference can downgrade the overall network performance. Macrocells that before could provide service might become useless now due to the presence of numerous femtocells. In this way, the possibility of having extensive femtocell deployments will be diminished or even neglected. Therefore, new approaches to solve this problem must be investigated.

In the co-channel deployment of CDMA femtocells, the self-optimization of the transmitted power is a key factor to avoid cross- and co-layer interference. In this kind of systems, all users share the same bandwidth, being the radiated power by a user seen as interference by all the others. Therefore, decreasing the transmitted power, the noise rise and the interference can be minimized.

In the co-channel deployment of OFDMA femtocells, although important, the self-optimization of this power is not the only way to avoid interference. In this kind of system, sophisticated sub-channel assignment techniques will also help to mitigate cross- and co-layer interference. However, the joint allocation of both power and subchannels will provide a better result.

In the following two sections (Section 8.6.4 and Section 8.6.5), power and subchannel self-organization techniques are depicted.

8.6.4 Power Selection

The self-optimization of the radiated power by the femtocells will play one of the most important roles in successfully deploying a femtocell layer. In order to mitigate the cross- and co-layer interference, femtocells must dynamically tune their radiated power (control/data channels), according to the changing conditions of the environment (passing users, channel state, etc.).

The self-optimization of the transmitted power will help to:

- adapt the femtocell coverage to the household structure;
- reduce the interference created towards macrocell users passing by;
- reduce the attempts of macrocell–femtocell handover by underlay macrocell users.

In the following, different techniques for the self-organization of the radiated power by a femtocell are presented.

Self-Organization of the Radiated Power Taking the Presence of Close Cells into Account

Claussen *et al.* have proposed an attractive approach for mitigating cross-layer interference [32]. This technique has been devised to minimize cross-layer interference in UMTS scenarios, but it can be extended to co-layer interference and other technologies.

In downlink, this approach consists of a power control algorithm for pilot and data channels that ensures a constant femtocell coverage. Each femtocell C_i^{femto} sets its power P_i to a value that on average is equal to the power received from the closest macrocell C_j^{macro} at a target femtocell radius r, selected according to the features of the household (Figure 8.13). In this way, a constant femtocell radius is warranted independently of the physical distance from the macrocells. The maximum femtocell transmitted power P_i can be computed as:

$$P_i = min \begin{cases} \underbrace{P_j + G_j - L_j - Lp_j(d)}_{\text{macro power at femto radius}} - \underbrace{G_i + L_i + Lp_i(r)}_{\text{femto channel gain at femto radius}} \\ P_{i,max} \end{cases} \tag{8.1}$$

where P_j denotes the power transmitted by macrocell C_j^{macro}, $Lp_j(d)$ indicates the average macrocell path loss at the macrocell distance d, $Lp_i(r)$ represents the average femtocell path loss at the target cell radius r, and $P_{i,max}$ is the maximum power that can be radiated by femtocell C_i^{femto}. Moreover, G stands for the antenna gains and L for the equipment losses. Note that decibels are used in this formulation.

In reality, the power received from the strongest macrocell can be estimated based on average channel modelling, or using path loss measurements at the femtocell target

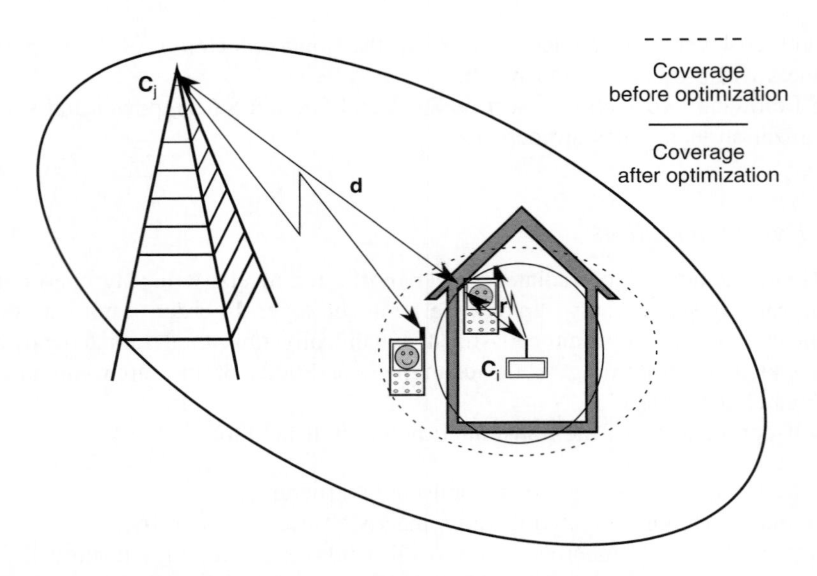

Figure 8.13 Self-optimization of the femtocell radiated power based on the received power from the nearest macrocell

radius. However, this would require femtocell knowledge of macrocell data (position, power, etc.). To avoid this, the power received from the strongest macrocell can be derived using the in-built sensing capability of the femtocell (listening mode) or user measurement reports.

In uplink, a maximum interference allowance is set for each macrocell C_j^{macro}. Afterwards, this budget is shared between all the existing femtocells C_i^{femto} under the macrocell. The maximum user transmitted power P_{UE} can be computed as:

$$P_{UE} = min \begin{cases} \frac{P_{j,budget}}{N} + Lp_j(t) \\ P_{UE,max} \end{cases} \quad (8.2)$$

where $P_{j,budget}$ denotes the maximum allowed interference in macrocell C_j^{macro} coming from the femtocells, N indicates the number of femtocells under the macrocell coverage, $Lp_j(t)$ represents the current path loss from the femto user to the macrocell, and $P_{UE,max}$ is the maximum power that can be radiated by subscriber UE_x^{femto}.

This technique can be further improved by sharing this budget only between femtocells with active UL connections, and by dynamically computing the budget according to the current interference conditions (noise raise) of the macrocell sector.

Self-Organization of the Radiated Power Taking the Handover Attempts into Account

Claussen *et al.* [33] have presented a method for coverage adaptation that uses information on mobility events of passing and indoor terminals. Each femtocell sets its pilot power to a value that maximizes its coverage and minimizes on average the total number of attempts of passing and indoor users to connect to the femtocell.

Let us define an unwanted event as those handovers or attempts to handover in which the user connects to the femtocell and immediately hands back to a macrocell or another femtocell. This fast hand back may be produced either by passing users moving across a residential area that continuously hands over from an open femtocell to another or to the umbrella macrocell, or by non-subscribers trying to connect to CSG femtocells. Wanted events, contrarily, are those allowed handovers that stay longer than a predefined time.

In this model, the femtocell counts the number of unwanted and wanted mobility events. If the number of unwanted events is larger than a given threshold n_1 per time t_1, the femtocell reduces its pilot power by a step of ΔP_1. If the number of unwanted events is smaller than a given threshold n_2 per time t_2, the femtocell increases its pilot power by a step of ΔP_2. These parameters must be tuned according to the scenario to achieve optimal performance. Note that after decreasing/increasing the pilot power, the mobility event counters are reset.

Using this technique, the femtocell coverage shrinks when the leakage of power from indoors to outdoors is large, causing unwanted mobility events, or increases when there are no passing users around the femtocell premises (Figure 8.14).

Self-Organization of the Noise Rise Threshold in the Femtocell

In [21], the authors propose to address UL interference from uncontrolled macrocell and femtocell users by tuning the noise rise threshold in the femtocells.

This uncontrolled interference occurs in the following scenarios:

- A macrocell user located in the femtocell premises and transmitting at high power in the UL can jam the UL communications of the femtocell due to the short path loss between this macrocell user and the femtocell. This is a common scenario in CSG femtocells.
- A femtocell user can generate a large UL noise rise at its own femtocell if this user is located too close to the femtocell and can not be powered down due to its limited dynamic range.

In this case, the femtocell will allow a greater UL noise rise in order to cope with this type of interference, thus the throughput of the UL femtocell users is unaffected. Note that this parameter needs to be modified only when the interference is strong, and afterwards, it can be returned to its normal value.

However, allowing greater noise rise thresholds will increase the UL interference caused at the macrocells, thus the performance of the UL macrocell users is degraded. In addition, when operating at large noise rise thresholds, bursty interference will create large fluctuations in the signal quality of the femtocell pilot that power control may not be able to cope with [34]. Therefore, a better solution could be to adapt dynamically the noise figure of the femtocell. This technique is introduced in the following section.

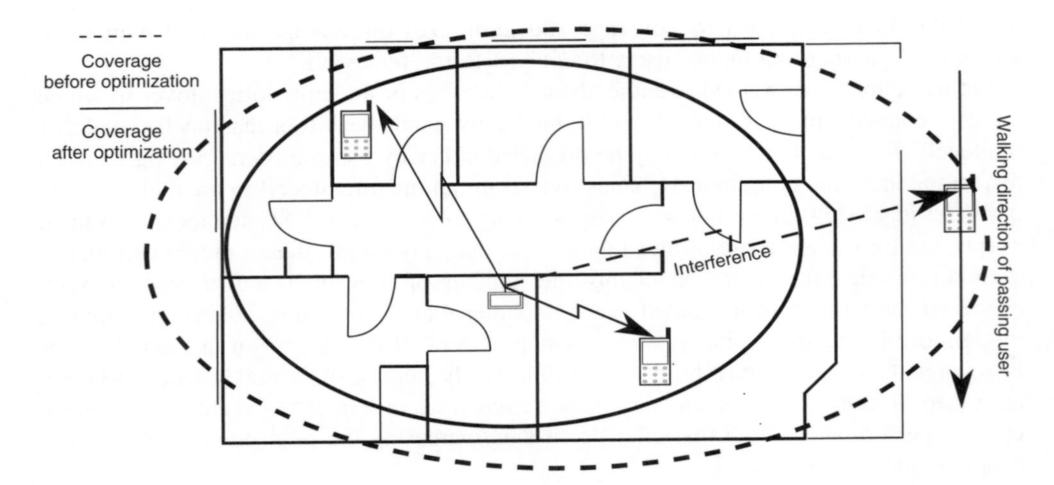

Figure 8.14 Self-optimization of the femtocell coverage power based on the number of unwanted mobility events

Self-Organization of the Receiver Gain in the Femtocell

In [21], the authors propose addressing UL interference from uncontrolled macrocell and femtocell users by tuning the receiver gain (noise figure) of the femtocells. In this way, the dynamic range of the femtocell is moved such that the interfering UL users do not block the femtocell UL communications. In this way, the interference is closer to the thermal noise level, leading to a lower noise rise operation. As a result, the interference is desensitized (attenuated) at the receiver, leading to a larger noise figure. Note that this parameter needs to be modified only when the interference is strong, and afterwards, it can be returned to its normal value. The effect of such attenuation could be to increase the UL power of the users connected to the femtocell, and therefore, the UL interference caused to the macrocells. As a safety mechanism the femtocell must limit the UL transmission power of its connected users in order to decrease the interference towards the surrounding macrocells.

8.6.5 Frequency Allocation

As introduced in Section 8.6.3, OFDMA femtocells can fight interference not only by optimizing its transmitted power, but also using sophisticated subchannel assignment algorithms. The target of these self-organization techniques is to allocate different subchannels to those connections that could suffer from interference due to the presence of others.

Figure 8.15 represents a downlink scenario where a loaded macrocell uses all the available subchannels to transmit information to its users. However, when a macrocell user, e.g. user 6, is located close to a femtocell, e.g. femtocell 2, it can happen that the macrocell user 6 and femtocell 2 employ the same subchannels. This is illustrated by the spectrum occupancy of links L_2 and L_6, which occupy subchannels 3 and 4 simultaneously, this results in a large DL interference for macrocell user 6. This problem

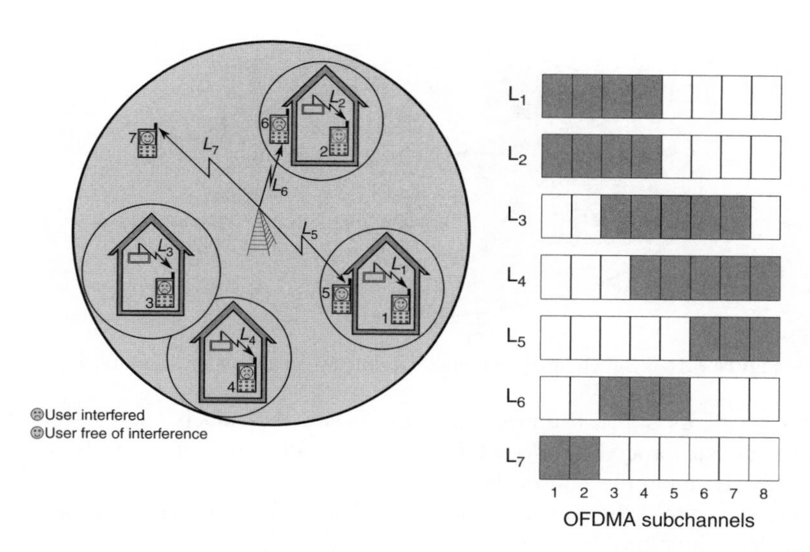

Figure 8.15 Downlink allocation of OFDMA subchannels in a macro/femtocells network with co-channels assignment

can be avoided by means of self-organization, monitoring the air interface and assigning different subchannels to user 2. For instance, the spectrum occupancy of users 1 and 5 illustrates how interference can be avoided by the femtocell whilst providing sufficient resources for a satisfactory user experience.

In the following, different strategies to mitigate interference in femtocell scenarios through the self-organization of subchannel allocation are presented.

Self-Organization of the subchannel Assignment Based on Broadcast Messages

In [35], López-Pérez *et al.* present a method for the distributed assignment of subchannels for OFDMA femtocell networks, based on the broadcast of information messages between neighbouring femtocells (Figure 8.16) (Section 8.5.2).

The idea is that each femtocell estimates the probability of usage of each sub-channel and distributes this information to its neighboring femtocells, sending a local broadcast message. Besides these sub-channel usage probabilities, the broadcast message also contains information about the power applied to each sub-channel, and the power of the pilot signal. Based on the information obtained from its neighbors over the broadcast messages, a femtocell prioritizes the usage of its sub-channels, i.e. according to the following quality indicator:

$$badness_j(k) = \sum_{i \in \mathcal{N}_j} p_i^{interf}(k) \cdot p_i^{usage}(k), \tag{8.3}$$

where $\mathcal{N}_j$ is the set of the neighbours of node j, $p_i^{usage}(k) \in [0, 1]$ denotes the probability of usage of subchannel k in femtocell F_i, which was reported by the last broadcast, and $p_i^{interf}(k) \in [0, 1]$ indicates the intensity (near/far femtocell) of the possible interference coming from F_i to F_j.

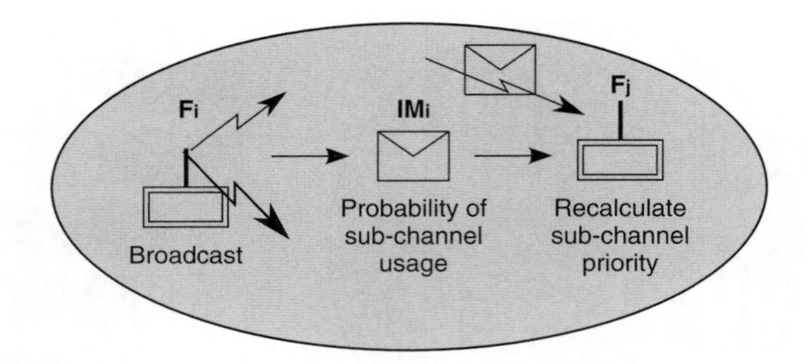

Figure 8.16 Subchannel self-organization by means of message exchange

The femtocell uses the badness value to update the subchannel assignment of its users. This update procedure is performed periodically, and the time between consecutive updates is randomly chosen from the interval of $[0, \ldots, 2T_{bc}]$ time units. This is done in order to avoid that several femtocells change their sub-channel allocation at the same time (coordination).

Between updates, the femtocell collects the messages broadcast by its neighbors. These are processed at the next update, in which the femtocell first recomputes the badness of each subchannel based on existing messages. Afterwards, the femtocell rearranges its sub-channel allocation so that the users get assigned to the sub-channels having the lowest badness values. Finally, it estimates its own sub-channel usage probabilities using the new assignment and broadcasts them to its neighbors.

To compute $p_i^{usage}(k)$ in femtocell F_i, a monotonically decreasing function is used. In this function the busy subchannels have a much larger probability of usage than the idle ones. Nevertheless, not all busy and idle subchannels have the same probability of usage (Figure 8.17). In this way, the femtocell indicates to its neighbouring femtocells which subchannel will be used or freed if a new user connects or disconnects.

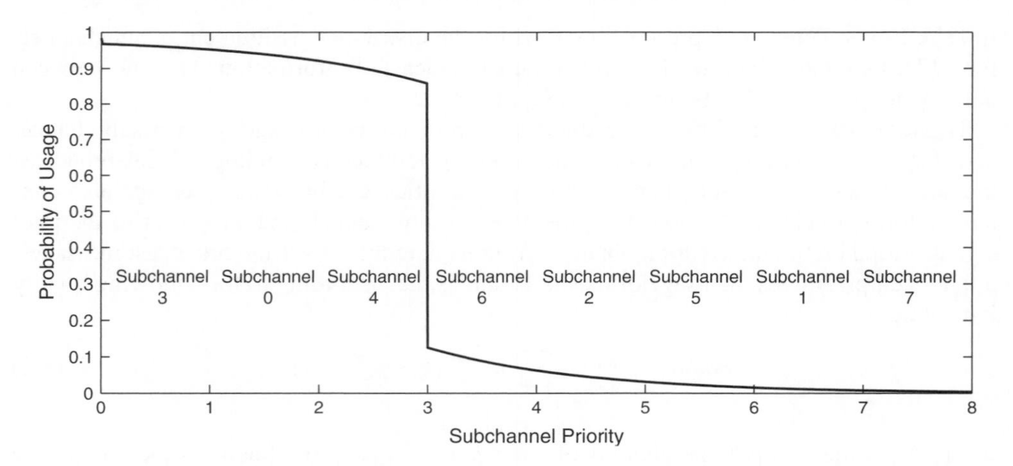

Figure 8.17 Probability of subchannel usage. In this case, there are eight available subchannels, and three of them are being used by the femtocell (subchannels 3, 0 and 4). If a new user appears, it will be assigned to sub-channel 6, while if a user disappears, sub-channel 4 will be freed

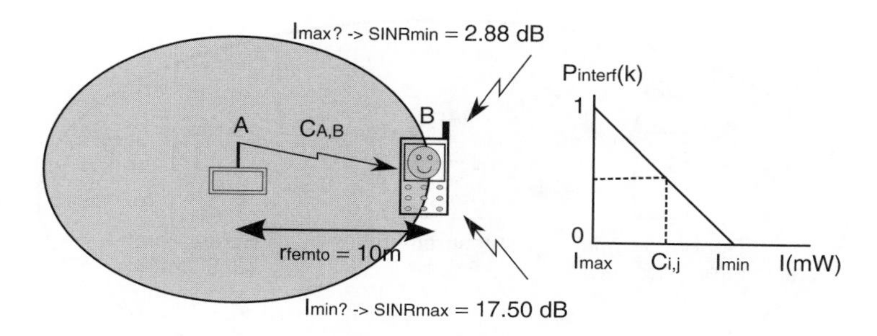

Figure 8.18 Subchannel self-organization by means of message exchange. In this case the femtocell radius is set to 10 m, and the SINR of the maximum and minimum RAB (modulation and coding scheme) is set to 20 dB and 0 dB, respectively

To compute $p_i^{interf}(k)$, the following model is used. Given a worst case scenario, where a femtocell A, whose cell radius is r_{femto}, provides coverage to a user B located in its cell edge (Figure 8.18), and considering the following:

- The maximum interference I_{max} that user B can suffer leads to a minimum SINR $SINR_{min}$, which is the SINR threshold of the minimum RAB defined in the system.
- The minimum interference I_{min} that user B can suffer leads to a maximum SINR $SINR_{max}$, which is the SINR threshold of the maximum RAB defined in the system.

Note that I_{min} and I_{max} can be calculated as follows:

$$I_{min} = \frac{C_{A,B}}{SINR_{max}} - \alpha^2, \quad I_{max} = \frac{C_{A,B}}{SINR_{min}} - \alpha^2 \tag{8.4}$$

where σ stands for the background noise density.

Then, by using the signal strength $C_{i,j}(k)$ of the subchannel k and the linear penalty function defined by Equation (8.5), the intensity of the possible interference (near/far) can be derived. Note that $C_{i,j}$, I_{max} and I_{min} must be in mW.

$$p_i^{interf}(k) = \begin{cases} 1, & \text{if } C_{i,j}(k) > I_{max} \\ \dfrac{C_{i,j}(k) - I_{min}}{I_{max} - I_{min}}, & \text{if } I_{min} < C_{i,j}(k) < I_{max} \\ 0, & \text{if } C_{i,j}(k) < I_{min} \end{cases} \tag{8.5}$$

It is to be noted that using the pilot signal power indicated in the broadcast message, and measuring the pilot signal strength of the sender, the receiving femtocell can estimate the path loss to the sender. Furthermore, using this path loss and the indicated power applied in each sub-channel, the receiver can estimate the received signal strength from the sender in each sub-channel.

Self-Organization of the Subchannel Assignment Based on Measurement Reports

In [35], López-Pérez *et al.* present a method for the distributed assignment of subchannels for OFDMA femtocell networks, based on the use of measurement reports sent by the

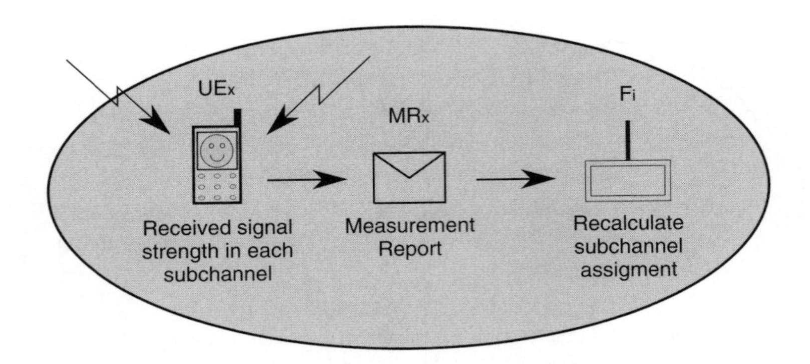

Figure 8.19 Subchannel self-organization by means of measurement reports

subscribers to the femtocells (Section 8.5.3). This approach has been initially designed for the femtocell layer, but it can be applied to two-layer networks, if the measurement reports contain information about the macrocells.

In this approach, a user UE_x sends an MR MR_x to its serving femtocell F_i on a regular basis T_{mr} (Figure 8.19). A MR MR_x indicates the received signal strength suffered by user UE_x in each sub-channel k. Then, femtocell F_i eventually updates its subchannel allocation according to all MRs received. This update event happens after a random distributed time after the last update event in F_i. In this way, that several femtocells change their subchannel allocation at the same instant is avoided.

When an update event happens, femtocell F_i gathers the information of all received MRs and builds an interference matrix W_i. The dimensions of W_i are $M_i \times K$, where M_i denotes the number of users connected to femtocell F_i, and K indicates the number of subchannels. Furthermore, $w_{m,k}$ represents the received signal strength or interference suffered by user m in subchannel k.

Once the interference matrix W_i is built, F_i computes its new subchannel allocation using the following optimization procedure, whose target is to minimize the sum of the overall interference suffered by the users of the femtocell.

$$min \sum_{m=0}^{M_i-1} \sum_{k=0}^{K-1} w_{m,k} \cdot \gamma_{m,k} \tag{8.6a}$$

subject to:

$$\sum_{m=0}^{M_i-1} \gamma_{m,k} \leq 1 \qquad \forall k \tag{8.6a}$$

$$\sum_{k=0}^{K-1} \gamma_{m,k} = 1 \qquad \forall m \tag{8.6b}$$

$$\gamma_{m,k} \in \{0, 1\} \qquad \forall m, k \tag{8.6c}$$

where $\gamma_{m,k}$ is a binary variable (Equation 8.6d) that is equal to 1 if user m is using subchannel k, and 0 otherwise. Assumption (8.6b) ensures that a subchannel is assigned to at most one user, while assumption (8.6c) ensures that all connected users have only one subchannel.

This optimization problem can be solved efficiently using backtracking, since its solution space is small due to limited number of users that can connect to a femtocell at a time, e.g. four.

In the following, the performance of these two self-optimization techniques (broadcast messages and measurement reports) is briefly depicted in order to highlight some related issues. Note that for comparison, two more subchannel assignment techniques are presented:

- A worst case assignment where the femtocell always assigns the first free subchannel starting from $k = 0$.
- A random assignment where the femtocell randomly selects a free subchannel from the available ones.

The scenario used in this experimental evaluation consists of an ideal free space area of 300×300 m, with a wide deployment of 130 femtocells. Although, this scenario can be considered as not realistic, it represents a worse case scenario since the signals of interfering femtocells are not attenuated by the presence of obstacles. Moreover, users are generated according to a Poisson process (intensity λ), and they stay in the system for a certain time, which is exponential distributed (mean ψ). Only the DL case is considered in this simulation. Note that the users are normally distributed within the coverage area of the femtocells. Figure 8.20 and Table 8.1 show the scenario and parameters of the experimental evaluation, respectively.

Figure 8.21 illustrates the cumulative distribution function of the SINR experienced by femtocell subscribers in a femtocell scenario. This figure shows that the message-broadcast and the measurement-report based methods perform better than does random assignment. This fact shows that using self-organization, the performance of the system can be enhanced. Moreover, the fact that the method using measurement reports outperforms the one using message broadcasting suggests that utilizing information collected at the user positions is important for the avoidance of the interference. Note that the information collected at the femtocell position does not accurately estimate the interference circumstances of a user. Imagine a femtocell user located indoors but close to an open window, the interference suffered by this user coming from a nearby macrocell is not the same as that felt in the position of the femtocell, which is located far away from the window and shielded by the house walls. Therefore, it is recommended that information collected at the user position (measurement reports) should be used when devising self-organization algorithms for tuning the RF parameters of the femtocells.

8.6.6 Antenna Pattern Shaping

As can be derived from above, the leakage of power from indoors to outdoors due to femtocell deployments, will not only increase interference, but also the core network signalling due to unwanted mobility events. To avoid this leakage of power, two different techniques that can be combined have been presented in Section 8.6.4, taking into account the influence of the closest macrocell or the number of unwanted mobility events.

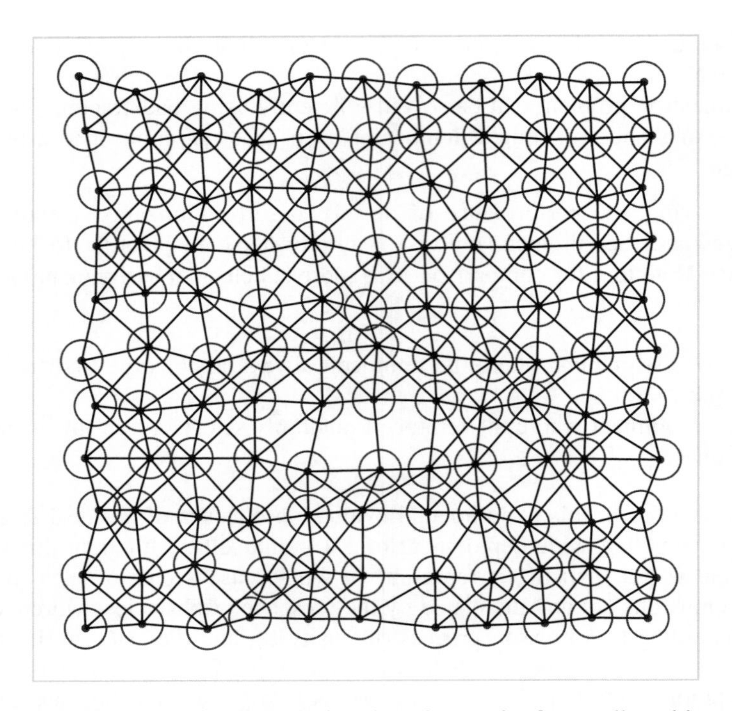

Figure 8.20 Simulation scenario. The black points denote the femtocell positions, the circumferences indicate the femtocell coverage areas, while the links between femtocells represent a neighbourhood relationship between those femtocells. Femtocell A is a neighbour of B if the received signal strength coming from A is larger than the sensitivity of the antenna of B. Note that in this scenario, the neighbourhood relationships are symmetric, if A is neighbour of B, then B is neighbour of A. However, the existence of asymmetric relationships will not affect the performance of the proposed algorithms

However, the performance of these techniques can be further improved by tuning not only the output power, but also the antenna pattern of the femtocell (Figure 8.22).

When using a single omnidirectional antenna, the femtocell can only self-optimize its radiated power in order to minimize its impact on the macrocell layer and other femtocells. However, it could happen that in order to reduce the number of unwanted mobility events, a femtocell has significantly to reduce its power. In this way, the indoor coverage provided by the femtocell may be compromised, thus resulting in an inadequate subscriber performance. To solve this issue, multiple antenna elements can be installed in the femtocell to create different antenna patterns that will be used to adapt the femtocell coverage to the scenario. Nevertheless, using multiple antenna elements in the femtocell might be inconvenient due to the tight size and price constraints required to commercialize a femtocell successfully. Therefore, these multiple femtocell antenna elements must be of reduced volume and cost. Moreover, the system handling the array of antenna elements must be of low complexity. As a result, simple antenna switching systems are preferred over complex beam forming.

Table 8.1 Simulation parameters

Parameter	Value
Scenario size	300×300 m
Femtocells	100
Carrier	2.3 GHz
Bandwidth	5 MHz
Duplexing	TDD 1:1
DL symbols	19
UL symbols	18
Preamble symbols	2
Overhead symbols	11
Frame duration	5 ms
Subcarriers	512
Pilot subcarriers	48
Data subcarriers	384
Subchannels	8
Femtocell radius	10 m
Femtocell TX Power	10 dBm
Femtocell antenna gain	0 dBi
Femtocell antenna pattern	Omni
Femtocell antenna sensitivity	−108 dBm
Femtocell noise figure	4 dB
User antenna gain	0 dBi
User antenna pattern	Omni
User noise figure	7 dB
User body loss	0 dB
Expected user/hour (λ)	1500
Mean holding time (ψ)	600 s
Broadcast message frequency	10 s
MR frequency	10 s
Thermal noise density	−174 dBm/Hz
Path loss model	COST 231Hata

In [36], Claussen *et al.* present a femtocell architecture, where four low-size low-cost antennas are installed in the femtocell access point. Nevertheless, in this architecture, no more than two antennas are used simultaneously at any time in order to keep the impedance mismatch low. The use of one or the combination of two antennas can generate 10 different antenna patterns.

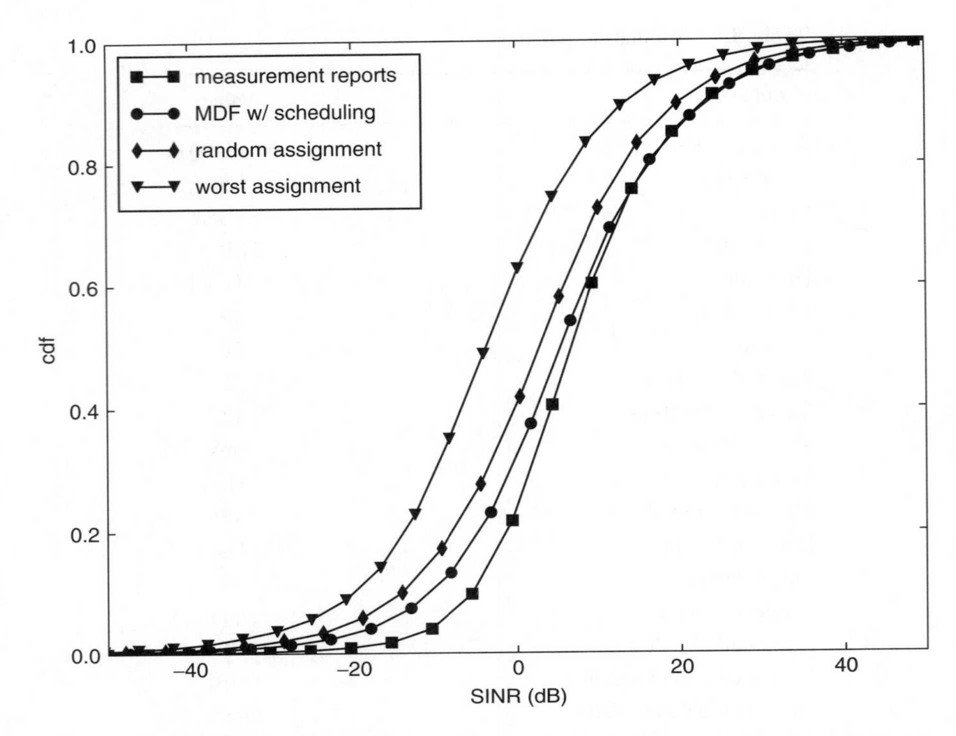

Figure 8.21 Cumulative distribution function of the SINR experienced by femtocell users

The target of the self-optimization procedure here is to select the antenna pattern and pilot power that maximize the indoor coverage, and minimize the number of mobility events. For that, during operation, the femtocell:

- counts the number of unwanted and wanted mobility events, in a similar way to the one presented in Section 8.6.4;
- periodically collects information about the coverage performance of all defined antenna patterns performing path-loss measurements.

These measurements can be derived using the in-built sensing capability of the femtocell and information on its transmit power and antenna gain, or from measurement reports coming from the femtocell end-users.

Then, if the number of unwanted events is larger than threshold n_1 per time t_1, the femtocell reduces its pilot power by a step of ΔP_1, but if the number of unwanted events is smaller than threshold n_2 per time t_2, the femtocell increases its pilot power by a step of ΔP_2. Furthermore, in this case, every time the pilot power decreases or increases triggered by the mobility event counters, the antenna pattern selection is re-evaluated and the combination of the best antenna pattern and pilot power are chosen.

In this way, the antenna pattern of the femtocell can be dynamically shaped in such a way that the indoor coverage is maximized and the number of mobility events is

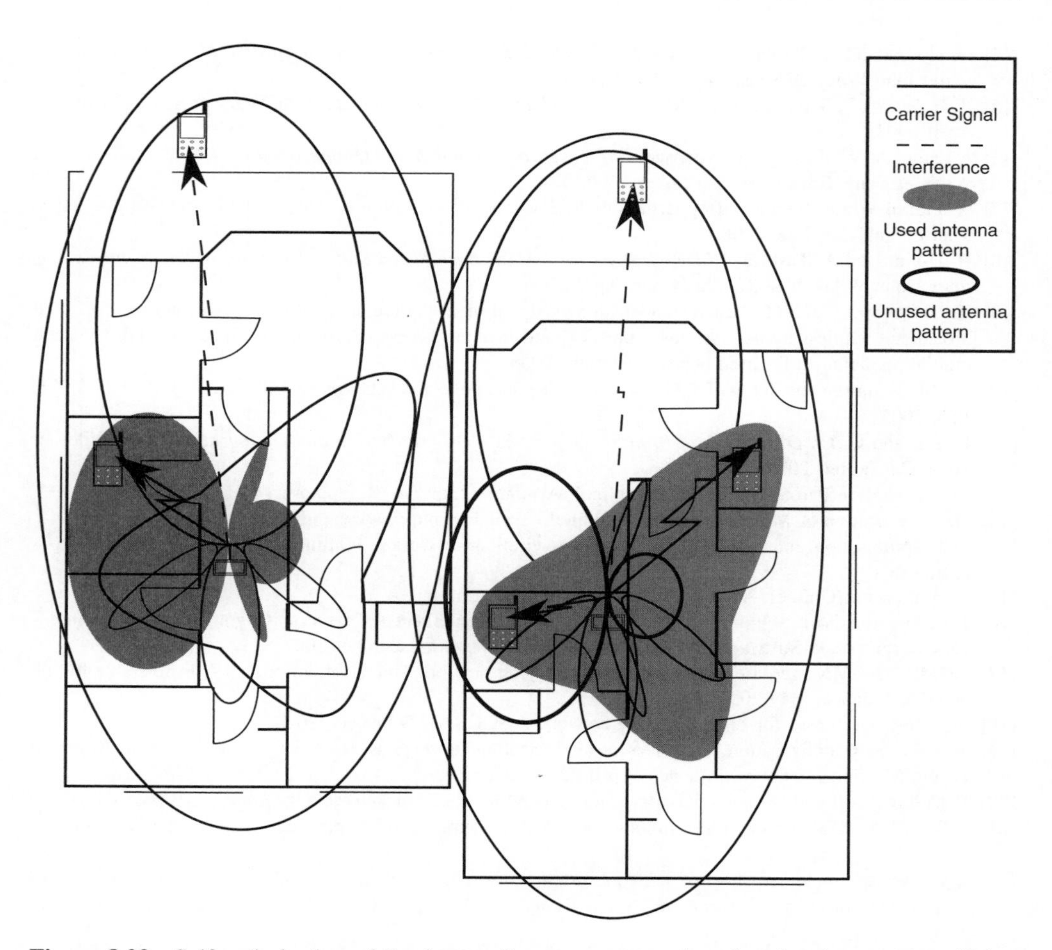

Figure 8.22 Self-optimization of the femtocell antenna pattern based on handover events. Using multiple antenna elements, multiple antenna patterns can be generated. The femtocell will select the antenna pattern that minimizes the attempt of passing users to connect to the femtocell. In this way, interference will be also minimized

minimized. This is always done taking the circumstances of the scenario (household shape, position of subscribers and passing by users, etc.) into account, since the self-optimization algorithm works on the basis of measurements collected from the environment.

References

[1] 'ETSI GSM specifications,' Series 01–12.
[2] A. Eisenblätter, 'Frequency assignment in GSM networks: Models, heuristics, and lower bounds,' PhD dissertation, Technische Universität Berlin, Fachbereich Mathematik, Berlin Germany, 2001.
[3] A. Eisenblätter, H.-F. Geerdes, T. Koch, A. Martin, and R. Wessäly, 'UMTS radio network evaluation and optimization beyond snapshots,' *Mathematical Methods of Operations Research*, vol. 63, no. 1, pp. 129, February 2006.

[4] H. Holma and A. Toskala, *WCDMA for UMTS: Radio Access for Third Generation Mobile Communications*. John Wiley & Sons, Chichester, July 2004.
[5] B. H. Walke, P. Seidenberg, and M. P. Althoff, *UMTS: The Fundamentals*. John Wiley & Sons, Chichester, April 2003.
[6] J. Laiho, A. Wacker, and T. Novosad, *Radio Network Planning and Optimisation for UMTS*, 2nd ed. John Wiley & Sons, Chichester, December 2006.
[7] A. Parkin-White, 'Femtocell opportunities in 3G network evolution scenarios,' Femtocell Europe 2008. Manson analysis, June 2008.
[8] H. Holma and A. Toskala, *HSDPA/HSUPA for UMTS: High Speed Radio Access for Mobile Communications*. John Wiley & Sons, Chichester, April 2006.
[9] IEEE Std 802.16e-2005, 'IEEE standard for local and metropolitan area networks: Air interface for fixed broadband wireless access systems – physical and medium access control layers for combined fixed and mobile operation in licensed bands,' February 2006.
[10] 3GPP, 'Requirements for evolved UTRA (E-UTRA) and evolved UTRAN (E-UTRAN),' 3GPP TR 25.913, June 2005.
[11] H. Lui and G. Li, *OFDM-Based Broadband Wireless Networks Design and Optimization*. John Wiley & Sons, Chichester, 2005.
[12] 3rd Generation Partnership Project. [Online]. Available: http://www.3gpp.org
[13] Next Generation of Mobile Networks. [Online]. Available: http://www.ngmn.org
[14] Self-Optimisation and self-ConfiguRATion in wirelEss networkS. [Online]. Available: http://www.fp7-socrates.eu
[15] FemtoForum. [Online]. Available: http://www.femtoforum.org
[16] J. Reed and J. Neel, 'Future wireless standards and the emergence of wimax,' Cognitive Radio Technologies: Services and Software for Intelligent Wireless Networks, October 2007.
[17] NGMN, 'NGMN use cases related to self organising network, overall description,' A Deliverable by the NGMN Alliance, May 2007.
[18] Socrates, 'Use cases for self-organising networks,' SOCRATES Project, 2008.
[19] ——, 'Self-configuration, -optimisation and -healing in wireless networks,' Wireless World Research Forum Meeting 20, Ottawa, Canada, April 22–24 2008.
[20] S. Carlaw, 'IPR and the potential effect on femtocell markets,' in *FemtoCells Europe*. ABIresearch, 2008.
[21] 3GPP, 'Universal mobile telecommunications system (umts); fdd home node b (hnb) rf requirements,' 2009.
[22] J. Edwards, 'Implementation of network listen modem for wcdma femtocell,' in *Cognitive Radio and Software Defined Radios: Technologies and Techniques*. London: 2008 IET Seminar, September 2008, pp. 1–4.
[23] Alcatel-Lucent, 'HNB to HNB handover architecture,' 3GPP TSG RAN3 63bis, March 2009.
[24] S. Srinivasa and S. Jafar, 'The throughput potential of cognitive radio: A theoretical perspective,' in *Signals, Systems and Computers, 2006. ACSSC '06. Fortieth Asilomar Conference*. Pacific Grove, CA: Electr. Eng. & Comput. Sci., Univ. of California Irvine, Irvine, CA, October 2006, pp. 221–225.
[25] D. Cabric, S. Mishra, and R. Brodersen, 'Implementation issues in spectrum sensing for cognitive radios,' in *Thirty-Eighth Asilomar Conference on Signals, Systems and Computers*., vol. 1, November 2004, pp. 772–776.
[26] 3GPP, 'Self-configuring and self-optimizing network use cases and solutions (release 8),' 3GPP TR 36.902 V1.0.0 (2008-02), February 2008.
[27] E. Dahlman, S. Parkvall, J. Sköld, and P. Beming, *3G Evolution HSPA and LTE for Mobile Broadband*. Elsevier Ltd, 2007.
[28] Samsung, 'Overview of SON and femtocell support,' IEEE C802.16m-08/1252, December 2008.
[29] Aircent, 'Challenges in deployment of UMTS/HSPA femtocell,' White paper, February 2008.
[30] Ericssson, 'Introduction of automatic neighbour relation function,' 3GPP R3-072014, September 2007.
[31] M. Amirijoo, P. Frenger, F. Gunnarsson, H. Kallin, J. Moe, and K. Zetterberg, 'Neighbor cell relation list and physical cell identity self-organization in LTE,' in *IEEE International Conference on Communications Workshops, 2008. ICC Workshops '08*., New Orleans, USA, May 2008, pp. 37–41.
[32] H. Claussen, L. T. W. Ho, and L. G. Samuel, 'An overview of the femtocell concept,' *Bell Labs Technical Journal*, vol. 3, no. 1, pp. 221–245, May 2008.

[33] H. Claussen, L. T. W. Ho, and L. G. Samuel, 'Self-optimization coverage for femtocell deployments,' in *Wireless Telecommunications Symposium*, ser. 24–26, California, USA, April 2008, pp. 278–285.

[34] Y. Tokgoz, F. Meshkati, Y. Zhou, M. Yavuz, and S. Nanda, 'Uplink interference managment for HSPA+ and 1×EVDO femtocells,' in *IEEE Globecom*, Hawaii, USA, December 2009.

[35] D. López-Perez, A. Ladanyi, A. Juttner, and J. Zhang, 'OFDMA femtocells: A self-organizing approach for frequency assignment,' in *Personal, Indoor and Mobile Radio Communications Symposium (PIMRC)*, Tokyo, Japan, September 2009.

[36] H. Claussen and F. Pivit, 'Femtocell coverage optimization using switched multi-element antennas,' in *IEEE International Conference on Communications*, Dresden, Germany, June 2009.

9

Further Femtocell Issues

Guillaume de la Roche and Alvaro Valcarce

In the previous chapters of this book, several technical topics related to femtocells were presented. For example, the different architectures were introduced, and the problem of interference and how to solve it with self-organization were widely studied. However, there are still other issues that need to be explained, and new solutions have to be found.

Too few femtocell trials were performed to be able to affirm that one solution is better than another, that is why this chapter mainly provides guidelines and presents different options, and their advantages and drawbacks.

The issues presented below combine both technical and commercial challenges. The questions we will answer are listed from the most technical to the most business related, and are summarized next:

- How can femtocells timing accuracy be ensured at a low cost?
- How can security and authentication be ensured?
- Is femtocell location necessary and what are the options?
- How is femtocell access defined and controlled?
- Why should new applications be developed?
- Are there health issues?

9.1 Timing

In wireless systems, the crystal oscillator ensures the precision of the internal clock. The internal clock is responsible for:

- the accuracy of the absolute timing to ensure frame alignment between receiver and transmitter and to avoid Intersymbol Interference (ISI);
- the spectrum accuracy to maintain frequency alignment between the receiver and the emitter.

Femtocells: Technologies and Deployment Jie Zhang and Guillaume de la Roche

The synchronization of clocks is further necessary to reduce interference. For example in WiMAX, the frames have to be perfectly synchronized in time. Moreover, if the clock is not accurate enough, errors will occur when creating the frame in the spectrum domain, and thus frequency shifts of the subcarriers could be observed, making it impossible to handover from cell to cell. This is called Intercarrier Interference (ICI). Finally, a good timing accuracy is required to cope with Doppler effects due to moving users. Therefore, a major challenge for femtocells, and also a condition for making them succeed, is that their clocks are accurately synchronized.

Many solutions have been proposed for synchronizing clocks, but this becomes challenging when dealing with the fact that femtocells have to be manufactured at a low cost. Indeed, the main drawback is that accurate oscillators are very expensive, hence it has been reported [1] that the crystal represents the highest individual cost item in a femtocell. That is why some new solutions have to be found, in order to ensure timing synchronization at a reduced price.

9.1.1 Clock Accuracy Requirements

The accuracy of a clock is usually measured in parts-per-billion (ppb) or parts-per-million (ppm). These units represent the maximum variation obtained over a high number of oscillations. For example a watch crystal has a typical error of 20 ppm, giving a maximum error per day equal to $0.00002 \times 24 \times 60 \times 60 = 1.7$ seconds.

In the 3GPP specifications, the requirements defined for a NodeB ask for a precision of 50 ppb. However in Release 6 it has been relaxed to 100 ppb for indoor base stations, and later in Release 8 is reduced to 200 ppb for Home nodeB with certain standards. Some typical accuracy requirements for femtocells recommended by 3GPP are summarized in Table 9.1. Even if it is reasonable for macrocell base stations to afford expensive and accurate oscillators, this is not the case for FAP, which need to be manufactured at low prices. Therefore cheap and easily implementable solutions are still required in this field.

9.1.2 Oscillators for Femtocells

Piezoelectricity is the ability of certain materials (like quartz) to create an oscillating electrical potential when mechanical pressure is applied. The resonance of this material

Table 9.1 Clock accuracy requirements

Standard	Frequency accuracy	Time/phase
GSM	100 ppb	N/A
CDMA	100 ppb	1 μs time
CDMA2000	100 ppb	3 μs time
WCDMA/FDD	200 ppb	N/A
WCDMA/TDD	200 ppb	2.5 μs phase
TD-SCDMA	100 ppb	2.5 μs phase
WiMAX TDD	8 ppm	5 μs time
WiMAX FDD	8 ppm	N/A

can be used to create a signal oscillating at an accurate frequency. Cheap crystals usually have a precision of about 20 ppm. However, the main drawback of such material is that the oscillation frequency changes with the temperature. Furthermore these changes do not repeat exactly upon temperature variation, i.e. resonators exhibit an hysteresis in the frequency variation [2]. That is why in femtocells some more advanced oscillators must be used, in order to compensate for the errors due to the variations in temperature.

Temperature Controlled Crystal Oscillators

A Temperature Controlled Crystal Oscillator (TCXO) is a type of oscillator that compensates for temperature changes to improve stability. In a TCXO, the signal from a temperature sensor is used to generate a correction voltage that is applied to a voltage-variable reactance, also called *varactor*. The varactor then produces a frequency change equal and opposite to the frequency change produced by temperature, as represented in Figure 9.1.

TCXOs are used in many applications, which is why they are the cheapest accurate oscillator components. Because when using a TCXO there are always delays between the measurement of the temperature change and the generation of the frequency correction, the

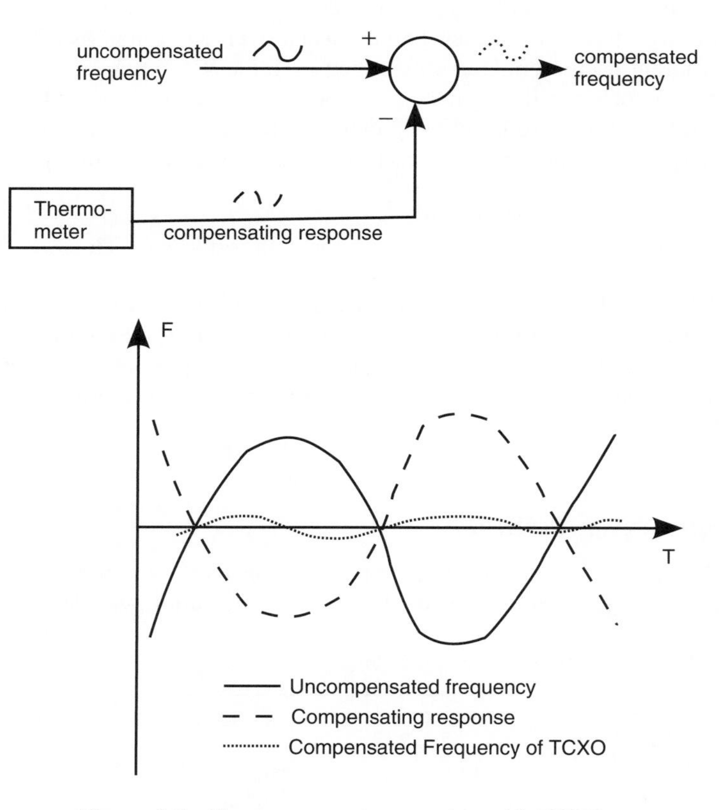

Figure 9.1 Temperature compensation with TCXOs

compensated frequency is not perfectly stable, as represented in Figure 9.1. Hence another kind of oscillator, called an Oven Controlled Oscillator (OCXO), has been proposed for improved accuracy.

Oven Controlled Oscillators

An OCXO is an oscillator enclosed in a temperature-controlled chamber. Inside this temperature-controlled chamber, also called an oven, the temperature is maintained to a fixed value. That is why with such oscillators the frequency variation is minimized. OCXOs offer the best possible stability for a crystal. The only oscillators that are more accurate that have been developed are those based on atomic clocks. However, such oscillators are expensive compared with TCXOs. Moreover they consume more energy and are larger. Hence they are unlikely to be embedded inside femtocells.

Hybrid TCXO–OCXO Oscillators

Some hybrid oscillators have also started to be manufactured. These are based on a TCXO embedded inside a simplified oven. In this kind of oscillator, the temperature inside the oven does not have to be perfectly stable, because the errors are compensated by the TCXO. That is why they are cheaper than the usual oven controlled oscillators. The main features of different oscillators are summarized in Table 9.2.

As explained before, oscillators have to be accurate in time in order to ensure a good performance of the system. To build femtocells at a low cost, expensive oscillators cannot be used so there are two possible approaches to ensuring clock accuracy:

- change the crystals from time to time;
- propose methods to synchronize the oscillator with a reference clock.

The idea of asking customers to bring back their femtocells to have their oscillator changed is inconvenient for both operators and customers. Even if in the future it is expected that the price of the hybrid oscillators will decrease, it is important to research into new solutions that will synchronize the clocks, in order to reduce as much as possible the price of FAPs.

9.1.3 Timing Synchronization

Timing synchronization allows manufacturers to use cheaper oscillators like TCXOs inside their femtocells [3]. To compensate for the accuracy deviation, external accurate clocks

Table 9.2 Types of crystal oscillators

	TCXO	OCXO	Hybrid
Price	Lower	Higher	Medium
Accuracy	Lower	Higher	Medium

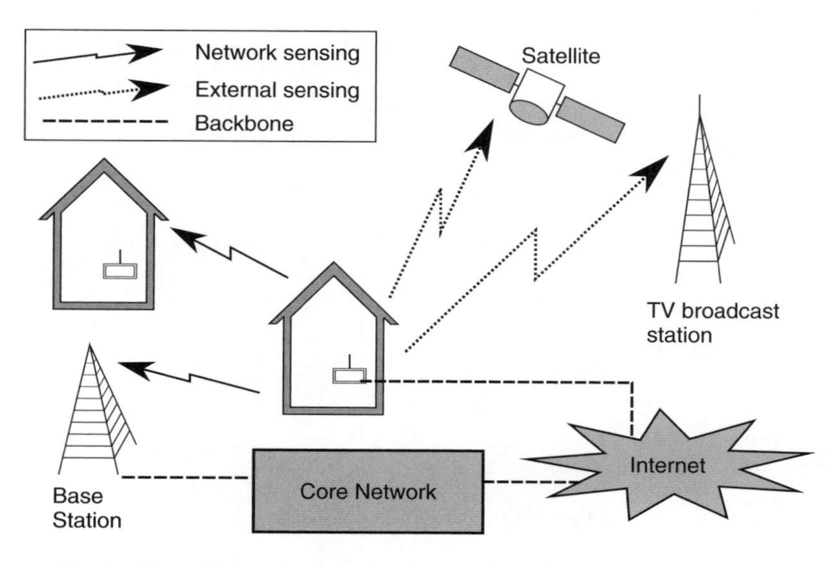

Figure 9.2 Three main possible femtocell synchronization techniques

are used to synchronize the oscillator. This is done on a regular basis, depending on both the accuracy of the oscillator in the FAP, and the accuracy of the external source used as a reference.

As represented in Figure 9.2 there are three mediums linked to the femtocell, that can be used to perform the synchronization of the clock:

- the backbone connection;
- the sensing of the other cells in the network (both macrocells and femtocells);
- the sensing of external sources like Global Positioning System (GPS) or Television (TV) signals.

These methods are analysed in the following paragraphs.

Synchronization via Backbone

In this approach, the FAP can use the Asymmetric Digital Subscriber Line (ADSL) connection to synchronize via backbone to a clock on the network of the operator. However such synchronization could suffer delays because the timings on the Internet can vary in unpredictable manners depending on the traffic. That is why some protocols, like the IEEE 1588 [4], specify the PTP (Precision Time Protocol) for synchronization in a network. The aim of such a protocol is to synchronize different clocks with varying precision and resolution and stability.

The IEEE1588 protocol supports heterogeneous systems and works with two steps:

- establish a master/slave hierarchy of the available clocks;
- synchronize each slave clock with its associated master clock.

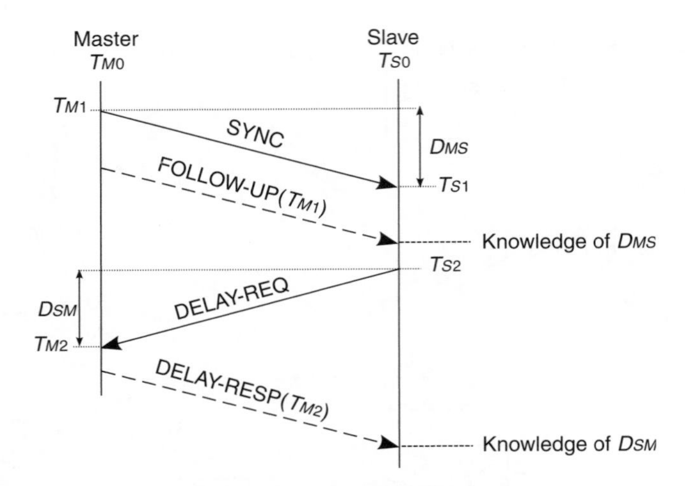

Figure 9.3 IEEE1588 algorithm [4]

In order to synchronize each slave, several steps are necessary in which messages are exchanged between the slave and the master as represented in Figure 9.3:

- The master clock periodically sends (depending on the stability of the slave clocks of the system) a *SYNC* request at time T_{M1} (T_{M1} is stored).
- When the slave clock receives a *SYNC* message it stores the arrival time T_{S1}.
- After a certain delay the master clock sends a *FOLLOW-UP* message containing the value T_{M1}.
- The slave clock periodically sends a *DELAY-REQ* message and stores the departure time T_{S2}.
- When the master clock receives the *DELAY-REQ* message, it sends a *DELAY-RESP* message containing the time T_{M2}, which is the arrival time of the DELAY-REQ message.

After this exchange of messages, the slave clock can estimate the downlink transmission delay *DMS* (delay from master to slave), which is equal to the difference between T_{S1} and T_{M1}:

$$D_{MS} = T_{S1} - T_{M1} \tag{9.1}$$

Moreover, the uplink transmission delay D_{SM} (delay from slave to master), which is equal to the difference between T_{M2} and T_{S2}, can be computed:

$$D_{SM} = T_{M2} - T_{S2} \tag{9.2}$$

Under similar uplink and downlink conditions (symmetric links), the one way delay D_W is the mean of the uplink and downlink delays:

$$D_W = \frac{D_{MS} + D_{SM}}{2} \tag{9.3}$$

Finally the offset of the slave clock with respect to the master clock is:

$$Offset = D_{MS} - D_W = \frac{D_{MS} - D_{SM}}{2} \tag{9.4}$$

The slave clock can finally adjust its new timing by adding or removing the corresponding *Offset* value.

This synchronization method does a fine adjustment of time, and can be easily implemented to synchronize timing of femtocells with the core network. However, in some cases like ADSL, the link is asymmetric and the delays D_{MS} and D_{SM} are not equal. That is why some improvements to the previous algorithm have been proposed in [5]. This approach uses block burst transmission to estimate the asymmetric ratio R of the communication and consider a more accurate offset:

$$Offset = D_{MS} - \frac{D_{MS} - D_{SM}}{R + 1} \tag{9.5}$$

However, it is important to note that synchronization using an Internet Protocol (IP) network is not always accurate, due to the fact that delays on the network, depending on traffic, can vary a lot.

IEEE1588 is not the only protocol that can be used to synchronize via the backbone. For example Netwrok Time Protocol (NTP) is also a good alternative. NTP uses User Datagram Protocol (UDP) on port 123 as its transport layer. NTP is based on Marzullo's algorithm and it is one of the oldest Internet protocols. However it is still in use to ensure time distribution on the web. The last version currently in use is the version NTPv4 [6].

Synchronization via the backbone, using IEEE1588 or NTP, suffers two main disadvantages:

- latency due to delays in the IP network;
- IEEE1588 and NTP can be bandwidth demanding.

These disadvantages are greater when the femtocell is located several hops from the timing server. In this case the precision is lower and the synchronization has to be performed on a more regular basis. That is why synchronization via the backbone is only possible in locations where the mobile operators are sure that there is a high Internet bandwidth available. To ensure efficient synchronization of the network, other options can be proposed.

Synchronization via Sensing of the Network

To avoid using the backbone connection as a reference, a good approach could be to listen to neighbouring cells, in particular the surrounding macrocell. Indeed, the condition for low price is not requested by the macrocells, which is why they are equipped with accurate oscillators, and also very often with GPS receivers to synchronize them. This is why the timing accuracy is high in macrocells, and an efficient synchronization solution would be for the FAP to listen to the nearest macrocell to synchronize its clock. However, it also suffers from a major drawback in locations where the macrocell coverage is poor. Indeed, femtocells have initially been developed to increase indoor radio coverage when the macrocell reception is poor, or when there is no macrocell reception, which is the case for example in rural areas where the first femtocells have been deployed. In such scenarios, a sensing of the macrocells to synchronize the clock will suffer major drawbacks:

- If the macrocell coverage is poor, some delays or some errors will occur, making such sensing difficult.
- If there is no macrocell coverage, it will be impossible to synchronize the femtocell.

It could be argued that the second disadvantage is not really valid, because if there is no macrocell coverage at the femtocell location there is no need for the femtocell to be synchronized. However, if the femtocell and macrocell do not overlap, it does not necessarily mean that synchronization is not required, due to the fact that it is possible that a handset can see both the macrocell and the femtocell. For example, if the clock is not accurately synchronized it is possible that a subscriber, who enters his home, will not be able to handover from one cell to another. Moreover, if the frequency shift of the femtocell is too high, it could happen that the mobile would not be able to decode the different channels of the femtocell. That is why in rural areas, where the macrocell coverage is usually lower, other solutions should be proposed.

Note that in the case of successful femtocell deployment, it is possible that many cells will overlap each other. It could happen, for example, in dense urban scenarios. A good solution for the synchronization could be, instead of only listening to the macrocells, to also use the neighbouring femtocells.

In such a scenario, even if the macrocell reception is too poor, femtocells distribute the timing between each other in order to fine tune their clocks. They listen to neighbouring timings, and compute an average depending on certain priorities to be defined. Many similar approaches called gossip algorithms have been proposed in sensor networks in order to distribute the information among the different nodes [7]. Such methods are efficient especially when there is a high number of nodes, that is why they are not expected to be used during the first femtocell deployments.

Synchronization via External Sources

Because in the first deployments the number of neighbouring cells will be low, femtocell manufacturers also try to use other references to synchronize the clocks. The main reason is that the sensing of their network via backbone can suffer delays and inaccuracy, while the sensing of the neighbouring cells via wireless is not always possible. Hence the idea of using GPS or TV signals has also been proposed.

- GPS is often used for navigation purposes. Thanks to a constellation of satellites it allows the system to determine not only the location of the receiver, but also its accurate timing. This is why GPS receivers are widely included in base stations. For example, in the USA it is mandatory for CDMA2000 to have a GPS receiver. Indeed, GPS signals are used as a synchronization resource for most of the cellular networks.
  GPS do not increase the price of femtocells much because they are mass produced. Moreover (see below on location), some femtocells will have to include a GPS receiver for localization purposes, which is why its usage for timing is quite normal. However, the use of GPS signals indoors poses some important technical challenges, due to the

fact that GPS reception is poor inside buildings. It means that in most scenarios, if the femtocell is not located close to the windows, synchronization will not be possible. A solution to overcoming this problem would be to add an external antenna connected to the femtocell, but this will make the installation more complicated. Is is also interesting to note that some enterprises currently focus on the manufacture of more advanced receivers called a-GPS, that should be able to decode the timing information with low received power. However, these are not expected to be used during the first femtocell deployments.

- In many countries, most locations nowadays have good TV reception. The main advantage of such signals is that, since they are broadcast at low frequencies, they penetrate well inside buildings compared with GPS signals, and indoor reception is good enough to allow femtocells to be located anywhere inside the building without an external antenna. That is why some companies have started to include a simplified TV receiver, just capable of decoding timing, in femtocells [8]. Such receivers also have the advantage of being manufactured in very large quantities, so can be offered at a low price. They are indeed cheaper than GPS receivers. However, unlike GPS that can be used worldwide, the usage of TV receivers is area dependent, because the kinds of TV signal are specific to each region. In locations with new mobile numerical TV networks like DVB-H for example, TV time synchronization can be very accurate. However, in less developed countries, with poor traditional TV reception, an accurate synchronization is more difficult to provide.

9.1.4 Choosing a Solution

As explained before, ensuring an accurate timing at a low cost is one of the most important challenges in femtocell networks. Different solutions have been proposed but it is hard to claim that one is really better than the others. Moreover, they all suffer different kinds of drawback. In general, the more expensive the oscillator is, the more often it has to be synchronized. Nowadays, OCXO oscillators are very expensive and can not be integrated inside femtocells. Hybrid TCXO–OCXO oscillators are also too expensive and there are still a few years to wait until they can be made at a low cost. That is why the current solution is to use TCXO oscillators and synchronize them on a regular basis. The main characteristics of the different synchronization methods are summarized in Table 9.3. The choice of one solution or another will depend mainly on the location, and the specific requirements of each operator. Another important issue to solve for the operators is to ensure good security when using their network.

Table 9.3 Synchronization techniques

	Backbone	Neighbouring cells	GPS	TV
Advantages	Easy implementation	Cheap	Available everywhere	Accurate
Drawbacks	Low accuracy	Requires good macro coverage	Requires external antenna	Location dependent

9.2 Femtocell Security

Most potential femtocell customers are very concerned about their privacy. For example, it is well known that with Wireless Fidelity (WiFi), the use of Wired Equivalent Privacy (WEP) encryption keys is easy to crack. With femtocells, especially if they are used in open access, any mobile user is allowed to connect to femtocells, which need to be secured, in order to protect the private information of the femtocell subscriber, and to avoid illegal use of the femtocell by unauthorized users.

9.2.1 Possible Risks

In femtocell networks, there are three main concerns related to security that have to be addressed:

- Network and service availability: since the link between the FAP and the core network is IP based, there are risks of Denial of Service (DoS) attacks. These occur when, for example, a hacker connects to the link between the FAP and the core network, which could overload the processing capacity of the network, and avoid legitimate subscribers connecting to their FAP. Some other examples could be to degrade or disrupt the network quality by sending unauthorized messages to the core network. Such attacks have to be prevented in order to ensure the availability of the network to subscribers.
- Fraud and service theft prevention: some attacks occur when unauthorized users connect to a femtocell and make illegal use of it. For example, in CSG mode, it is important to prevent non-subscribers from accessing a femtocell. Another example would be a user that authenticates as another femtocell subscriber in order to avoid billing. With such a fraud all the calls would be paid by the regular femtocell subscriber.
- Privacy and confidentiality: since in femtocell networks the user's data travels over the Internet, it is subject to the same security issues as usual IP communications. Moreover, femtocells are also expected to be plugged into the home network as home gateways. In this case it is, therefore, important to protect all the data accessible from the femtocell gateway.

As represented in Figure 9.4, there are three main locations where security risks could occur:

- on the Internet;
- in the FAP;
- on the wireless link.

All these threats to the femtocell network lead operators and manufacturers to include secure solutions inside their femtocells. First Internet Protocol Security (IPsec) is commonly used as a solution to ensure the link between the FAP and the femtocell gateway (risk 1). Then some approaches to ensure a secure authentication have been proposed to avoid hackers controlling the FAP (risk 2).

Finally, some solutions can be proposed to protect the wireless link between the user and the FAP (risk 3).

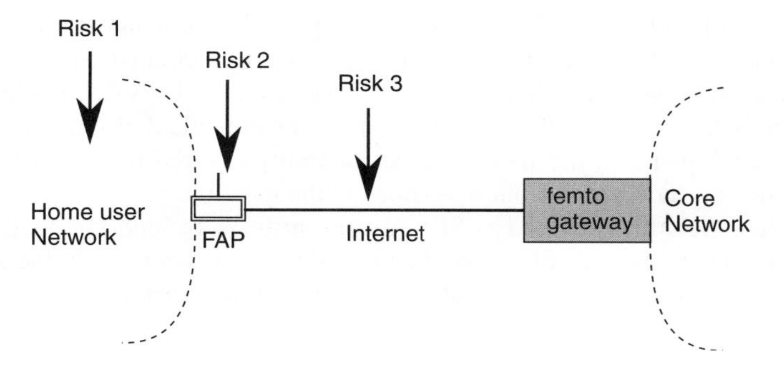

Figure 9.4 Security architecture

9.2.2 IPsec

IPsec is an Internet protocol, whose aim is to ensure security and authentication on the Internet. It operates on the third layer of the Open Systems Interconnection (OSI) model. The IPsec standard is defined by Internet Engineering Task Force (IETF). With IPsec, packets are divided into two parts: an IP header, and the data. The protocol can operate in two modes:

- Transport mode: only the data to be transferred is encrypted, and the header is not modified.
- Tunnel mode: all the packet (header and data) is encrypted and encapsulated into a new packet with a new header.

As represented in Figure 9.5, the IPsec is built to protect the link between femtocell users and the core network, which is why the tunnel is built between the FAP and the Femto Gateway (FGW).
IPsec is based on three main protocols:

- A security association protocol. This creates the security keys that will be used for encryption and to authenticate the two entities at the extremities of the tunnel. The creation of the keys is based on the secret shared concept, and a complex algorithm is responsible for sharing the secret between both entities. Different Extensible Authentication Protocol (EAP) protocols can be used to share the keys. These will be described in the next section.

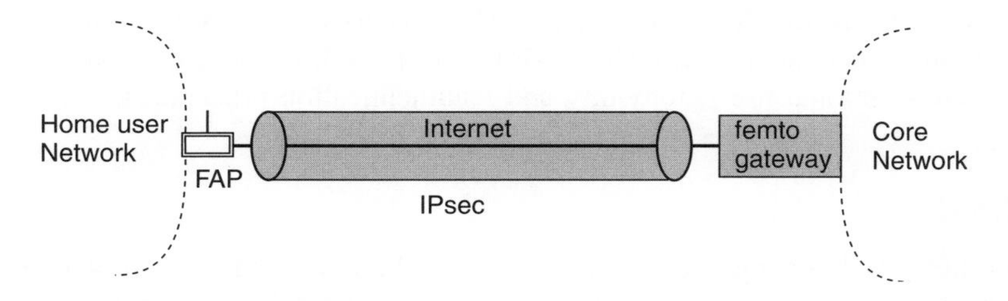

Figure 9.5 Security architecture

- Authentication Header (AH) is a protocol that provides authentication of the contents of the packet through the addition of a header that is calculated based on the content in the packet. It is based on checksums that depend on the keys defined by the security association. Which parts of the packet are used for the calculation, and the placement of the header depend on the mode (tunnel or transport) and the version of IP (v4 or v6). AH does not encrypt but only provides authentication.
- Encapsulating Security Payload (ESP) ensures privacy by encrypting the data. This algorithm uses the key to combine the data in order to encrypt it. Only the security association users know the keys and are able to decrypt at the other side of the IPsec tunnel.

9.2.3 Extensible Authentication Protocol

EAP is a universal authentication framework frequently used in wireless networks. Many implementations have been proposed depending on the technologies. Few of them have been applied to femtocells and are implemented in the FGW to ensure security and authentication.

EAP-TLS

EAP-Transport Layer Security (TLS) is the most well known EAP, and is implemented by all wireless equipments. It is based on the use of a Public Key Infrastructure (PKI) to create and manage the digital certificates. EAP-TLS was formerly called EAP-Secure Socket Layer (SSL), and the last implementation is detailed in [9]. In this protocol, a certificate authority links the public keys with their respective users. The certificate can be established automatically by software or manually by the users themselves. The user, the keys, the certificates and their validity are managed by the PKI.

EAP-SIM

EAP-Subscriber Identity Module (SIM) is an implementation for GSM using a SIM card. It is defined in [10]. In this approach the shared information between the two entities is contained on the SIM card.

EAP-AKA

EAP-Authentication and Key Agreement (AKA) is used for UMTS combined with a Universal Subscriber Identity Module (USIM) card [11]. It is based upon symmetric keys and it includes optional user anonymity and reauthentication procedures.

EAP-IKEv2

EAP-Internet Key Exchange version 2 (IKEv2) is the improved version of EAP-Internet Key Exchange (IKE) proposed in 1998. It is a very secure solution that has the following options:

- The public key is embedded in a certificate, and the corresponding private key is known only by one of the two entities (called asymmetric pair).
- Use of passwords known to both the FAP and the FGW.
- Use of symetric keys known by both the FAP and the FGW.

EAP-IKEv2 offers the possibility of choosing a different option for each direction. IKE is still widely used and the last implementation called IKEv2 is described in [12].

9.2.4 Femtocell Secure Authentication

A secure authentication is important for both operators and subscribers:

- Operators want to be sure that valid mobile users are allowed to connect to the core network.
- Femtocell subscribers want to be correctly identified in the core network.

Two main efficient technical solutions can be used for this purpose: a X.509 certificate based authentication, or an authentication using SIM card.

X.509 Authentication

X.509 certificates are usually used for authentication in IP-based networks. With such an approach, the sensitive information (i.e. the serial number) is stored in a specific hardware component called Trusted Platform Module (TPM). This element is a protected memory whose content can not be modified. With this approach, the identification of the FAP is defined at the manufacturing stage. When a customer purchases a FAP from an operator, the operator will associate this new customer to the serial number of the femtocell. The serial number information is given by the manufacturer directly to the operator, so that no other entity has access to this information. Later, when the customer uses his femtocell, its public key can only be used together with this serial number.

SIM Card Authentication

Another option to authenticate a FAP is to use a SIM card (GSM) or a USIM card (UMTS). In this case the protected information to authenticate a user is stored in a SIM/ USIM card, and this has to be installed inside the FAP (see Figure 9.6). In this approach, when a customer purchases a FAP from an operator, this operator will authenticate the user thanks to the information stored in both the SIM/USIM card and the Authentication, Authorization and Accounting (AAA) server.

Comparison of Both Approaches

SIM/USIM have been successfully used by operators for the authentication of handsets. However this solution suffers a drawback due to the fact that in femtocell networks, unlike traditional mobile networks, customer IP hosts (i.e. the FAP) are allowed to connect to the core network. This is a problem since nowadays many laptops are equipped with

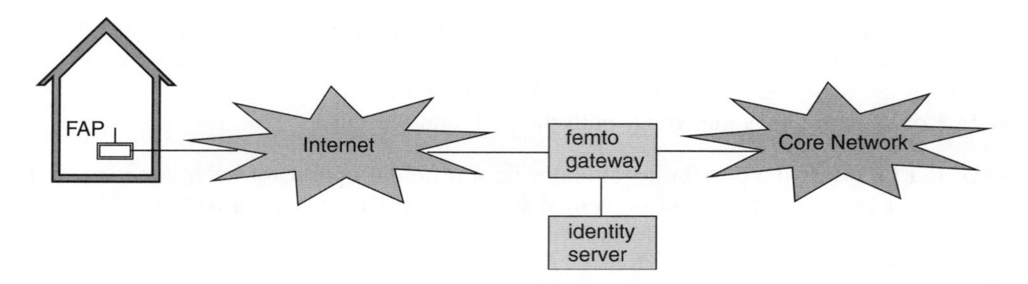

Figure 9.6 Architecture for X509 authentication

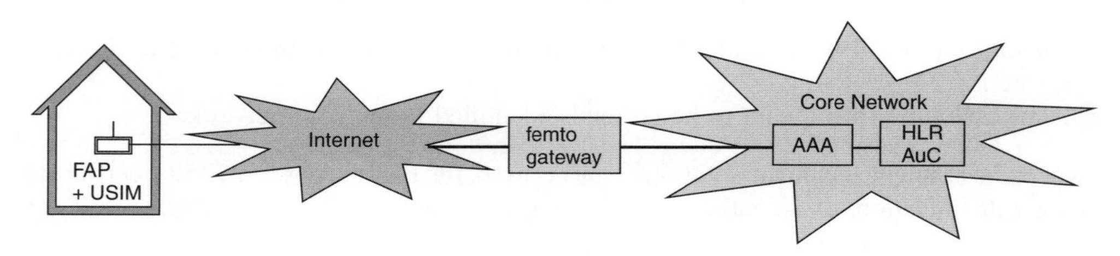

Figure 9.7 Architecture for USIM card authentication

SIM/USIM cards reader. With such laptops, any user could successfully authenticate as a FAP because it would be recognized by the FGW. Then, such a user could connect directly inside the core network with the possibility of accessing or modifying operator specific information. Moreover, with SIM/USIM card authentication, the software in the FAP could be modified to unlock the FAP, as is the case with mobile phones.

Finally, TPM authentication is not only very difficult to modify, but it also avoids performing the identification of the user inside the core network as represented in Figure 9.7. The main properties of both approaches are summarized in Table 9.4. It has been verified that both approaches have some advantages [13]. This is why the use of a hybrid model that combines both approaches needs to be investigated.

9.2.5 *Protection of the Wireless Link*

If the previously described authentication mechanisms help to make the femtocell network more secure, another solution to avoid unwanted users and to ensure privacy would be

Table 9.4 Secure authentication techniques

USIM card	X.509
Can be modified	Difficult to hack
Protection of core network required	No interaction with core network
Manufacturing and distribution of cards needed	No cards management
Possibility of using another FAP	Change of FAP made by operator

to protect the wireless channel itself. Hence approaches to optimize the radio coverage could be proposed to ensure that nobody is able to sniff the channel.

Such approaches can include the use of multi-antennas in the femtocell to preform beam forming, or the implementation of efficient power control algorithms to avoid radiating too much signal. An easy solution, for example, is to propose a sleep mode where the femtocell reduces its power when nobody is at home. Finally, another more radical option could be, in an enterprise where high security is required, to add special security materials to the walls to ensure that no RF radiation can penetrate the walls.

If customers are very concerned about the security of the calls or the assurance of their privacy, another important challenge is related to the provision of efficient emergency calls, which can only be performed if the FAP is able to locate itself.

9.3 Femtocell Location

Another important challenge related to femtocells is the ability to determine their location. The location estimation is necessary for emergency call services, but also for other reasons detailed later. According to a recent survey [14], 80% of the respondents said that it was important or extremely important that their cell location could be pinpointed on a map. The reason argued was that, if they pay for a femtocell, they deserve the same service in the case of emergency call as if they were using their landline.

Concerning the solutions to ensure the location of femtocells, it is possible that the decisions will be made depending on the regulations in the country where the femtocells are deployed.

9.3.1 Need for the FAP Location

Determining the location of the femtocell is necessary in the situations given below.

FAP Location for Emergency Calls

When a mobile user performs an emergency call (like 911 in the USA), the call is usually directed to the dispatch centre corresponding to the area where it was initiated. This helps emergency services to be more efficient when identifying the address of the caller and contacting the closest emergency services.

The location is easy to determine when using landline phone services, because each number is associated to a fixed address. Unfortunately it is not so simple when the calls are performed using femtocells. Moreover, since femtocell subscribers will have free calls using their mobiles, it is expected that, if femtocells are deployed, most of emergency calls will be done via femtocells. This is why, in some countries, severe regulations have been proposed, in order to ensure an accurate location of the femtocell when an emergency call is initiated. For example, in the USA [14], under a Federal Communications Commission (FCC) mandate, an emergency caller's location must be identified within 50 metres 67% of the time and 100 meters 95% of the time for handset based location technologies. The required accuracy will depend on the regulations in each country.

Each country must also decide if, in the case of closed access femtocells, non-subscribers should be allowed to perform emergency calls, especially in situations where there is not sufficient macrocell radio coverage.

FAP Location for Network Planning

Knowing the location of the femtocell can help operators to plan their network. In some countries, different areas are associated with different parts of the spectrum in order to avoid different operator frequencies overlapping. If an operator has chosen to allocate one frequency band for macrocell users and another for femtocell users, it could be useful to estimate the position of the femtocells. With such knowledge the frequency planning of the femtocell could be done depending on the frequencies of the neighbouring macrocells.

FAP Location for Access Control

As explained before, some operators allocate different channels depending on the area. This is why, if a femtocell subscriber chooses to move his FAP to another location, it is important for the operator to be aware of the new position. Moreover, the operator can choose to forbid the use of the FAP in certain areas. For example if moved abroad, the femtocell's usage should be blocked in order to avoid interference with other operators. Finally, it may be possible that some applications embedded in the FAPs would be activated only in some selected regions, which is why localization could be helpful.

9.3.2 Solutions

Different solutions have been proposed to ensure an estimation of the FAP position.

- *GPS positioning:* some femtocell manufacturers have chosen to include a GPS receiver inside their equipment. As explained in Section 9.1, GPS suffers low reception inside homes. However, the FAP is not expected to be moved often and so it could be possible for it to store the last received GPS coordinate in order to give a good estimation of the position of the FAP.
- *Cell sensing:* the position could also be estimated using geolocalization methods. With such an approach the FAP senses the neighbouring macrocells, and uses triangulation methods, based on the received signal power or the time of arrival. With such information the FAP could estimate its position. However this solution is only possible if there are several macrocells in the surroundings of the femtocell.

  If there were numerous neighbouring femtocells, a collaborative geolocation algorithm could also be performed. With this method, the position, estimated thanks to the macrocells, is made more accurate by taking into account (with lesser importance) the information obtained via the sensing of the femtocells.

  Finally, it is important to note that cell sensing can be difficult to implement, because it requires many cells, and because triangulation techniques are not always efficient in femtocell environments such as urban areas, where many reflections and diffractions occur.

- *TV signal:* TV signals can be used not only for timing accuracy, but also to estimate the position of a FAP. The main advantage of such signals is that their levels are higher than GPS signals, and they use a wide range of frequencies, making them more efficient against fading. According to [15], it has been verified with measurements in the USA that using TV signals outperforms GPS, because of its cheaper price and higher accuracy.
- *Internet IP address:* it is possible to identify the location by the IP address of the Internet connection. However this information is not always reliable, unless the whole chain (Internet and mobile operator) work together to provide this information. In practice, IP addresses could be used when a unique operator offers a combined gateway incorporating the broadband router and the FAP.
- *Customer address:* The last solution is to associate the home address of the subscriber to its FAP at the sale point. The advantage of this approach is that the exact address is located. The drawback is that the operator should be informed each time the subscriber moves to another location.

Some decisions will have to be taken in order to solve the problem of femtocell location. These decisions will depend mainly on the regulations in the different countries, which will decide if a location is required for emergency calls, and at which accuracy.

9.4 Access Methods

As explained in Chapter 6, interference is a strong issue in femtocell networks:

- *Cross-layer* interference is caused by femtocells to the macrocell layer and vice versa.
- *Co-layer* interference occurs between neighbouring femtocells.

Interference can be reduced by optimizing the resources (e.g. radiated power or allocated frequencies) of each femtocell. This optimization of the parameters has to be done by the FAP itself, because in the case of a successful large deployment, the operators will prefer to minimize their tasks. This is why, as explained in the previous chapter, each FAP has to be self-configurable (when it is switched on in order to adapt its initial parameters), and most of the parameters have to be self-optimizable (in order to adapt the femtocell to the fluctuations of the channel due to the neighbouring users and cells).

Moreover, even if the femtocell network is deployed in a self-organizable manner, global interference is also strongly dependent on the method of access to the femtocell, which decides whether a given user can connect or not to the femtocell.

In order to describe the access control procedures, it is important to note that in a two-layer network, users can be classified into two categories, depending on the connectivity rights that they are given:

- A *subscriber* of a femtocell is a user registered in it. Subscribers are thus defined as the rightful users of the femtocell, and they are usually mobile terminals of the femtocell owner and close family or friends.
- A *non-subscriber* is a user not registered in the femtocell.

According to 3GPP, three access methods to the femtocells, have been proposed: *closed access*, *open access* and *hybrid access*. The closed access mode is also called Closed Subscriber Group (CSG), and the list of subscribers of a femtocell is also called the CSG list. The rights given to each category of users are given in Table 9.5.

On the one hand, in closed access mode, as represented in Figure 9.8, only subscribers can connect to their femtocell. On the other hand, open access femtocells (Figure 9.9) are accessible by everyone. Since both approaches have some drawbacks, hybrid access also has been proposed (Section 9.4.3). In this approach, subscribers have a preferential access to their femtocell, and non-subscribers have the right to connect with limited access to the femtocell resources. In the following, the principles, advantages and drawbacks of the different approaches are described.

Table 9.5 The access modes as defined by 3GPP

	Closed access	Open access	Hybrid access
Subscriber	Access	Access	Preferential access
Non-subscriber	No access	Access	Limited access

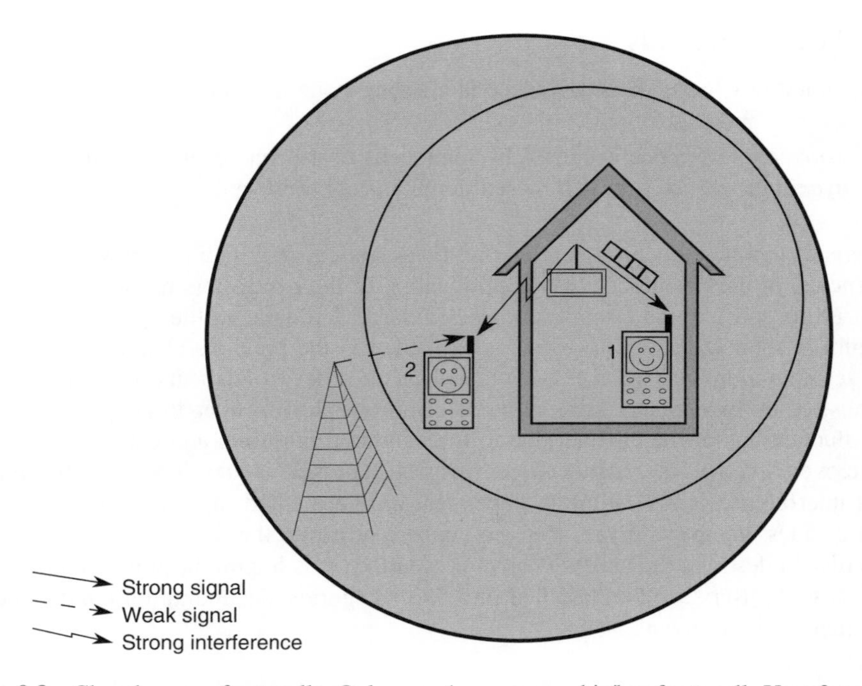

Figure 9.8 Closed access femtocells. Only user 1 can access his/her femtocell. User 2 can only connect to macrocells

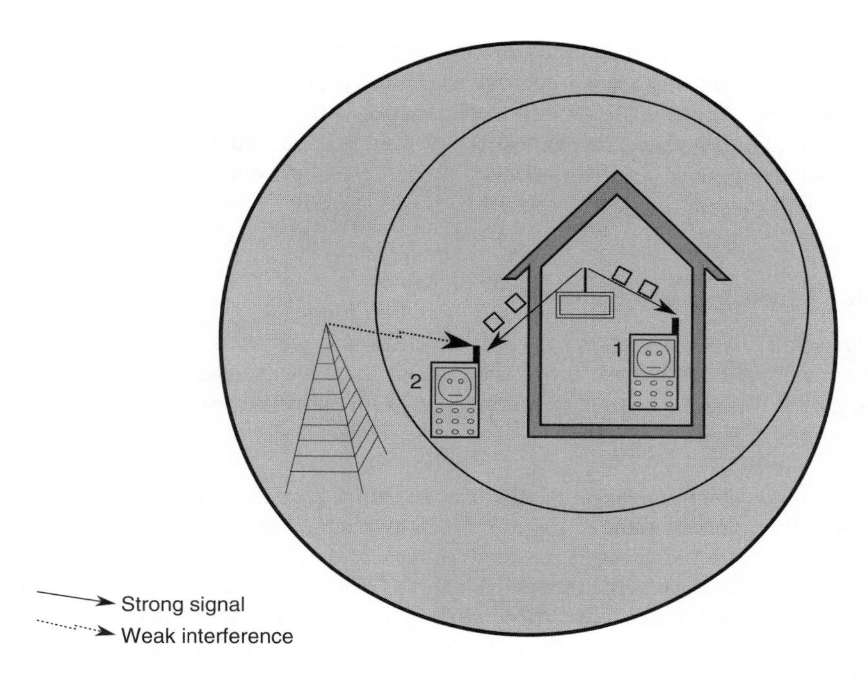

Figure 9.9 Open access femtocells. Both users 1 and 2 can connect to the femtocell

9.4.1 Closed Access

In closed access, only femtocell subscribers are allowed to connect to their femtocells. In this case, the list of registered users is decided by the femtocell owners and set up by the operator.

The list of authorized users resides in the core network and it is therefore not possible to modify this information in real time. The current CSG femtocell deployment is based on a solution where the CSG list is managed by the femtocell owner via a protected web page. This page allows the customer to login and to add or remove guest users by entering their mobile numbers. Then, on a regular basis the operator can update the CSG database in the core network. As will be explained later, non-registered guest users create a large amount of interference and hence this list should be updated as regularly as possible. Since this task has a cost, a daily update seems to be a reasonable frequency for an operator.

CSG Femtocells Deployment

The first large commercial femtocell deployments are to occur in areas where radio coverage from macrocells is poor but broadband connectivity is sufficiently deployed. This corresponds to certain areas as for example the middle of North America, where the first commercial deployments started at the end of 2008. However, in these deployments interference is not an important issue due to the low population density.

It is expected that most femtocell deployments will be aimed at the home market, starting in rural areas where the macrocell coverage is poor, and developing progressively in cities, where there is a higher number of potential customers. In cities, interference avoidance will play an important role for CSG femtocells.

Technical Challenges and Solutions

In urban scenarios with closed access femtocells, non-subscribers connect through the strongest macrocell. As a result, cross-layer interference is generated. For example, when a passing user walks along a street, where many CSG femtocells are deployed, the following occurs:

- The downlink communication of this user is jammed by the FAPs.
- The uplink communication of the femtocells is jammed by this user.

The closer to the femtocells the mobile user is, the worse the quality of the communication is, due to this interference. The most challenging case is the scenario where a non-subscribed outdoor user enters a home where a FAP is located. He then becomes a guest non-subscriber and produces interference in the whole house. A solution to this is to authorize non-subscribers to connect for a short period of time. This is in fact equivalent to a hybrid access model, where the restriction of the guest users resides in the length of time for which they are allowed to use the femtocell.

Co-layer interference comes up between neighbouring femtocells, especially in dense deployments. In most cases users install their femtocells in random positions within their homes. Therefore, subscribers are sometimes severely jammed by neighbouring cells, thus being unable to connect. For example, in a high building where each floor contains many flats, many femtocells are visible to each other, resulting in a large amount of interference.

To reduce the negative impact of the femtocells on the overall network, operators and manufacturers seek new solutions to minimize both the cross-layer and co-layer interference, hence interference cancellation and avoidance techniques for femtocell networks are currently a main research challenge.

The ideal solution for mitigating interference would be to adapt the shape of the best server area of the femtocell to the exact shape of the building. However this is unrealistic because both the FAP position and the building structure are unknown. The alternative solution is to adapt the signal level of the femtocell indoors, so that it is larger than that of any other cell. This can be done through a learning process, where automatic control algorithms are implemented. For example, in [16], the femtocell listens to neighbouring cells, and adapts its radiated power considering a security margin so that it is higher than the other cells.

It is also possible to change the power depending on the time of day, or on the presence of subscribers. For example a sleeping mode can be implemented, in which the femtocell reduces its power when no user is connected, and where the signal level is increased depending on the number of simultaneous users. If automatic power control algorithms are implemented, FAPs might increase their radiated power while trying to achieve a better service quality. Moreover, this may also introduce more interference to other macrocells and femtocells. This is why any power control algorithms should be carefully implemented, and adapted depending on the type of scenario.

Another solution for optimizing the shape of the cells, and thus reduce interference, is the use of sector antennas. This has been proposed [17] to minimize the overlapping of coverage areas. Furthermore, the use of several radiating elements to perform beam forming can be used. However it is important to note that sector antennas and multiple antennas can increase the price of the femtocells if not correctly engineered, which is why it is important to find simple solutions that can be manufactured at a low cost.

Finally, and as described in previous chapters, another solution for reducing interference is to use more frequency resources. This is why the use of OFDMA femtocells seems promising. OFDMA allows the allocation of orthogonal frequency and time resources to different users. In order to be efficient, frequency subchannels and time slots have to be properly assigned by the femtocells themselves (see Chapter 8). For example, the automatic subchannel allocation of WiMAX femtocells has been investigated [18]. It should also be noted that LTE femtocells, also called Home eNodeB (HeNB)s in 3GPP, are highly regarded by most of the operators.

Business Model

Closed access is the favourite method for customers of home femtocells according to some surveys [19]. The main reason is that most of the potential customers are only interested in paying for a femtocell if they are sure they will have full control over the list of authorized users. Moreover, femtocells will probably allow all types of user to perform emergency calls by law. This implies that some resources might be released in order to assure that non-subscribers can also make emergency calls. This resembles a hybrid model, where non-subscribers have a limited access that allows them to be authorized to make emergency calls.

The billing of both the FAPs and the calls also has to be decided. Some operators will prefer to sell the FAP at a fixed price and then propose a preferential price for calls. Other approaches would be to include the rent of the femtocell in the monthly price, with or without a preferential price for the calls. However, it is expected that to attract customers, operators will offer special rates or free calls to the femtocell subscribers. Moreover, the first billing solutions that have been implemented base the price of the whole call on the cell where the call was initiated. Some more advanced solutions should be proposed to change the billing during the call duration.

Finally, the problem of femtocell guest users must also be solved in the case that a roaming agreement exists between the operators of the subscriber and the guest's.

9.4.2 Open Access

All users (subscribers and non-subscribers) are allowed to connect to open access femtocells. There is thus no distinction between these two groups, who are referred to as users. Open access femtocells can be deployed in three main different scenarios:

Open Access Deployments

If the first femtocell deployments are mainly in closed access mode, different scenarios will appear in the future. This will mainly be due to the fact that, if FAPs are manufactured at low prices, they could be advantageous for planned indoor deployments compared with other approaches like Distributed Antenna System (DAS) or picocells. The main open access femtocell scenarios are:

- *Open access home deployments* could occur in dense areas where interference is high. Indeed, such an approach would allow an improvement in the performance of the network, by solving the problem of passing or guest users, thus reducing both co-layer and cross-layer interference. However, open access reduces the performance for subscribers and increases substantially the amount of handovers between cells due to the movement of outdoor users. This has a negative effect on the operator because the signalling in the network increases and calls can drop due to failure in the handover procedure.
- *Enterprise deployment* may also be targeted in the future for the industry, SMEs or even larger companies. This case is different from previous scenarios in that the size of the environment requires the use of several femtocells. Femtocells can be either deployed by the operator or self-installed by the end-customer. If they are deployed by the operators, they can be planned, thus improving system performance. Moreover, if femtocells are self-configurable, such a deployment would be interesting for the operator who would not have to care about maintenance. In this scenario femtocells will be open access so that all the employees can connect to them.
- *Hotspot deployment* is also planned to occur in the future in a similar manner to the widely deployed WiFi hotspots. They could be deployed in public areas such as airports, parks and train stations in order to improve coverage. Moreover, the idea of outdoor femtocells deployment, also referred as metrozone [20] has been investigated. However such deployment will not occur in the near future until femtocells are totally self-configurable in order to reduce negative impacts on the macrocell layer. Moreover the problem of deploying an outdoor ADSL backbone still has to be solved before this can be implemented.

Interference in Open Access

In order to improve performance, the ideas for reducing interference proposed for closed access (power configuration, multiple and sector antennas, or OFDMA) are also necessary. To illustrate this, several system-level simulations have been performed using the simulator presented in Chapter 5. The test environment was a residential area where WiMAX femtocells and a macrocell have been deployed. The simulator is based on a deterministic radio coverage calibrated with real measurements [21]. Numerous Monte Carlo snapshots were run to obtain statistics of cell throughputs, the histograms of which are displayed in Figure 9.10.

It was verified here that the overall network throughput of open access outperforms that of closed access. This is due to the fact that, when non-subscribers connect to the femtocells, resources are released from the macrocells allowing more users to connect.

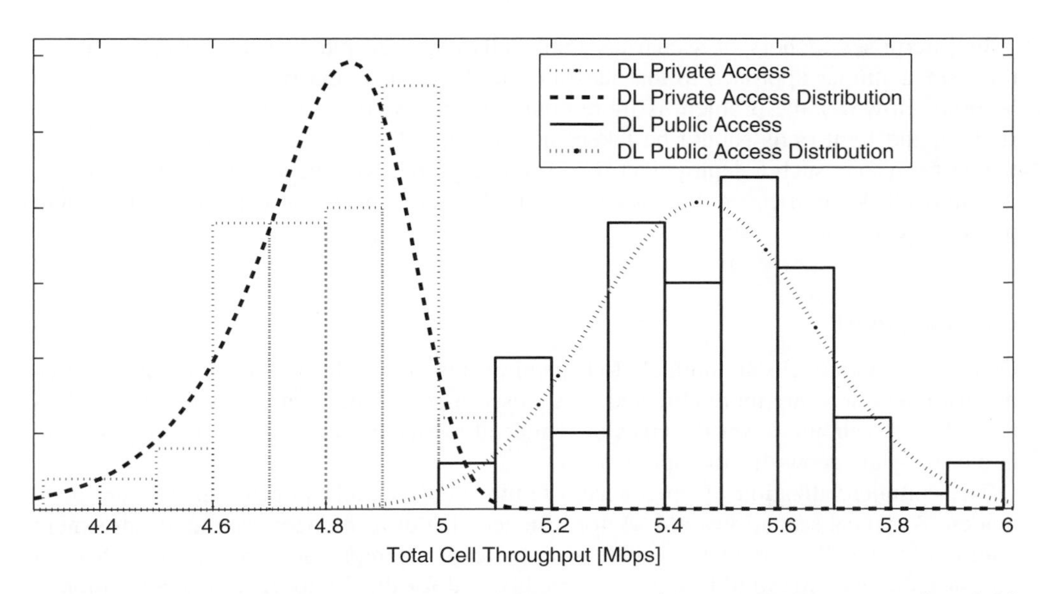

Figure 9.10 Total throughput in a residential area covered by a WiMAX macrocell and 25 femtocells

Furthermore it is clearly seen that a CSG approach reduces the overall network capacity by almost 15%. Such open access deployment is thus interesting for the operator, because it increases the capacity of the overall network, and allows the operator potentially to save energy by reducing the power radiated by its macrocells.

Handovers in Open Access

Mobile users in an open access femtocell network also cause a high number of handovers, which reduces the overall network performance. Hard handover is the most commonly supported handover in femtocells [22]. This is due to delays in the backbone connection, which do not allow the use of soft handovers where the communication is held in parallel in different cells. Therefore, a passing user will always handover between femtocells in the street, increasing substantially the signalling in the network.

Furthermore, the chances for an unsuccessful handover also increase, especially if the neighbouring list is not properly updated. Regardless of this, solutions have been proposed in which a femtocell-centric sensing of the environment is used to obtain parameters of the surrounding environment and to update the list of neighbours. Unfortunately, if femtocells are massively deployed, there could be confusion regarding the cell identities of surrounding femtocells. Even if the different standards (HSDPA, WiMAX (Wireless Interoperability for Microwave Access), LTE (Long Term Evolution), ...) could support enough cell identities for a femtocell network, the search time might be prohibitive for a successful handover and calls might still be dropped. Hence, before open access femtocells are deployed, research is required in order to propose solutions that allow the network to support more handovers, and to avoid confusion between large numbers of cells.

In enterprises or hotspot scenarios, femtocells could be planned by the operators. In this case a similar approach to that used during the deployment of WiFi networks could be used. However the problem is more complex because the neighbouring outdoor base stations and femtocells should be taken into account. This is why, in order to optimize the performance, such a deployment requires a very good knowledge of the environment. A femtocell deployment tool as described in Chapter 1 can be used for the deployment of such scenarios.

Business Model

Open access femtocells are unlikely to be deployed in homes due to the preferences shown by customers, who are more attracted by a closed model. Moreover, in such scenario, it is not clear who should cover the costs of femtocell maintenance, because such deployment is more advantageous for the operators.

The commercialization of open access femtocells is mainly targeted at the enterprise market. A typical scenario would be, for instance, that of a company having an agreement with the femtocell operator to provide low cost calls to employees when in the office. In this case the operator would propose a monthly rent for the femtocells. Such a solution is efficient in ensuring Fixed Mobile Convergence (FMC) [23], because the employees can perform seamless switch from the indoor network (using landline billing) and the outdoor network (using mobile operator billing).

In hotspot scenarios, femtocells are used to increase the coverage and capacity of the network. In this case, they would be deployed by the operator.

9.4.3 Hybrid Access

The main characteristics of both closed and open access are given in Table 9.6.

Concerning the home market, both closed and open access methods suffer some important drawbacks:

- Closed access femtocells suffer and cause high interference.
- Customers are unlikely to accept paying for open access femtocells.

Table 9.6 Closed vs open access

Closed access femtocells	Open access femtocells
Higher interference	More handovers
Lower network throughput	Higher network throughput
Serves only indoor users	Increased outdoor capacity
Home market	SMEs, hotspots
Easier billing	Security needs

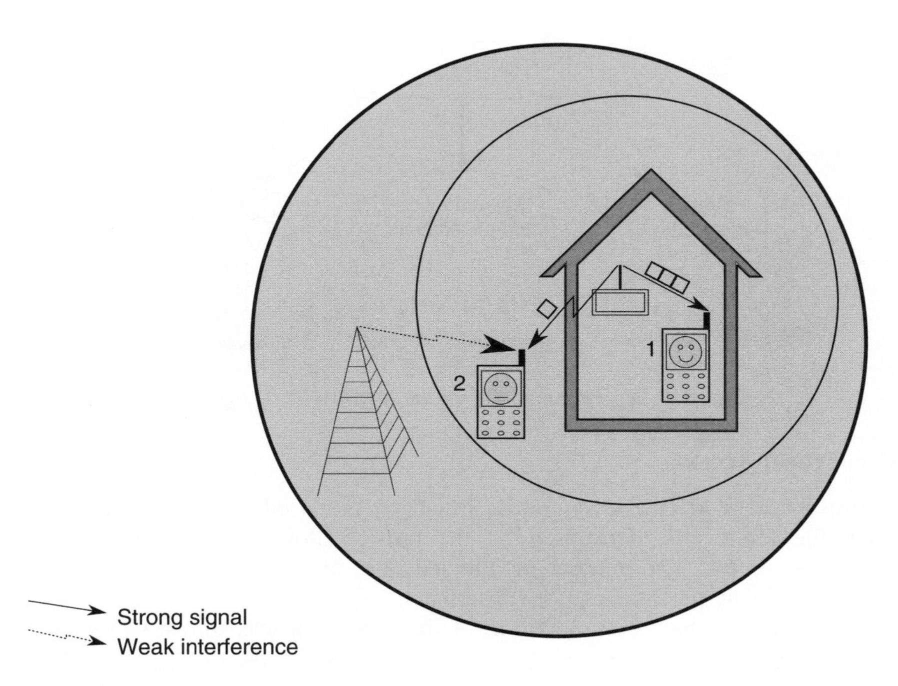

Figure 9.11 Hybrid access femtocells. User 1 connects with a preferential access, and user 2 with a limited access

There is thus a need for a compromise. That is why a new access method called hybrid has been proposed, based on the principles represented in Figure 9.11 and combining the two following features:

- a CSG access mode for the users in the CSG list, offering a preferential access,
- an open access mode for the non-subscribers, offering a limited access.

There are many possible solutions to defining an hybrid model, i.e. to define how resources are shared between subscribers and non-subscribers. As represented in Figure 9.12, three important factors should be taken into account.

- Time dependency: should the number of shared resources be changed over time or not?
- Treatment of non-subscribers: should a finite amount of femtocell resources be assigned to all non-subscribers or to individual users?
- Sharing of the resources: should all resources be assigned equally to all users (shared), or should a fixed amount be booked for subscribers (restricted)?

Some ideas for developing new hybrid models are described below for CDMA and OFDMA.

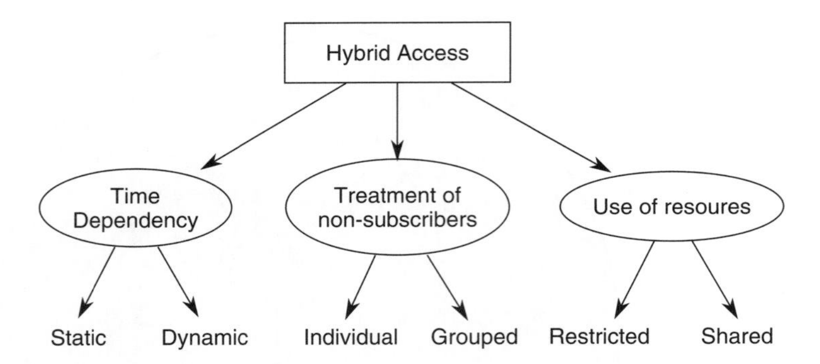

Figure 9.12 Factors defining the hybrid access algorithm

CDMA Hybrid Access

An adaptive access to HSPA femtocells has been proposed and analysed [24]. In this approach the femtocells, depending on the traffic, allow a certain number of non-subscribers to access. The number of allowed non-subscribers is carefully adapted, to avoid decreasing the performance of subscribers.

Using this method, based on system level simulations, it was shown that when several non-subscribers connect to the femtocell, the performance of the overall network is vastly increased.

However, in CDMA all users share the same frequencies and hence all users (subscribers and non-subscribers) interfere with each other. Therefore each time a new user connects to a femtocell it reduces the performance of connected users. In such a model there is no preferential access for subscribers, that is why this model is more interesting to non-subscribers. A solution to this problem could be for the operator to have more than one frequency band available, and allocate one band for subscribers and the other for non-subscribers. This would reduce the impact on subscribers. However it is unlikely to occur due to frequency restriction problems.

In such an adaptive access method, the main challenge is to find the number of authorized non-subscribers that maximizes the performance of the overall network and minimizes the reduction of performance of subscribers.

OFDMA Hybrid Access

OFDMA, as explained previously, can help to reduce interference by allocating both time slots and subcarriers between the different users. In OFDMA systems, subchannels contain a series of subcarriers, which can be adjacent or not, in order to exploit frequency diversity. In such a hybrid model, it is possible that some subchannels of the femtocell are released and used by non-subscribers. Depending on the rate of preferential access given to the subscribers, the number of subchannels allocated to subscribers will be adapted. If operators own only a small bandwidth, it is also possible with OFDMA to subdivide subchannels over the time domain.

Due to its two degrees of freedom (time and subcarriers) OFDMA is an ideal mode for hybrid models. However the number of resources to share with non-subscribers must be carefully balanced depending on the scenario and the time of the day. In general it is more efficient to dynamically change this number of resources in order to adapt better to the variations of the channel.

Commercial Challenges

Hybrid models, making a compromise between closed access and open access, are probably the most efficient in terms of performance. However, a business model for the deployment of such femtocells must be proposed. A first solution would be to consider that they could be deployed under the same model as are CSG femtocells, especially if the chosen model always gives preferential access to subscribers. A fairer solution would be to propose such femtocells at a lower cost for customers, because they also increase the performance of the macrocell layer. Another business model could be to adapt the price of the monthly bill, depending on the number of non-subscribers allowed to use the femtocell during the month.

9.5 Need for New Applications

As explained in Chapter 2, a major competitor for the femtocells is the use of dual mode handsets, also called Unlicensed Mobile Access (UMA), providing outdoor communications via GSM/UMTS and indoor communications via WiFi. These two competitors (femtocell and UMA), can both ensure the FMC, being the main advantage of the femtocells in that they can be used with any mobile phone. Therefore it is important that operators and manufacturers work together, in order to propose specific applications that give a greater advantage to femtocells over UMA.

9.5.1 Evolution of Consumer Interest in Femtocells

Most femtocell manufacturers are currently focusing not only on the technical challenges as described in this book, but also on the development of new femtocell services. For example, IPaccess [25] has summarized the consumers' propositions in historical order as follows:

- Indoor radio coverage: this was the initial goal of femtocells, i.e. ensuring a good radio coverage indoors, where the macrocell coverage was not sufficient.
- Cheap calls at home: for users already having a good indoor radio coverage, the concern is to have indoor calls at a lower price. It is also the main concern for enterprise femtocell deployment.
- Mobile data: femtocells are not only aimed at voice, but also data services. Data services like video and web started to be widely used with UMA, which is why femtocells have to offer advanced data services.
- Femtozone services: this kind of service is both voice/data, and is used automatically when the user is in the range of the femtocells. The operators are today very concerned about deploying such services.

- Connected home services: a recent concern for customers is the concept of connected home. This is mainly the case with the scenario of gateway femtocells, were the services at home (like computer, phone, TV, printer, or camera) are all connected to the same *box*. In this case the phone can access mobile services via the femtocell.

Femtozone and connected home services are thus new concepts, and there are currently few propositions in this area. However, it is important to note that the success of femtocell technology depends mainly on this new kind of application.

9.5.2 *Development of New Applications*

Some new applications have currently been deployed and demonstrated at conferences, in order to advertise femtocells and show their advantage compared to UMA. However, the contributions are still not sufficient and more innovative ideas need to be developed.

Femtozone Services

To develop new applications, two main specific concepts can be used: the presence concept and the virtual number concept.

- *Presence applications* are based on the fact that customers always have their mobile phone with them. In such a case it is possible to initiate some automatic services each time the subscriber enters/leaves the range of their femtocells.

  For example, when a user enters his home, it could be possible to automatically upload all the new pictures in the mobile onto the home server.

  Another example could be for a mother to set up her child's mobile, so that she will receive a Short Message Service (SMS) when the child arrives at home or leaves the house.
- *Virtual numbers* can be added in the home. The calls would be managed only by the femtocell and not by the operator. Such numbers can be used to create groups of users inside the house. For example, a unique mobile number could be used to reach all the users inside the house.

Connected Home Services

The notion of connected home services aims at using the femtocell as a link between the mobile and the home network. Another idea would be to use the mobile to control some equipment like the TV. With connected home services, the mobile becomes a controller that helps to manage, via the FAP, all connected equipment inside the house. Such services could also help to diversify the role of the mobile, thus making more revenue for the operators.

Femtocell API

Femtocell applications can be implemented in different locations:

- in the FAP itself for most of the applications,
- in the handset that interacts with the FAP.

For those implemented in the FAP, the applications should be developed by the manufacturer. Hence, as already suggested [26], a good option could be that the operator provides an Application Programming Interface (API), in order to allow external companies easily to develop their software. This was the case for example with the iPhone from Apple, where an API is used with success by many third party companies. Another example is the Android operating system for mobile devices released by Google in 2008.

The API could be provided by femtocell manufacturers with a development kit and documentation, a software development kit, some easy form of testing and an online application store. The proposition of such API could be a good approach to help femtocells overcome UMA, by proposing more innovative services and applications.

9.6 Health Issues

One of the concerns with the use of femtocells is the health issues associated with the employment of radio frequency radiation. This is a major issue in most Western countries because many customers are restrained from using wireless technologies due to the fear of the potential dangers of radio waves. This is hence a major obstacle in the deployment of the technology that increases the digital divide.

9.6.1 Radio Waves and Health

Since this concern has direct influence in the commercial success of femtocells, the Femtoforum has produced a document related to health issues in femtocells [27]. This document was published to reassure the customers by explaining that the femtocells comply with with RF exposure requirements.

A large amount of research has been and is currently being undertaken in order to try to find a link between RF radiation, and illnesses like cancer.

It has been shown that high doses of radiowaves can increase the temperature of body cells and thus also tissues. However this has been legislated for in most countries and, as long as the law is obeyed, it should pose no danger. It is mainly due to this that:

- No diseases have been related to low temperatures increase.
- The radiowave amplitude has to be extremely high in order to raise body tissue by just one degree celsius. Hence, strict radiation levels have been defined.

Since it is not easy to link radio waves to body malfunction, researchers have tried to perform statistical surveys that check the potential effects of RF exposition. However these typically reveal that those who are not exposed to radiation can also feel sick, and hence most of the evidence is inconclusive. Therefore the main protection that can be nowadays guaranteed to the users, is to ensure that the maximum radiated power respects the existing law.

9.6.2 Power Levels Due to Femtocells

The RF exposure due to a radio device, at distance r and direction (θ, ϕ), is often quantified thanks to the power density S, expressed in W/m^2:

$$S = \frac{RGP}{4\pi r^2} \tag{9.6}$$

$R(\theta, \phi)$ is the normalized radiation pattern in the direction (θ, ϕ), i.e. R is equal to unity in the direction of maximum radiation, G is the antenna gain and P the imput power of the antenna.

The International Commission on Non-Ionizing Radiation Protection (ICNIRP) is an international organization responsible for fixing the maximum Non-Ionization radiation (NIR) limits. NIR limits correspond to the maximum authorized RF radiation generated by RF transmitters. According to ICNIRP [28], the NIR limit for the general public in the range of 2 GHz – 300 GHz is defined with a power density $S = 10\,W/m^2$. This limit has also been validated by the World Health Organization (WHO), and is used in all countries in the world.

All FAP, like WiFi access points, use a maximum radiated power equal to 0.1 W. With such power, and using Equation (9.6), the power density received from a FAP at a distance $r = 10\,cm$ in the direction of maximum radiation, is equal to $1.3\,W/m^2$. This means that a user located very close to his femtocell will receive a power about eight times smaller than the authorized limit. Moreover, it is important to note that, in an indoor environment, the received power at a distance r usually decreases by a proportion equal to $1/r^2$. This means that the femtocell user, who will in reality be always located at a larger distance than 10 cm, will receive a RF radiation very far below the ICNIRP limit.

In fact, the main RF radiation that a user will receive is due to their mobile phone. This is due to the fact that, during a call, the mobile is located close to the head of the caller, and, as explained bellow, the power decreases as $1/r^2$. This is why, in general, the radiation from the mobile is higher than that from the FAP. Moreover, the power radiated by a mobile phone is always adapted to ensure a reliable communication. This means that, in poor radio coverage areas, a mobile phone will have to emit more power when connecting to an outdoor macrocell, which is why, when using a femtocell, which ensures a good indoor coverage, the power level radiated by the mobile will be reduced very much, thus radiating fewer RF waves to the caller's head.

Finally, femtocells if they are widely deployed, could allow operators to reduce the power of their macrocells, which would also reduce outdoor RF waves. In this way the overall level of radio signal, both indoor and outdoor, may be reduced.

References

[1] D. Chambers, 'Crystal Frequency Oscillators in Femtocells,' http://www.thinkfemtocell.com, October 2007.

[2] J. A. Kusters and J. R. Vig, 'Hysteresis in quartz resonators-a review,' *IEEE Transactions on Ultrasonics, Ferroelectrics and Frequency Control*, vol. 39, no. 3, May 1991.

[3] 'Femtocell: choosing your oscillator,' Rakon Technical Report, 2008.

[4] N. I. of Standards and Technology, 'Ieee1588 website,' http://ieee1588.nist.gov/, 2008.

[5] S. Lee, 'An enhanced ieee 1588 time synchronization algorithm for asymmetric communication link using block burst transmission,' *IEEE Communications Letters*, vol. 12, no. 9, pp. 687–689, September 2008.

[6] D. L. Mills, 'Network time protocol version 4 reference and implementation guide,' GSMA, Tech. Rep., June 2006.
[7] N. Marechal, J.-B. Pierrot, and J.-M. Gorce, 'Fine synchronization for wireless sensor networks using gossip averaging algorithms,' in *IEEE International Conference on Communications, ICC 2008*, May 2008, pp. 4963–4967.
[8] 'In-building timing and location for femtocells,' Rosum FemtoSync Technical Report, October 2008.
[9] B. Aboba and D. Simon, 'RFC2716: PPP EAP TLS Authentication Protocol,' IETF, Tech. Rep., October 1999.
[10] H. Haverinen and J. Salowey, 'RFC4186: Extensible Authentication Protocol Method for Global System for Mobile Communications (GSM) Subscriber Identity Modules (EAP-SIM),' IETF, Tech. Rep., January 2006.
[11] J. Arkko and H. Haverinen, 'RFC4187: Extensible Authentication Protocol Method for 3rd Generation Authentication and Key Agreement (EAP-AKA),' IETF, Tech. Rep., January 2006.
[12] H. Haverinen and J. Salowey, 'RFC5106: The Extensible Authentication Protocol-Internet Key Exchange Protocol version 2 (EAP-IKEv2) Method,' IETF, Tech. Rep., February 2008.
[13] 'Challenges in Deployment of UMTS/HSPA Femtocell,' Aricent White Paper, February 2008.
[14] 'Connecting when it counts: the role of femtocells in emergency calls,' Intrado Technical Report, 2008.
[15] M. Rabinowitz and J. Spilker, 'A new positioning system using television synchronization signals,' *IEEE Transactions on Broadcasting*, vol. 51, no. 1, pp. 51–61, March 2005.
[16] K. Utsumi, H. Sasai, T. Niiho, M. Nakaso, and H. Yamamoto, 'Performance of macro- and co-channel femtocells in a hierarchical cell structure,' in *IEEE 18th International Symposium on Personal, Indoor and Mobile Radio Communications (PIMRC)*, Sept 2007, pp. 1–5.
[17] V. Chandrasekhar and J. G. Andrews, 'Uplink Capacity and Interference Avoidance for Two-Tier Femtocell Networks,' *IEEE Transactions on Wireless Communications*, February 2008.
[18] D. Lopez-Perez, G. de la Roche, A. Valcarce, A. Juttner, and J. Zhang, 'Interference avoidance and dynamic frequency planning for wimax femtocells networks,' in *11th IEEE Singapore International Conference on Communication Systems (ICCS)*, Nov 2008, pp. 1579–1584.
[19] S. Carlaw, 'Ipr and the potential effect on femtocell markets,' in *FemtoCells Europe*. ABIresearch, 2008.
[20] K. Graham, 'Femtocells beyond the home and office,' Femtocell Europe Conference, London, June 2008.
[21] A. Valcarce, G. De La Roche, A. Jüttner, D. López-Pérez, and J. Zhang, 'Applying FDTD to the coverage prediction of WiMAX femtocells,' *EURASIP Journal on Wireless Communications and Networking*, Feb. 2009.
[22] 'Radio Resource Management,' Aricent White Paper, January 2008.
[23] Ericsson. (2008, Sep.) Press Release. [Online]. Available: http://www.ericsson.com/ericsson/press/releases/20080911-1250616.shtml
[24] D. Choi, P. Monajemi, S. Kang, and J. Villasenor, 'Dealing with Loud Neighbors: The Benefits and Tradeoffs of Adaptive Femtocell Access,' in '*IEEE Global Telecommunications Conference (Globecom)*', Dec. 2008, pp. 1–5.
[25] ip.access, 'Femtocell briefing paper: Introducing 3G femtocells successfully to the home,' in *Femtocells Europe*, 2008.
[26] D. Chambers, 'Using a Femtocell API to create innovative new applications?' http://www.thinkfemtocell.com, January 2009.
[27] 'Femtocells and Health,' Femtoforum White Paper, 2008.
[28] 'Radiofrequency Radiation Measurements Public Wi-Fi Installations in Hong Kong,' Office of the Telecommunications Authority, March 2009.

Index

Femtocells: Technologies and Deployment Jie Zhang and Guillaume de la Roche